James K. Young

A Practical Treatise on Orthopedic Surgery

designed for the use of students and practitioners

James K. Young

A Practical Treatise on Orthopedic Surgery
designed for the use of students and practitioners

ISBN/EAN: 9783337089764

Printed in Europe, USA, Canada, Australia, Japan

Cover: Foto ©berggeist007 / pixelio.de

More available books at **www.hansebooks.com**

A PRACTICAL TREATISE

ON

ORTHOPEDIC SURGERY.

DESIGNED FOR THE USE OF

STUDENTS AND PRACTITIONERS.

BY

JAMES K. YOUNG, M.D.,

INSTRUCTOR IN ORTHOPEDIC SURGERY, UNIVERSITY OF PENNSYLVANIA; ATTENDING SURGEON
ORTHOPEDIC DEPARTMENT, UNIVERSITY HOSPITAL; FELLOW OF THE AMERICAN
ORTHOPEDIC ASSOCIATION; FELLOW OF THE COLLEGE OF PHYSICIANS OF
PHILADELPHIA; MEMBER OF THE PHILADELPHIA COUNTY
MEDICAL SOCIETY, ETC., ETC.

*ILLUSTRATED WITH TWO HUNDRED AND EIGHTY-FIVE
WOODCUTS.*

PHILADELPHIA:
LEA BROTHERS & CO.
1894.

TO

D. HAYES AGNEW, M.D., LL.D.,

In Memoriam,

JOHN ASHHURST, Jr., M.D.,

WILLIAM G. PORTER, M.D.,

DE FOREST WILLARD, M.D., Ph.D.,

AND

A. SYDNEY ROBERTS, M.D.,

MY SURGICAL TEACHERS,

TO WHOM I SHALL ALWAYS REMAIN INDEBTED FOR PERSONAL KINDNESSES
AND PROFESSIONAL ASSISTANCE, AND FOR WHATEVER SURGICAL
KNOWLEDGE, JUDGMENT AND SKILL I MAY HAVE ACQUIRED
FROM THEIR INTELLIGENT INSTRUCTION.

PREFACE.

—

In the following pages the author has endeavored to provide students and practitioners with a guide to Orthopedic Surgery in accordance with the most approved knowledge of the present day. Though based largely upon personal experience, this volume owes not a little to the existing literature of its subject, old as well as new. Systematic treatises on general and special surgery, monographs, and articles in periodicals have been faithfully consulted for material at once valuable and available for a work of the present scope and purpose. Much attention has been devoted to the mechanical part of the subject, to pathology, which it is believed will be found thoroughly modern, and to treatment, which is, of course, the ultimate object of the work.

222 S. Sixteenth St.,
Philadelphia.

TABLE OF CONTENTS.

ORTHOPEDIC SURGERY.

—

CHAPTER I.

INTRODUCTION AND CLASSIFICATION.

, THE word ORTHOPEDY (from ὀρθός, straight, and παῖς, child)—Fr., *orthopédie;* Ger., *orthopädie;* It., *ortopédia*—according to its derivation and its earlier use implies the art of removing deformities in young children. In the present day its meaning has been extended to apply to the treatment of deformities in persons of all ages. From its derivation, again, it might be taken to embrace the rectification of a great variety of abnormal conditions in which deformity is a conspicuous feature—the reduction of dislocations, the removal of tumors, etc.; but in modern practice the application of the term is limited to certain kinds of deformities, especially those of a chronic and progressive character.

Orthopedic surgery may be defined as that department of surgical science which includes the preventive, mechanical, and operative treatment of chronic and progressive deformities.

As a special branch of medicine the sphere of its influence reaches out in three directions. By its employment of gymnastics it pursues the field of hygeia; by its operative procedures it allies itself with, but does not invade, the realm of general surgery; and by its therapeutic prevention and cure of deformities it advances in the path of practical medicine.

Orthopedics is, therefore, a legitimate special section of general surgery.

Although the effective relief of deformities has only been accomplished within a comparatively recent period, their treatment was not neglected by the ancients. Hippocrates displayed an intimate acquaintance with club-foot in his treatise "On Articulations," and described a method of correcting infantile deformity which may still be employed with advantage. As medicine as a science advanced, the radical cure of hare-lip, described by Galen, Celsus,

2

and the Arabian physicians, became a common operation. The plastic art was well understood, and section of the sterno-cleido-mastoid muscle, for the relief of wry-neck, was frequently performed.

Ambroise Paré, Severinus Arcæus, and others contributed to the knowledge and treatment of deformities, and Thilenius (1784), Sartorius (1804), and Delpech (1816) advanced the art in a scientific direction, and prepared the way for the brilliant discovery of subcutaneous tenotomy of Stromeyer in 1830, and its establishment as a principle in operative medicine.

As a special branch of surgery, orthopedics dates its existence from the time of Audry (1741), who coined for it a name, and as an established specialty it obtained its first firm foothold upon the profession from the time of the discovery of Stromeyer. Its advances have been accomplished by the energy of such French surgeons as Delpech, Bouvier, Bonnet, Malgaigne, Guérin, and Pravaz; such German surgeons as Heine, Dieffenbach, and Langenbeck; such English surgeons as Little, Tamplin, Lonsdale, Brodhurst, Adams, and others; and such American surgeons as Bigelow, Detmold, Mütter, Brown, and others. Like many other great discoveries in surgery, subcutaneous tenotomy was at first recklessly and extravagantly employed, until fifty years ago orthopedics became an almost despised and rejected art. In this country, under the influence and energy of Davis, Sayre, and Taylor, it was rescued from its obscure position and established upon a satisfactory basis as a special branch known as the American School of Orthopedic Surgery.

Previously and subsequently to Stromeyer's discovery, mechanical appliances formed the fundamental part of the treatment of deformities. In the modern methods of orthopedic practice mechanical appliances and subcutaneous tenotomy supplement each other. While the introduction of anæsthetics into surgical practice and the perfection of the antiseptic method have greatly enlarged the field and precision of orthopedic operations, they have not diminished the mechanical ingenuity which manifests itself as such an important element in this branch of surgery. The tendency has rather been toward mechanical extravagance.

Modern ideas demand that the orthopedic surgeon of to-day should be an educated surgeon in every sense of the word. Carefully trained in clinical and operative surgery, thoroughly skilled in mechanical principles, he must be equally prepared in all three branches of his special art : the treatment and prevention of orthopedic diseases, the application of apparatus, and the performance of operations—in other words, must be a physician, a mechanician, and a surgeon. In this respect the orthopedist most resembles the ophthalmologist, who must treat disease, refract, and operate. As in ophthalmic practice refraction forms the greater part of the work, so also in orthopedic surgery, measurement and the application of mechanical appliances will demand the greatest attention.

Every student of medicine upon graduating, or before entering upon the practice of his chosen profession, should be instructed in the fundamental principles and practice of the orthopedic art. Particularly should he be instructed in the use and application of such appliances as he can manufacture for himself.

The increasing interest in this subject is well indicated in the establishment of special dispensaries, practical courses, and clinical professorships in all the more important colleges and post graduate schools in this country. In many of these, moreover, as the University of Pennsylvania, the New York Orthopedic Hospital, and the Boston Children's Hospital, special machine shops are attached, in which, under the direct supervision of the surgeon, the mechanical appliances are made. The orthopedic art is thus elevated in importance as a special branch of surgery, and the orthopedic surgeon is advanced in dignity and reputation.

Orthopedic affections may be looked at in any one of three ways, and accordingly a topographical, a pathological, or an anatomical-pathological arrangement may be adopted.

The first, or topographical arrangement, is the one usually employed by systematic writers, and the deformities are taken up *seriatim* as they affect the different portions of the body : the head, neck, trunk, upper extremity, lower extremity, etc. Such an arrangement has been used by Reeves.

The second, or pathological arrangement, offers certain advantages, since it gives a clue to the cause and nature of the affections. Thus we have the division into acquired deformities and congenital deformities, and a subdivision of the acquired deformities into three classes as they arise from causes directly, indirectly, or both directly and indirectly, affecting the articulations ; a subdivision also of the congenital deformities into congenital distortions and congenital malformations was preferred by Little, and a somewhat similar one has been more recently employed by Schreiber.

The third, or anatomical-pathological arrangement, is the one here presented as being the most scientific and satisfactory. The subject is divided into six parts, the affections being classed as they are deformities dependent on—I. Lesions of bone; II. Lesions of synovial membrane; III. Lesions of cerebro-spinal system; IV. Impaired nutrition, or diathesis; V. Embryonic disease or disturbances of development; VI. Accident or traumatism. These six classes are divided, and are again subdivided into the individual affections. It is not presented as a perfect arrangement, but it is the one the writer has found most convenient in teaching, and is a slight modification of that used by Dr. A. Sydney Roberts, his predecessor. The subject of "acquired club-foot," while properly classed in the table under Paralyses, will, for convenience, be considered under the generic head, "Club-foot," under Congenital Embryonic Disease.

CLASSIFICATION OF DISEASES BELONGING TO THE DEPARTMENT OF
ORTHOPEDIC SURGERY.

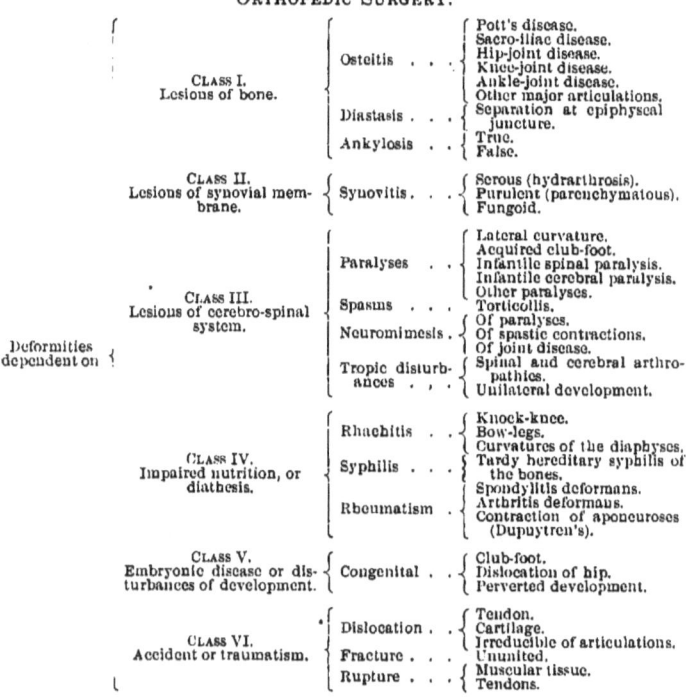

Deformities dependent on			
CLASS I. Lesions of bone.	Osteitis	{	Pott's disease. Sacro-iliac disease. Hip-joint disease. Knee-joint disease. Ankle-joint disease. Other major articulations.
	Diastasis	{	Separation at epiphyseal juncture.
	Ankylosis	{	True. False.
CLASS II. Lesions of synovial membrane.	Synovitis	{	Serous (hydrarthrosis). Purulent (parenchymatous). Fungoid.
CLASS III. Lesions of cerebro-spinal system.	Paralyses	{	Lateral curvature. Acquired club-foot. Infantile spinal paralysis. Infantile cerebral paralysis. Other paralyses.
	Spasms	{	Torticollis.
	Neuromimesis	{	Of paralyses. Of spastic contractions. Of joint disease.
	Tropic disturbances	{	Spinal and cerebral arthropathics. Unilateral development.
CLASS IV. Impaired nutrition, or diathesis.	Rhachitis	{	Knock-knee. Bow-legs. Curvatures of the diaphyses.
	Syphilis	{	Tardy hereditary syphilis of the bones.
	Rheumatism	{	Spondylitis deformans. Arthritis deformans. Contraction of aponeuroses (Dupuytren's).
CLASS V. Embryonic disease or disturbances of development.	Congenital	{	Club-foot. Dislocation of hip. Perverted development.
CLASS VI. Accident or traumatism.	Dislocation	{	Tendon. Cartilage. Irreducible of articulations.
	Fracture	{	Ununited.
	Rupture	{	Muscular tissue. Tendons.

CHAPTER II.

Pott's disease of the spine is a progressive inflammatory lesion of the vertebral bodies or inter-vertebral disks, leading to their partial or complete destruction, usually terminating in ankylosis with the characteristic posterior deformity. It is so called from the accurate description of the disease given by Sir Percival Pott in 1783, although it was well known to Hippocrates, Galen, and the earlier writers on medicine, and also described by Camper and Sévérin.

Synonyms. English, *Caries or Ostcitis of the Spine; Vertebral Arthritis; Vertebral Tuberculosis; Angular Curvature; Posterior Curvature.* Greek, *Kyphosis.* French, *Mal de Pott; Cyphose.* German, *Spitzbuckel; Winkelförmige; Kinkung der Wirbelsäule.* Most of these are too precise, pathologically considered, or obviously contradictory—as the term "angular curvature," which from a geometrical point of view would correspond to a round triangle, or a square circle—hence it would seem best to retain the extensively employed "Pott's disease," or, if a more scientific designation is required, to substitute *spondylitis*, as being the least open to objection.

Frequency. Authors differ in their opinions in regard to the influence of sex in the etiology of this disease; those who believe in a traumatic origin consider it more frequent in boys from their presumed greater liability to injury. Schreiber makes the statement that "boys are, in general, much more likely to become affected than girls," and quotes the statistics of Nebel, who, out of 54 cases, found 31 male and 23 female. On the other hand, Fisher found a greater frequency among females; out of 500 cases treated at the National Orthopedic Hospital, London, 261 were females and 239 males. Mohr found out of 137 cases, 69 were males and 68 were females. Gibney found out of 2455 cases, 1329 males, 1126 females. Taylor, out of 412, 234 males and 178 females. Bradford and Lovett, at the Children's Hospital, Boston, out of 294, 152 males and 142 females. From the combined statistics collected from these sources there were 3989 cases, of which 2145 were males and 1844 were females. Allowing in this estimate for the preponderance of females over males in the population, Treves considers that it would make the disease appear equally common in both sexes, and correct the erroneous impression of a greater prevalence among males. Dr. A. Sydney Roberts expresses the same belief, and in this the writer concurs. The frequency of this disease in surgical practice is exhibited in the 2292 orthopedic cases treated at the Orthopedic Dispensary of the Hospital of the University of Pennsylvania, 346 or 19 per cent. of which were cases of this affection. From collected statistics it would appear that the relative frequency with which the disease attacks different portions of the vertebral column is: 1, dorsal; 2, lumbar, and 3, cervical.

The relative frequency as regards other joints is shown in the following: Pott's disease, 346; hip-joint disease, 298; knee-joint disease, 67; ankle-joint disease, 19; shoulder-joint disease, 2; elbow-joint disease, 6; wrist-joint disease, 2.

In the 1444 cases of deformity treated by Hoffa, 142 were cases of spondylitis, or 9.83 per cent., and these were taken from 67,919 cases of all kinds, of which Pott's disease formed 0.21 per cent.

The erect position of the human body is a factor in the production of this affection. It is asserted that spinal caries does not occur in quadrupeds. Davy[1] suggests that caries "is possibly one of the penalties we pay for walking in the upright position," while Albrecht[2] assumes that the upright position is the chief cause. Mohr, in 56 autopsies of osteitis of the spine, found the affection most common in the dorsal region (33 in 56 cases), next in the lumbar region (27 times), and next in the cervical (12 times).

In one hundred consecutive cases taken in order of attendance from my case-books, the following relative frequency was observed: 14 cervical, 50 dorsal, and 36 lumbar.

Redard[3] in a series of 100 cases observed at his service at the Dispensary Furtado-Heine, found the following: 6 cases in the cervical region; 5 cases in the cervico-dorsal region; 62 cases in the dorsal region; 5 cases in the dorso-lumbar region; 20 cases in the lumbar region; 2 cases in the lumbo-sacral region.

In regard to the individual vertebræ affected, Billroth and Menzel found the order of frequency as follows: First and second cervical;

[1] Schreiber: Allg. u. spec. Orthop. Chir., 1888, p. 93.
[2] Brit. Med. Journ., 1885, ii. 8-10.
[3] Traité de Chirurg. Orthopédique, 1892, p. 232.

sixth, fourth, and eighth dorsal; fourth and fifth lumbar; tenth and ninth dorsal; and the third cervical.

Etiology. Pott's disease occurs at all periods of life; it is seen in infancy, youth, adult life, and extreme old age; Bryant has even described a case occurring in the fœtus. By far the great number, however, occur from three to fourteen years. Gibney found 87 per cent. under fourteen years of age, 7 per cent. between twelve and fourteen years of age, and 4 per cent. over twenty-one years of age.

The following table from Hoffa illustrates well the relative frequency at all ages:

Period.	Mohr: 72 cases.	Drachman: 161 cases.	Taylor: 375 cases.
1 to 5 years	29 per cent.	41 per cent.	60.3 per cent.
6 " 10 "	22 "	36 "	18 "
11 " 15 "	20 "	13.7 "	6.4 "
16 " 20 "	16.7 "	5 "	
After 20 "	11 "	43 "	

The disease may be limited to one vertebra, or five or more may be affected. Thus, in the 81 cases collected by Bouvier, in 31 cases one or two vertebræ were affected; in 26 cases three, four, or five; in 24 cases more than five.

Age is, therefore, a predisposing cause, while sex appears to exercise but little or no influence. It is particularly frequent among the scrofulous, or those suffering from the condition known as strumous diathesis—a condition which, irrespective of external physical appearance or hereditary antecedent, renders the system peculiarly prone to chronic catarrhs, chronic inflammations of bones, glands and skin, retrogressive in character, occurring without adequate cause, and singularly liable to tubercular infection; those in whom there is, in other words, a constitutional predisposition to caseation, or to a tuberculosis of irritated parts. This is particularly well shown in the association of Pott's disease with strumous and tubercular affections in other parts of the body, such as "white swelling," caries or necrosis of bone, phthisis, etc., and in the antagonism, pointed out by Treves, which exists between such strumous disorders, by which two such affections are seldom manifest at the same time.

Scrofulosis, or the tubercular diathesis, and tuberculosis are, then, very important factors in the etiology of this affection, the former

being the predisposition, the latter the actual infection.   In 185 cases
examined by Gibney, an hereditary tuberculous taint was found in 76
per cent.   Again, while the diathetic condition is important as a predisposing etiological factor, a history of traumatism is usually presented as a direct exciting cause; with this predisposition to chronic
inflammation present in the system, a slight injury, or an undue use

FIG. 1.

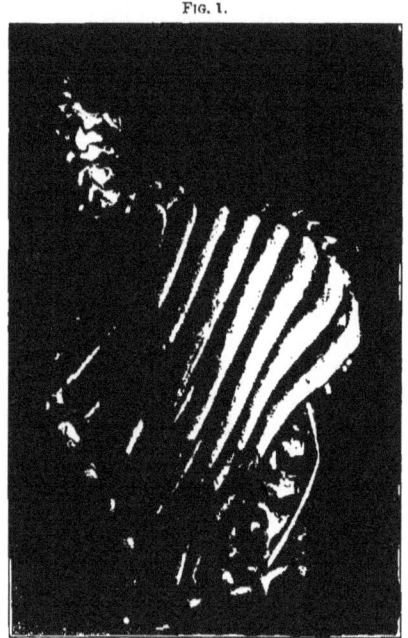

Photograph from specimen of dorsal Pott's disease, showing kyphosis and deformity of thorax.

of or pressure upon certain parts, may initiate the disease.   Thus, Taylor in 845 cases found a history of preceding trauma in 53 per cent.
It has, however, been the writer's experience to meet many cases of
extensive disease in which no such cause could be elicited; in which
definite relationship between trauma and disease was so obscure that
he could give but little credit to the former as an exciting etiological
factor.   It may be, then, that the various exanthemata, whooping-
cough, and other depressing diseases of childhood are more directly
responsible, or that the disease may occur without appreciable (exciting) cause.

Pathology. The pathological lesion peculiar to Pott's disease is a destructive osteitis terminating in interstitial absorption or caries, affecting the cancellous structure, and especially the anterior portion of the body; it may involve only one, or several, vertebræ. As a tubercular lesion, it does not differ from the tubercular osteitis occurring in the epiphyses of long bones, and consists essentially of a softening or medullization of the bone tissue, the various steps of the

FIG. 2.

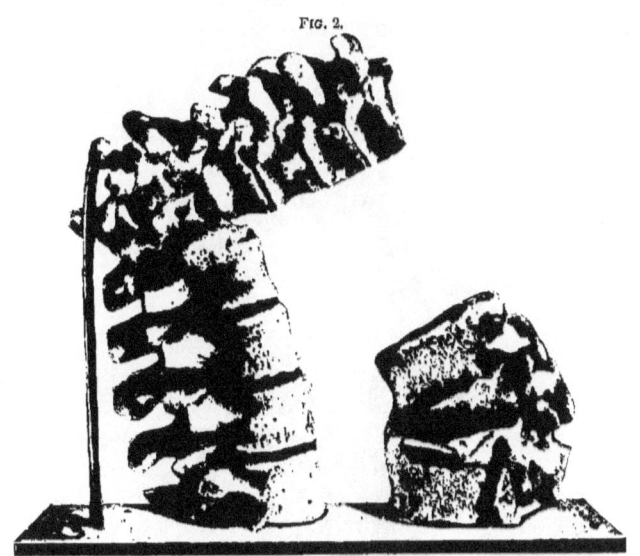

Photograph from specimen of dorso-lumbar Pott's disease, with section of vertebræ showing absorption of bodies.

process (congestion, formation of granulation tissue, and degeneration and softening of the new formations, with pus-formation, caseation, and bone absorption) following each other in slow succession. The macroscopic appearance of an area of tubercular osteitis is that of a large spot of hyperæmia surrounding an opaque or grayish centre, with possibly a spot of cheesy degeneration in its centre. Microscopical examination reveals numerous tubercles, and in them are discovered the tubercle bacilli. The bacillus tuberculosis is rarely or never present in the pus discharged from abscesses or sinuses. Instead of being limited to the bodies of the vertebræ the intervertebral fibro-cartilages and adjacent soft structures may be involved, or, in exceptional cases, it may be confined to the inter-vertebral sub-stances.[1] In that class of cases in which the disease terminates in interstitial absorption (caseation without suppuration), designated as dry caries, or *caries sicca*, many vertebræ are generally involved, and the marked angular deformity is replaced by posterior curvature. In the majority of cases suppuration occurs, and true caries exists; abscesses form and find exit in various situations according to the location of the affected vertebræ; frequently large sequestra are entirely cut off by areas of granulation tissue, producing the so-called *caries necrotica*.

FIG. 3.

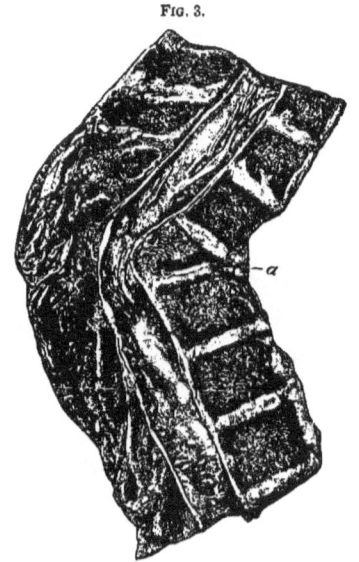

Pathological illustration of Pott's disease. (KRAUSE.)

The pathological changes in the membranes and cord in the paraple-gia which accompanies Pott's disease have been studied by Michaud, Courjon, Charcot, and Echeverria. According to these writers, the disease is seldom the result of direct pressure, but ordinarily begins as a pachymeningitis externus, with extension by contiguity or irrita-tion. The result of this is a thickening of the membrane, compres-sion of the cord, and the establishment of a compression-myelitis,

[1] Reeves: Bodily Deformities and their Treatment, Philadelphia, 1885, p. 126.

which is the cause of the paraplegia. Compression may also be induced by the direct pressure of the vertebræ, obliteration of the canal, caseous deposits, sequestra, or abscess. When the violence of the process has exhausted itself, the vertebral bodies and arches become consolidated and protected against sudden dislocation by a formative or osteoplastic osteitis which, beginning early by the development of osteophytes from and in front of the bodies of the vertebræ, has locked together and fused them into firm, bony ankylosis.

Symptomatology. Taken together the symptoms of well-established Pott's disease are so characteristic that a diagnosis is possible almost at a glance, and yet few diseases in their incipiency present such variations under so many phases as does this affection. There is present in the majority of cases a premonitory stage, often unobserved, the symptoms of which are included in the comprehensive term *malaise*—a condition between vigorous health and debilitating sickness, a want of energy, an irritability, a lowering of all the vital functions, an inactive calm which precedes and premonates the coming storm.

Among the first and most important symptoms is a state of spasm or muscular rigidity of the spine. This is an early, always present, and persistent sign, due either to reflex muscular spasm similar to that constantly found associated with joint disease elsewhere, or an unconscious, automatic effort of the patient to avoid motion or prevent and diminish jar in the affected vertebræ; is a reflex, tetanoid spasm, accompanied by a specific atrophy, especially of the erector spinæ group of muscles, and exhibits itself in the peculiar attitudes assumed, in the diminished normal flexibility of the spine, and in the slight lateral deviations of the column sometimes met with in this disease. The attitudes assumed by sufferers from this affection are characteristic of the different parts affected. In the cervical region the most common attitude is one of wry-neck. In the cervico-dorsal region the neck is pushed forward and the chin elevated, the shoulders are drawn up, the spine below the diseased area being straight, or in a condition of lordosis. In the dorsal region the spine is curved forward above and below the seat of disease, the shoulders are elevated, the body is shorter than normal, and the rigid spine gives a military attitude. In lumbar caries when the psoas is irritated, or a psoas abscess is present, the patient stands upon one leg, the thigh of which is flexed, the body bent forward and one hand resting upon the knee. Children with cervical caries, when fatigued, grasp the head with the hands about the sides of the face; in dorsal caries they rest the hands upon the hip, or in sitting on a chair, upon each side, or leaning forward rest both hands upon the thighs.

Though cases are recorded where pain is entirely absent throughout the entire course of the disease, it is usually a prominent, distressing symptom; the fallacy of the diagnostic value of local pain

FIG. 5.

FIG. 4.

FIG. 6.

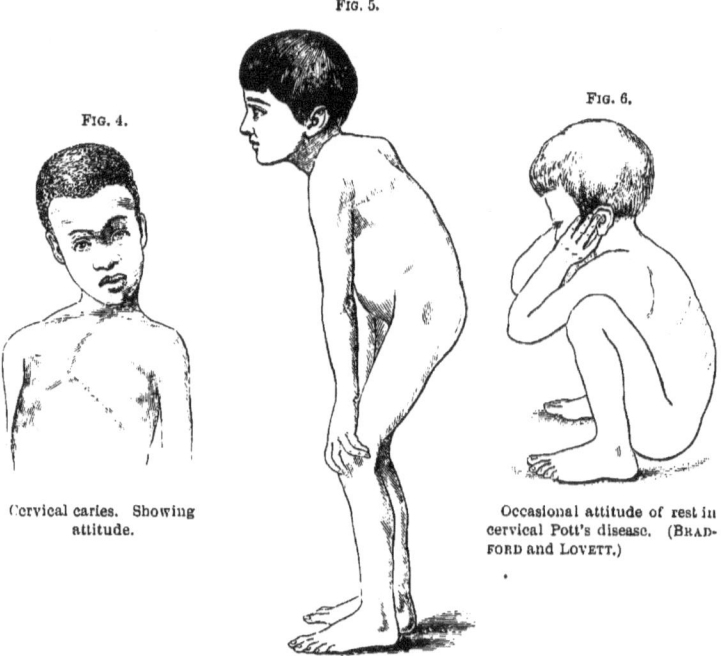

Cervical caries. Showing attitude.

Occasional attitude of rest in cervical Pott's disease. (BRAD-FORD and LOVETT.)

Attitude of rest in dorsal caries. (HOFFA.)

is now fully recognized. When present it is usually deep-seated, dull, subacute, and intermittent, and as a rule experienced at the peripheral distribution of the irritated nerves, either at or below the affected spot, very rarely above it, excepting in some cases of cervical caries. It may be subacute, intense, and lancinating, accompanied with hyperæsthesia, or may only amount to an irritation. Thus, a stiff neck, laryngeal irritation, gastralgia, pulmonary, intestinal, gastric, or cystic troubles, are frequently peripheral symptoms of a spinal caries.

FIG. 7.

Disease of the cervical vertebræ.

FIG. 8.

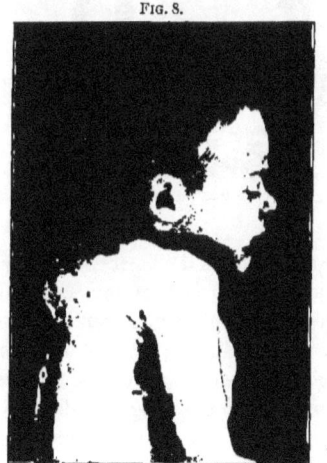

Upper dorsal Pott's disease.

Torticollis, occipital neuralgia, bronchitis, pneumonia, cystitis, and gastralgia are diseases frequently treated for this affection, and in one remarkable case lateral lithotomy was performed for supposed stone in the bladder, the autopsy of which exhibited lumbar caries. The writer has had related to him a case of caries supposed to be torticollis, where an attempt to suddenly correct the deformity resulted in sudden death. "Night cries" are of infrequent occurrence, but a peculiar "grunting" sound is frequently emitted by sufferers from cervico-dorsal caries.

FIG. 9.

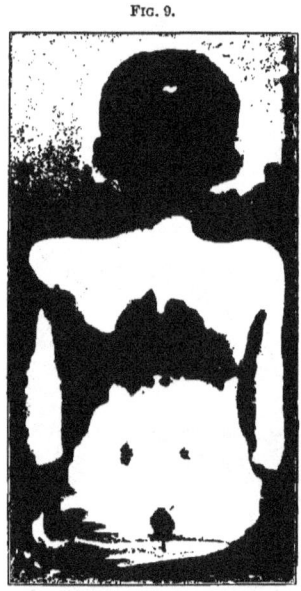

Mid-dorsal Pott's disease.

The posterior angular curvature is the most conspicuous, and often, especially in hospital practice, the first symptom which directs attention to the real seat of the disease; it results from the breaking-down of the vertebral bodies, the giving way of the anterior support of the column, and the projection backward of the spinous processes. As pointed out in speaking of the pathology of this affection, when but one or few bodies are involved, it is sharp and angular; when many, a long and gradual curve, the latter cases being usually of the *caries sicca* variety; when *angular* and *median* it is a positive sign of caries. From the anatomical construction of the vertebræ the deformity will reach its greatest degree of development, and be most conspicuous, in the dorsal region.

Angular curvature may be, and frequently is, absent in the cervical and lumbar regions. Slight lateral deviation is present in certain cases, especially the dorsal or dorso-lumbar regions, dependent upon undue muscular contraction or unilateral destruction of the vertebræ.

FIG. 10.

Lumbar Pott's disease.

The development of the deformity is usually gradual, but cases of rapid or even sudden appearance have been recorded, and indicate an active and progressive disease, or some sudden and unwonted action. With the occurrence of the angular projection in the dorsal region, the configuration of the anterior portion of the chest is likewise altered, amounting to a lateral flattening and "pigeon-breast."

The disease may run its entire course without suppuration, but in the majority of cases abscesses form a constant and important complication. In the former, caries sicca is the pathological lesion, and a "residual abscess" results, as pointed out by Paget.[1] As to the frequency of abscess formation, statistics vary. Nebel, in 54 patients suffering from Pott's disease, found 24 abscesses, and according to Taylor, abscesses developed in 14 per cent. of cases of kyphosis. In 61 post-mortem sections upon spondylitic patients, Mohr found 30 abscesses, and in 82 sections Nebel found 24 abscesses. The relative frequency of abscess in the different regions is well shown in the 183 cases of spondylitis reported by R. W. Parker,[2] in which about 8 per cent. of the cervical, 30 per cent. of the dorso-lumbar, and 70 per cent. of the lumbo-sacral cases suppurated. In general features the suppurations that accompany are identical with the cold abscesses that result from caries elsewhere, and with the formation the usual apyretic course may be interrupted by an elevation of temperature, even reaching 105 degrees, as recorded by Schaffer. As a rule, however, there is but little constitutional disturbance—a slight rise of evening temperature, slight rigors, and perspiration. Locally, however, abscesses may occasion much disturbance from pressure and distention of surrounding parts, or may assume great size and remain stationary for long periods, especially in children, without occasioning inconvenience; these collections of pus follow the fascias in the direction of least resistance, and usually open at some distance from the seat of origin ; both the direction and termination will be determined by the region of the spine affected. The anatomical importance of the fasciæ in protecting important parts and organs is evident. Treves, Jacobson, Hilton, and others, have called especial attention to the great importance of the cervical fascia, and Koenig, Heuke, Soltmann, and others, have pointed out the importance of the fasciæ generally in determining the direction and pointing of abscesses from whatever cause.

Abscesses in the cervical region, when they open externally, pass between the longus colli and scaleni muscles to appear posterior to the sterno-cleido-mastoid ; they may, however, open into the posterior wall of the pharynx as a post- or retro-pharyngeal abscess, or may burrow beneath the deep fascia into the thorax and form a mediastinal abscess, discharging finally into the trachea, œsophagus, or through an intercostal space. In rare instances, as in a recent case of the writer's, the pus may penetrate the pleura and form an empyema.

[1] Clin. Lect. and Essays, London, 1887.
[2] Brit. Med. Journ., January 12, 1884, p. 58.

In the dorsal region they burrow posteriorly to open on the back or side a short distance from the spine, or gravitate beneath the

FIG. 11.

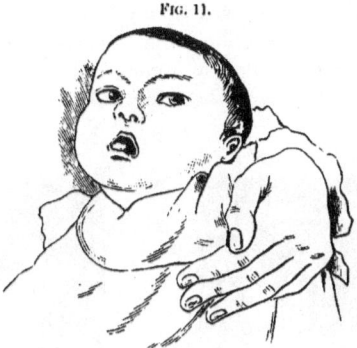

Retro-pharyngeal abscess: attitude and characteristic expression. (BRADFORD and LOVETT.)

ligamentum arcuatum internum within the sheath of the psoas magnus muscle, and beneath Poupart's ligament, to appear externally in Scarpa's triangle as a psoas abscess.

FIG. 12.

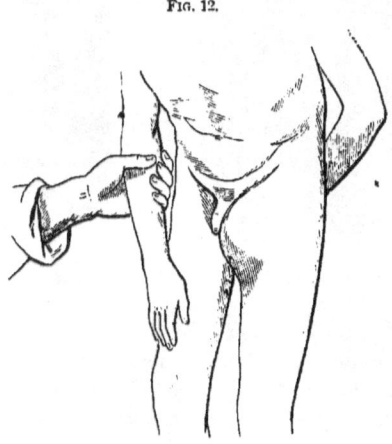

Psoas abscess. (HOFFA.)

In the lumbar region, owing to the peculiar anatomical conditions, purulent collections vary much in their course and exit. It will be

3

observed in examining a sectional diagram of the lumbar region that the sheath of the psoas muscle and the lumbar fascia are the most important structures in this connection. The sheath of the psoas is a thin, fibrous membrane derived from the iliac fascia, attached above to the ligamentum arcuatum internum, laterally by a series of arched processes to the intervertebral substance and prominent margins of the bodies of the vertebræ, becoming below continuous with the

FIG. 13.

FIG. 14.

Photograph of a case of bilateral lumbar abscess, showing sinuses.

Same case as former, showing posterior deformity and lateral sinuses.

iliac fascia. The lumbar fascia divides into three layers, enclosing the quadratus lumborum, multifidus spinæ, and erector spinæ muscles, and giving attachment to the internal oblique. Its anterior and middle layers are attached to the transverse processes, and its posterior layer to the spinous processes. Above, its anterior layer is attached to the lower border of the last rib, forming the ligamentum externum. The posterior surface of the psoas muscle is separated from the quadratus lumborum by the anterior layer of the lumbar fascia—a very thin fascia—but the greater part of the muscle is firmly supported behind by the erector spinæ muscle. In front, the quadratus lumborum is very thin and offers but little resistance to the exit of the pus. The sheath of the psoas and the lumbar fascia, thin as they are, nevertheless are essential factors in determining the direction, and with the direction the prognosis, of lumbar abscesses. So important do these fasciæ appear to the writer that

FIG. 15.   FIG. 16.

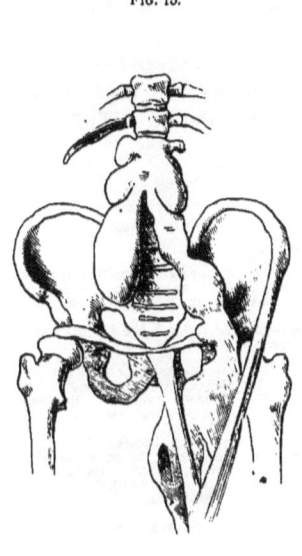

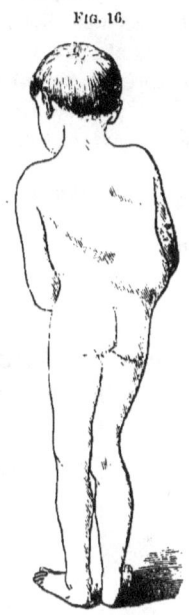

Lumbar and psoas abscess. (HOFFA.)   Lumbar abscess. (HOFFA.)

he has suggested a division of lumbar abscesses into *external abscess* and *internal abscess*, their relation to the sheath of the psoas fascia and the anterior layer of the lumbar fascia determining their position.[1] They may pursue the course and terminate similarly to the psoas abscess; may burrow backward and laterally along the middle layer of the lumbar fascia, separating the quadratus lumborum from the internal oblique, through the internal oblique and between the external oblique and latissimus dorsi, to appear at the outer border

---

[1] Trans. Amer. Orthop. Assoc., 1891, p. 177.

of the erector spinæ muscles, constituting lumbar abscess; may gravitate beneath the internal iliac muscles over the posterior brim of the pelvis, perforating the great sacro-sciatic foramen as a gluteal abscess; or, if the abscess find exit upon the surface of the vertebræ anterior to the attachment of the psoas fascia, it will terminate by burrowing along the great vessels, or become an iliac abscess, to again terminate as a gluteal abscess. After becoming an iliac abscess, the pus may rupture into the intestines, bladder, vagina, or rectum, and Owen has recorded a case of fistula-in-ano which had resulted from a spinal abscess.

These represent the usual classical courses, but cases are reported in which the pus has travelled great distances and discharged into the viscera or external parts remote from the seat of origin.[1]

TABLE OF SPINAL ABSCESSES.

Variety.	Course.	Exit.
Cervical	*a.* Anterior.	Into posterior wall of pharynx.
	b. Burrow beneath deep fascia into thorax as mediastinal abscess.	Into trachea, œsophagus, or through an intercostal space.
	c. Laterally between the longus colli and scaleni muscles.	Posterior to the sterno-cleido-mastoid.
Dorsal	*a.* Burrow posteriorly.	On the back or side a short distance from the spine.
	b. Within psoas sheath.	Beneath Poupart's ligament in Scarpa's triangle.
Lumbar	*a.* Enter psoas sheath.	As psoas abscess.
	b. Burrow between the fasciæ of the quadratus lumborum and abdominal muscles, through the internal oblique.	Posteriorly beneath the external oblique and latissimus dorsi at the outer border of the erector spinæ muscle.
	c. Gravitate beneath the internal iliac muscles over the posterior brim of the pelvis, perforating the great sacro-sciatic foramen.	As gluteal abscess.
	d. May be directed to the iliac region along the aorta and external iliac arteries.	As gluteal abscess.

The paraplegia of the lower extremities which so frequently accompanies and complicates, especially in the cervico-dorsal region, must be distinguished from that which results from compression of the spinal cord in cases of extreme distortion; it involves generally only the motor area of the cord, and occurs in the acute stage of caries from reflex paralysis due to some obstruction to the blood supply, or to communication of the inflammatory action by contigu-

[1] *Vide* Treves, Internat. Encyclop. of Surgery, vol. iv. p. 931.

ity from the seat of the disease to the peri-meningeal areolar tissue and membranes of the spinal cord—a pachymeningitis, or meningo-myelitis. The symptoms are those of a compression myelitis—gradual loss of power, with increased reflexes, exaggerated patellar reflex and increased ankle clonus in the early stage, followed by complete loss of power, contracture of muscles, atrophy of paraplegic parts, and loss of sensation in the later stages. The general health is but little affected, and the bladder and rectum are not disturbed, except when the lumbar portion of the cord is involved, or toward the end in very severe cases. The average duration before the appearance of paralysis is about three years, although it has appeared as early as four and a half months, and as late as eleven years. In the 59 cases reported by Taylor and Lovett, the duration was never over three years, except in 1 case, where it lasted for six years. Recurrence was observed in 6 cases, 4 cases having two attacks, and 2 having three attacks. It occurs in any number of consecutive cases in about the same proportion (being somewhat modified by early treatment), but much more frequently complicates the disease in the upper than in the lower part of the canal, for obvious anatomical reasons. Thus, in the analysis of 295 patients suffering from Pott's disease, Gibney[1] found that paralysis occurred in 62 in the course of the affection, 59 of these complicating 189 cases of disease in the upper dorsal and cervical regions, and only 3 complicating 106 cases of the affection in the lower portion of the canal. Recovery is the rule under efficient treatment, even when sensation has been lost, and when it occurs is generally complete.

Diagnosis. The importance of an early diagnosis cannot be overestimated; to this end the examination, both oral and physical, should be most thorough and painstaking. The entire back should be exposed in good light, and the flexibility of the spinal column tested, either standing, by the method of Adams (by placing the heels together, the lower extremities extended, and the body flexed as far forward as possible) or prone upon a hard couch. In children the latter position is the better. The surgeon places his left palm upon and fixes the pelvis, and grasping the feet with the right hand, flexes the knees, and ascertains the amount and range of flexibility. In small children a high degree of flexibility should be present. In obscure cases this examination should include the inspection of the

---

[1] Journal of Nervous and Mental Diseases, April, 1878, p. 254.

pharynx, accurate measurements, and electrical and other reactions of the extremities, examination of the major articulations and the tem-

FIG. 17.

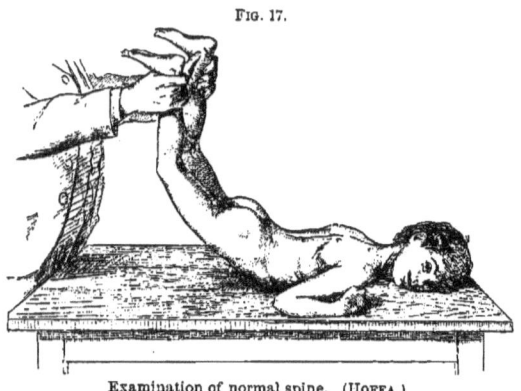

Examination of normal spine. (HOFFA.)

perature. Finally, it should consist essentially of a differential study (1) of the spinal deformity, (2) of the cord and nervous symptoms, and (3) of the abscess, or of all three conditions if present.

FIG. 18.

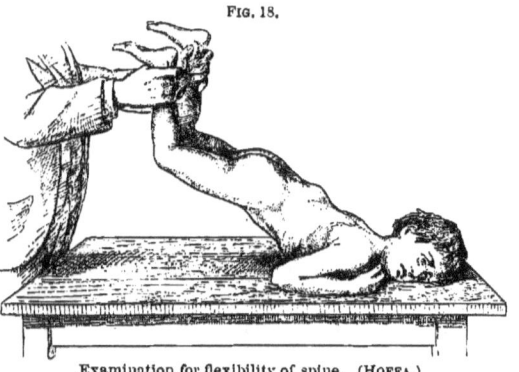

Examination for flexibility of spine. (HOFFA.)

The kyphosis is characterized by its angularity, median position, and rigidity; so marked are these peculiarities that confusion with lateral curvature is not likely to ensue, except in those cases of Pott's disease where there is marked lateral deviation. These, as pointed out before, are chiefly muscular, and are accompanied by an exacerbation of all the symptoms, which with the rigidity of the spine would serve to distinguish them. Aneurisms of the thoracic and abdominal aorta eroding the vertebral bodies give rise to symptoms characteristic of caries, as in the cases recorded by Quincke[1] and Roberts;[2] but in addition, the symptoms of aneurism would be associated, and the extensive curve, localized pain, and late period of life at which they occur, would render the diagnosis patent. The same considerations would eliminate carcinoma, sarcoma, and other malignant growths. Rheumatoid arthritis of the spine, or spondylitis deformans, is characterized by its occurrence late in life, stiffness and arching of the spine, and the absence of kyphosis, abscess, and muscular spasm. From functional spinal debility, rhachitic curve, etc., with which it is sometimes confused, a differential diagnosis is readily made, and need not be dwelt upon. In young children muscular rigidity may be present in rhachitic curves to a marked degree, but the curves are longer, less angular, more flexible, and *always* associated with other evidences of rickets.

The cord and nerve symptoms will, in the majority of instances, be found associated with marked kyphosis, or at least rigidity of the spine. In those rare cases in which paraplegia occurs without angular deformity, the latter symptom, and possibly a purulent collection, will assist the diagnosis. It is, however, in those neuromimeses, or so-called "hysteria of the spine," in which the greatest difficulty is encountered, and especially when associated with "hysterical paraplegia;" here the hysterical simulation closely resembles the genuine disease, but the pain is localized posteriorly and apparently acute; there is no reflex spasm; the muscular rigidity yields under gentle, firm pressure; and the paraplegia is usually sudden. It occurs usually in young neurotic women of pronounced brunette type, and there is usually associated ovarian tenderness, the globus hystericus, and other evidences of this condition. From hysterical, hyperæsthetic or neuromimetic spine the same conditions would aid in differentiation; but the absence of bony deformity, the exaggerated localized tenderness and pain, and, as a rule, the absence of real muscular spasm will render the nature of the affection patent. In severer cases, and after railway injury, or ligamentous traumatism from any cause, excessive muscular spasm is induced by flexion. In a recent case of this character without medico-legal interest, in an

[1] Dis. of Arteries: Ziemssen's Cyclop. of Med., vol. vi. p. 434.
[2] Pott's Disease: Cyclop. of the Dis. of Children, vol. iii. p. 1032.

eleven-year-old girl, the history of injury and subsequent suffering, the absence of deformity, the excessive spasm and atrophy of the erector spinæ group of muscles, rendered the diagnosis clear. From muscular rheumatism, lumbago, etc., it is distinguished by a history of associated rheumatic affections, the diffuseness and extent of the pain, and the later period of life at which the rheumatic affections occur.

Caries of the last lumbar vertebra is often mistaken for osteitis of the hip, and *vice versa*.[1] The differential diagnosis is at times exceedingly difficult to make, and for details the reader is referred to the chapter upon Hip-joint Disease. Sacro-iliac disease may be distinguished, if occurring in young adults, by the position of the swelling over the sacro-iliac joint, pain caused by pressure of the sides of the pelvis together, and the absence of lumbar spinal deformity. The differential diagnosis of lumbar Pott's disease from hip disease and sacro-iliac disease is well shown in the accompanying table :

	Coxalgia.	Lumbar Pott's disease.	Sacro-iliac disease.
Occurrence . . .	Children under 12 years.	A disease of childhood.	Young adults.
Pain	In knee or hip-joint.	Referred to peripheral distribution of lumbar nerves.	Localized over sacro-iliac joint.
Limb	Apparent elongation of limb, adduction followed by real shortening.	No elongation of limb.	Elongation only of limb.
Buttocks . . .	Early loss of gluteo-femoral fold	No change in buttocks.	No change during early stage. Late obliteration of gluteo-femoral fold.
Movements . .	Limited flexion and extension first sign.	No restriction of movements in hip-joints.	No restriction of movements when pelvis is fixed.
Pressure . . .	Pressure on sides of pelvis without pain.	Pressure on sides of pelvis without pain.	Pressure on sides of pelvis causes severe pain.
Deformity . . .	Swelling and induration about trochanter.	Angular median deformity of lumbar spine.	Swelling over sacro-iliac articulation.
Abscess . . .	Abscess sinuses lead to hip-joint.	Abscess sinuses lead to lumbar spine.	Abscess sinuses lead to sacro-iliac joint.

Infantile paralysis may be distinguished from the paraplegia of Pott's disease by the history, absence of rigidity or pain in the vertebral region, the muscular weakness and atrophy, and especially in cases of doubt, by the electrical reactions in the affected muscles, the

---

[1] *Vide* Gibney, Diseases of the Hip, 1884, p. 301.

description of which will be found in full in the chapter upon Infantile Spinal Paralysis.

The abscess accompanying Pott's disease may be diagnoticated from purulent collections from other causes and neoplasms in general. In the cervical region it is liable to be mistaken for simple abscess and adenitis. The former are usually acute, attended with fever, and superficial in character. In the latter the inflammation is circumscribed, deep-seated, and unattended with the characteristic spastic sensation.

In the dorsal region, chronic pleurisy with effusion or empyema, other inflammatory pulmonary affections, and malignant growths, are to be distinguished by the physical signs. In the lumbar region abscess must be distinguished from simple chronic abscess, abscess of lumbar glands, abscess from caries of the ilium, perinephritic and pericœcal abscesses. The condition of the psoas muscles will best indicate the presence or absence of pus within its sheath. In simple chronic abscess the symptoms of systemic disturbance are marked rigor, hectic fever, night-sweating, while those of vertebral caries are negative. Perinephritic and pericœcal abscesses are distinguished by the marked local symptoms, some disturbance of the organ about which the pus has collected, the condition of the psoas muscles, and the absence of all the characteristic manifestations of Pott's disease. When the abscess opens into and is discharged from the vagina, an attempt may be made to distinguish it from the blennorrhœa infantilis by the examination of the pus for bacilli tuberculosis (which, in all probability, cannot be found), and for the gonococci, which however, it must be remembered, cannot always be distinguished from other bacteria, particularly diplococci.[1]

The psoas abscess pointing below Poupart's ligament to the outer side of the femoral vessels may be mistaken for the abscess of hip disease, femoral hernia, cancerous tumor,[2] bubo, fatty, cystic, and other fluctuating tumors; varicose saphenous vein, aneurism, hydrocele of the inguinal canal, undescended testicle, etc.; the differential diagnosis of which may be rendered clear by the characteristic signs of each of these affections, and the absence of the positive signs of spinal caries. The difference between the abscesses of hip disease and spondylitis is well shown by a comparison of the tables of each.

Progress and Prognosis. Contrary to the generally accepted idea, many patients recover from this affection, there being a natural

---

[1] *Vide:* Vibert and Bordas, La Medicine Moderne, Nov. 13, 1890.
[2] Shaw : Holmes' System of Surgery, Am. ed., vol. iii., p. 313.

tendency thereto. The cures achieved in modern times may justly be attributed to early diagnosis, better knowledge of the etiology and progress of the disease, and the skilful adaptation of mechanical measures. The progress depends much upon the portion of the spine affected, and will be influenced by the amount of personal attention and care given.

Agnew,[1] though considering caries of the spine to be always a serious affection, does not, nevertheless, deem life seriously imperilled thereby. "Few children," says he, "who receive proper care perish from angular curvature; of the cases I have been able to follow I recall only two who have died from causes directly traceable to the spinal affection." The percentage of mortality in 269 cases collected from various sources was found to be about 27 per cent. They were as follows :

		Cases.	Deaths.
Billroth and Menzel .	.	61	23
Jaffé	.	82	22
Mohr	.	72	7
Nebel	.	54	19
		269	71

The deformity may be diminished in some cases by appropriate treatment, but once present it never entirely disappears. Nothing renders the prognosis more unfavorable than the occurrence of abscesses, especially when they exist for a long time and discharge at a point remote from the seat of disease. A fortunate result is most usual, however, when the abscess opens close to the affected vertebræ, and it has appeared to the writer that such cases do even better than where no abscess occurs. A cure may still follow after the abscess has opened and discharged for a considerable period, but abscesses which discharge for a long time are a drain on the vitality and a direct menace to life, owing to the ultimate amyloid and other degenerations of the internal viscera.

A "residual abscess" may become active at any, even a remote, time from the beginning of apparent cure of the vertebral lesion.

The paraplegia of Pott's disease, while a distressing and alarming complication, one which

FIG. 19.

JUNE 10 1867   MAR. 5 1879

A

B

Tracings from untreated case of spondylitis, at beginning (A) and twelve years later (B). (TAYLOR.)

[1] Principles and Practice of Surgery, 1881, vol. ii., p. 876.

materially lessens the prospects of cure, tends, as a rule, to spontaneous recovery, the average duration being a little less than one year. This is well shown in the analysis of fifty-eight cases by Gibney,[1] twenty-nine of which recovered, save one " still under treatment ;" again, Taylor and Lovett[2] report nineteen cases, of which seventeen recovered, one recovered partially; and one remained paralyzed ; and Sayre[3] has reported thirty-eight cases either partially or completely paralyzed, of which thirty-four recovered, and four remained under treatment. Relapses occur, but do not materially affect the prognosis. Exceptions to this favorable tendency are found in the paralysis in connection with caries, which is peculiarly apt to end in death, and where sensation and motion are both involved the restoration of the power of the limbs is only partial, or incurable paralysis persists. Even these cases are not hopeless, for cases of recovery after complete loss of sensation have been recorded elsewhere by the writer.[4]

A fatal issue may ensue from simple asthenia, the result of excessive suppuration, hectic pyæmia, amyloid degeneration of the internal viscera, intercurrent affections—tubercular or otherwise—rupture of an abscess internally, or from hemorrhage from perforation of a large bloodvessel.[5] An interesting case of the latter is recorded by Ashhurst,[6] where a psoas abscess caused ulceration and consequent rupture of a branch of the internal iliac artery, leading to rapid death. It may be safely said that more children perish from abscesses which open internally than from all other causes combined. Mohr found in 9 cases of fatal abscess perforation into the œsophagus in 2 ; pleura and lungs, in 2 ; pleura alone, in 1 ; peritoneum, in 1, and spinal canal, in 2. The prognosis as regards life, and the cause of death in persons cured of Pott's disease is well shown in the 31 specimens studied by Neidert,[7] at the Munich Pathological Institute. It was found that persons with slight deformity have as good a chance of life as normal individuals; persons with medium deformity die young of phthisis, while persons with severe deformity die of heart-failure or fatigue. The average age was forty-nine and a half years.

[1] Journal of Nervous and Mental Diseases, April, 1878, p. 254.
[2] Medical and Surgical Reporter, 1888, vol. ix., 136-139.
[3] Trans. Amer. Med. Assoc., Phila., 1879, vol. xxx., pp. 657-790.
[4] University Med. Magazine, vol. i., p. 348.
[5] Legouest: Gazette Hebdom., 1861, p. 76. Fuller: *Ibid.*, 1859, p. 524.
[6] Prin. and Prac. of Surg., second ed., 1878, p. 644.
[7] "Causes of Death in Deformities of the Vertebral Column." Inaugural dissertation, Munich, 1886.

Hypertrophy, with or without dilatation, of the right side was present in 24; muscular degeneration, in 4; stenosis of the mitral valve, in 2; acute miliary tuberculosis, in 1; phthisis, in 8; pneumonia, in 4, and carbuncle, in 1. In cervical caries death as sudden as in apoplexy may result where suitable support of the head has been neglected, as in two cases recorded by Little.[1]

The prognosis is more favorable in children than in adult cases. The tendency to cure is marked, varying with the resistance of the individual and the situation and extent of the disease. Of the 59 cases analyzed by Taylor and Lovett, 39 wholly recovered, 3 partially recovered, 5 died of intercurrent diseases, and in 12 cases the ultimate termination was unknown. Speaking in round numbers, three years may be said to be an average time for a course of treatment, and the patient should remain under observation from three to ten years.

Treatment. Equipped with modern improved mechanical means, one approaches the subject of treatment of spondylitis with greater confidence than earlier writers could possibly have done. In addition, practical surgeons appreciate the important rôle which improved hygiene, in its widest adaptation, plays. In considering this section of the subject, the general medical treatment will be considered, and only such mechanical measures as have been found in the hands of the writer capable of meeting the requirements in the majority of cases. Nature is engaged in a local germicidal warfare, and well-directed assistance may enable her to conquer.

Hygiene, generous and improved diet, sufficient sleep, proper clothing, salt baths, massage, etc., are important adjuvants, but *sunlight* and *fresh air* are the two most important agents required. In addition, measures should be taken to improve the general health by the use of tonics and alteratives; the selection of the particular drugs will depend upon the individual case and the judgment of the practitioner. Cod-liver oil and the hypophosphites are of the greatest value, and the writer has experienced the best results from the addition of strychnia and creosote, as in the formula made by Mulford & Co., of this city, as follows:

℞.—Quiniæ hypophos.	. 1 grain.
Ferri hypophos.	. 1/4 "
Calcii hypophos.	. 1/2 "
Sodii hypophos.	. 1/2 "
Potass. hypophos.	. 1/4 "
Manganii hypophos.	. 1/4 "
Strychniæ hypophos.	. 1/64 "
Creosote	. 1 gtt.

S.—For adult: One t. i. d. after food.

[1] Reynolds' System of Surgery, vol. iii., p. 320.

Recumbency as a means of treatment still has its advocates. Prolonged recumbency with withdrawal from sunlight and fresh air has a deleterious effect upon the health, and it has been suggested in a recent authoritative work that it also predisposes to tubercular meningitis.[1] Recumbency should be enjoined during the acute stage, and its evil effects may be obviated by the use of the "*stehbett*" of Phelps,[2] or what the writer most prefers, a light iron frame covered with canvas, to which the patient can be accurately fixed with straps and buckles, and which permits of his being lifted and carried readily about; such an oblong bed can be cheaply constructed of light gas-pipe or stout steel bars. An apparatus of this character permits also of extension, counter-extension, and fixation, a method of treatment of great advantage in cervical disease. Suspension as a mode of treatment is an old plan which has recently been brought into prominence; it was employed as early as 1826 by Prof. J. K. Mitchell;[3] independent of support, as a remedial agent in uncomplicated cases, it has no special value.

Complete suspension can only be employed temporarily in securing a better position of the trunk for the application of plaster-of-Paris or other fixative apparatus, and great care should always be exercised therein. It is now recognized that the deformity itself cannot be changed—the vertebral column above and below the gibbosity may be extended and straightened, but the diseased area remains fixed.[4] In fact, attempts to accomplish this have been followed by immediate paraplegia and even death.[5]

The machines and appliances employed are innumerable, but can all be classed in two groups : 1. The fixed jacket of plaster-of-Paris, and its modification in silicate of soda,[6] poroplastic felt[7] (Fig. 20), leather,[8] woven wire,[9] paper,[10] bamboo,[11] wood,[12] etc. 2. The spine brace.

The plaster-of-Paris jacket filled a long-felt want, and its vari-

[1] Treatise on Orthop. Surgery, New York, 1890, p. 58.
[2] Schreiber : Allg. u. spec. Orthop. Chir., 1888, p. 93. Bradford and Lovett.
[3] American Journal of the Medical Sciences, 1826.
[4] *Vide* Anders : Archiv f. klin. Chirurg., 1889, Band iii. p. 558.
[5] Mr. Willett : St. Bartholomew's Hospital Reports. Thomas Buzzard.
[6] Coover : Med. and Surg. Reporter, April 13, 1878.
[7] Cocking : Med. Press and Circular, 1879, N. s. xxviii. Walsham : Lancet, 1885, vol. i. p. 619.
[8] Agnew's Surgery, vol. ii. p. 880
[9] Roberts : Internat. Journ. of Surgery, vol. i., No. 4, p. 207.
[10] Med. Record, N. Y., 1887, xxxii. 647 ; N. Y. Med. Journ., 1886, xliv. 261.
[11] Slimshi : Tokio, Feb. 2, 1884, No. 305.
[12] Centralbl. f. Orthop. Chir. und Mechanik, Jan. 1889.

ous modifications have extended the range of its usefulness. The *technique* of its application is so well known that it need not be dwelt upon here. But the method as now employed must be briefly described. The best material for the rollers will be found to be fine

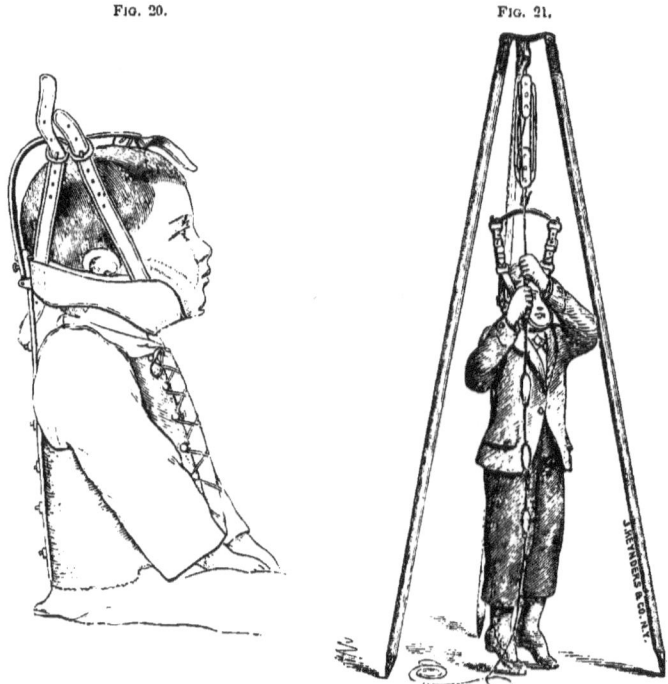

Fig. 20.   Fig. 21.

Felt jacket of Beely. (SCHREIBER.)   Tripod for suspension during application of the plaster jacket. (STIMSON.)

dental plaster, kept thoroughly dried, and cheese-cloth or butter-cloth. Stockinette in different sizes is more convenient than the knit shirts formerly employed, or an ordinary sleeveless woven shirt answers very well. In either case the garment must be a tight fit. The breasts must be protected with cotton, and in some cases a folded towel may be placed over the epigastrium as a "dinner pad," to be subsequently removed. The patient is to be suspended until only the toes touch the floor, and the bandages, thoroughly wetted by placing singly on end in warm water until the air-bubbles cease to rise from the submerged bandage, and squeezed nearly dry, are to be applied quickly from below upward and well rubbed between each layer. In this manner a thin, neat, and strong cast should be obtained. As soon as this is completed, the patient should be carefully placed in a recumbent position until the plaster has set, when the front should be cut down, the armholes trimmed, and the corset be bound and have the lace-hooks adjusted.

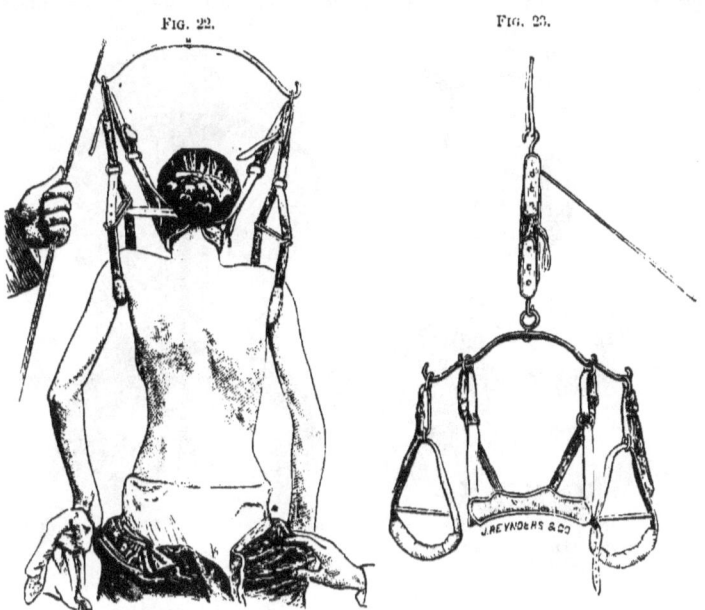

FIG. 22.

FIG. 23.

Patient suspended ready for the plaster. (STIMSON.)

Suspensory apparatus for application of the plaster jacket.

While possessing certain advantages of economy, requiring less special experience in its application, and being entirely beyond the control of the patient or attendants, there are certain positive disadvantages—the encircling of the body within a solid support; the inability to inspect the condition of the skin and note the progress of the affection; the formation of excoriations, ulcerations, and abscesses without the knowledge of the surgeon; the lack of cleanliness, etc.—which relegate it and its modifications to a secondary position. When split, furnished with lacings, and applied and removed at will, it loses part of its efficiency, but there is a gain in comfort and cleanliness. It is of decided value, however, for patients who are unable to bear the expense of even the cheapest apparatus; in such cases, with attention to detail, a cure may often be effected, and the writer has in public practice frequently proven this statement, especially when the disease was located in the lumbar region. In all cases of disease above the seventh dorsal vertebra, a chin-rest should be added to remove the superincumbent weight, and more

FIG. 24.   FIG. 25.

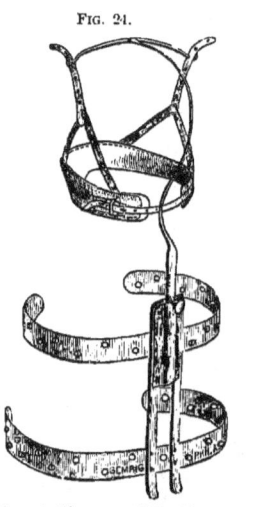 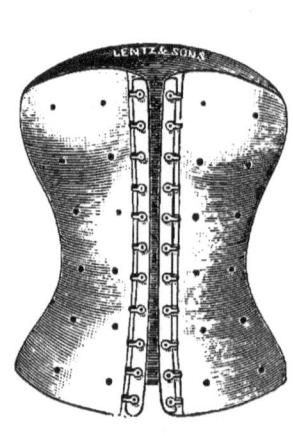

Sayre's "jury-mast" for disease   Perforated leather jacket for
of the cervical vertebræ.   Pott's disease.

particularly to extend and fix the diseased area. The "jury-mast" is the head-piece generally employed with the plaster jacket, but is open to the objection that, while it supports, it does not fix the head. This may be accomplished by attaching it to two uprights fitted to the back of the head; or, better, by the use of the chin-rest, to be described later under the spine brace.

In justice to this mode of treatment, it must be added that there are certain cases in which it is the best and most efficient; this applies particularly to the lateral deviation of the spinal column present in certain cases of Pott's disease and in lower lumbar disease. It is, moreover, important that the surgeon be familiar with both the plaster-of-Paris jacket and the spine brace, adapting each to the special requirements of the individual case. The leather corsets are prepared, over a counter-cast from the plaster-of-Paris jacket, from raw-hide, untanned leather, or saddler's skirting. As manufactured by surgical instrument factors they are ornate and durable, but entirely satisfactory leather jackets can be made by any practitioner, as suggested by Dr. Vance, of Louisville, Ky., and as repeatedly demonstrated by the writer to his classes. (Fig. 25.) The adjustable wooden corset is more difficult to manufacture, but may be made by almost anyone by attention to the details as given by the writer

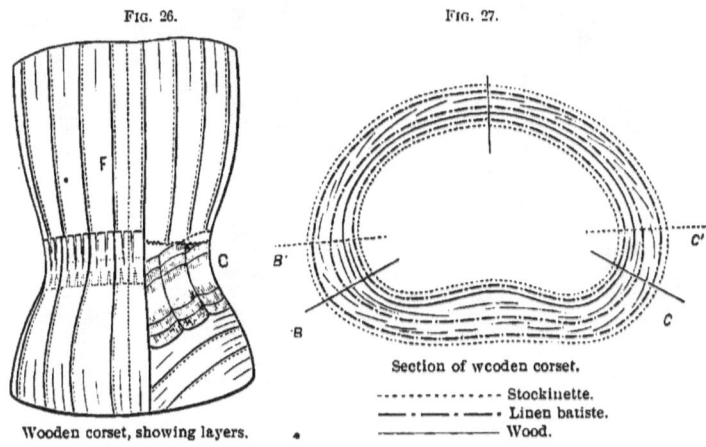

FIG. 26.   FIG. 27.

F

C

B'

C'

B

C

Section of wooden corset.

- - - - - - - - - - - - Stockinette.
— · — · — · Linen batiste.
——————— Wood.

Wooden corset, showing layers.

elsewhere[1] and as shown in the illustrations. It consists essentially of a wooden cuirass manufactured somewhat similarly to the felt and leather jackets, and composed of alternate layers of stockinette, wood, roller bandage, and linen, held intimately together with glue. (Figs. 26 and 27.)

The spine brace should act on the principle of a lever, the fulcrum being the diseased part of the spinal column, the weight at the waistband, and the power applied above to pull the part above the kyphosis back as far as possible. The writer has personally had the best results with the antero-posterior support as modified by Dr. C. Fayette Taylor, of New York, an apparatus constructed upon this principle. (Fig. 28.) It consists of a pelvic band, upon which are

[1] Annals of Gynecology and Pædiatry, August, 1891.

attached two uprights of the best annealed steel, admitting of easy manipulation and bending, connected with two transverse bars from which arise two shoulder-pieces. These uprights are separated

FIG. 28.   FIG. 29.

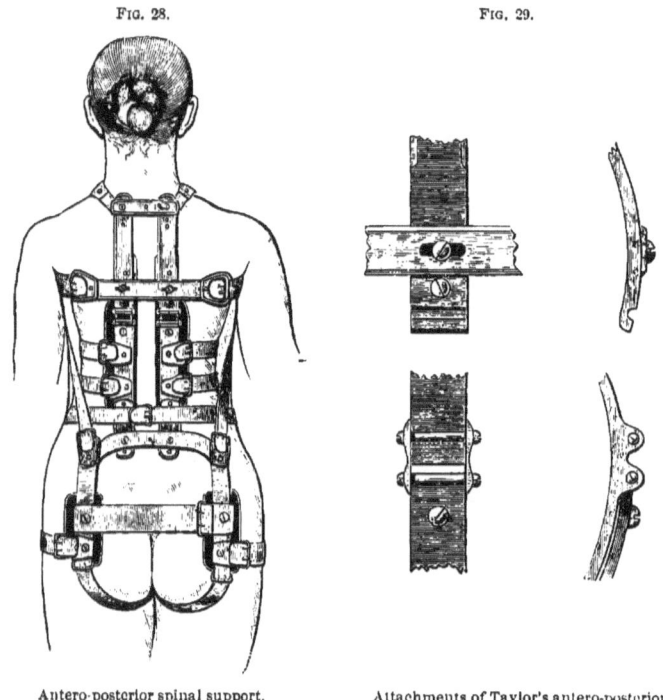

Antero-posterior spinal support.   Attachments of Taylor's antero-posterior
(TAYLOR.)   spine support.

sufficiently to rest, when applied, upon the transverse processes throughout the greater part of the spinal column and at the seat of deformity; extending some distance above and below are the pad plates (Fig. 29); the latter are pieces of softer steel fastened to the uprights by hinges and screws, admitting of ready removal and bending, and best covered with hard rubber or ground cork enclosed in canton flannel. The shoulder-pieces are provided with covered webbing straps, and the transverse bars and pelvic band with buckles. The apparatus is securely held in position by an apron of stout muslin and webbing straps. (Fig. 30.) The patient is placed prone upon a hard couch and the measurements are taken with a lead strip, carefully moulded to the inequalities of the transverse processes, from the anal fissure below to the upper border of the

FIG. 30.

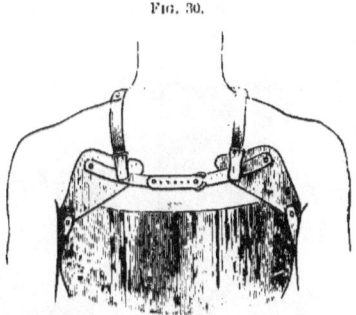

Chest-piece for spinal support. (TAYLOR.)

scapula above. (Fig. 31). This outline, carefully traced upon stiff cardboard, is cut out, and serves both as a plan for the manufacture of the instrument and a record for subsequent reference. The pelvic band is measured from one trochanter to the other. The upper parts of the uprights are bent backward, so as not to rest upon

FIG. 31.

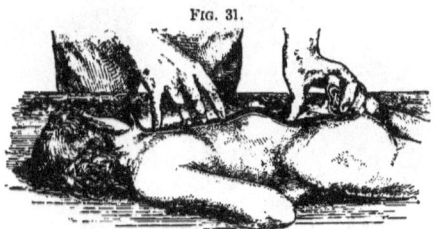

Measurement method in spine disease. (HOFFA.)

the surface when applied in the recumbent position, but make gentle traction backward. The brace should always be applied in the recumbent posture, the pelvic band be secured first, the axillary straps next, to the lower transverse bar, and, thirdly, the upper apron-strap to the buckles of the upper transverse bar. The middle straps of the apron are then secured over the uprights by means of safety-pins. In some cases of lumbar caries a swathe of plaster-of-Paris bandage (Fig. 32), muslin, or leather will add much to the comfort of the patient and increase the degree of fixation and pressure. Much ingenuity and variety may be displayed in the matter of pelvic bands, shoulder-straps, aprons, etc., but the principle of all is the same. (Fig. 33.) Great care must be observed to so regulate the pressure on either side of the deformity as to secure

FIG. 32.

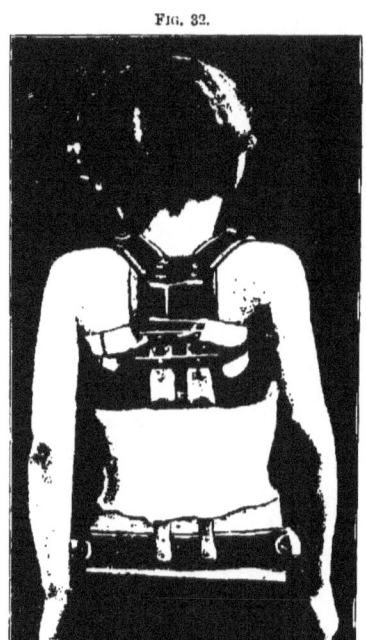

Antero-posterior support with straight band and plaster-of-Paris swathe.

FIG. 33.

Spine brace with triangular rubber scapular pads. (WHITMAN.)

fixation of the inflamed area without undue pressure. Dr. Judson has formulated the following rule :

The apparatus may be considered as having reached the limit of its efficiency if it makes the greatest possible pressure on the projection compatible with comfort and integrity of the skin.

The apparatus must be worn day and night, unless some special complication calls for its removal, and under no circumstances should the patient assume the upright position without the support. In all cases where the disease is above the seventh cervical vertebra the apparatus must be supplemented by a chin-rest, to fix the diseased area and support the superincumbent weight (Fig. 34) ; this consists of an ovoid steel ring, made to open on the side and secured when closed, arranged so that it can support the skin in a hard-rubber cup and exert pressure upon the occiput, and attached to the steel upright

FIG. 34.

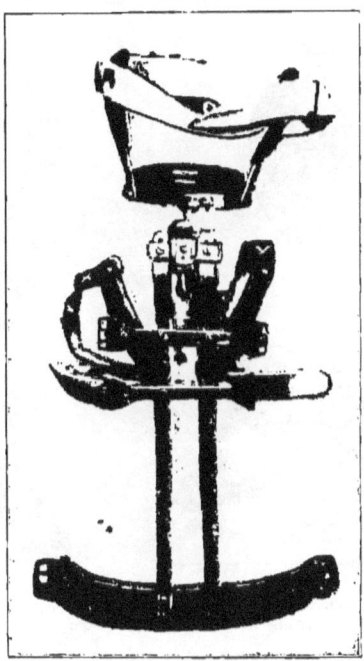

Antero-posterior spine brace, with chin-rest.

by a ball-and-socket joint, which allows of motion in all directions or can be fixed at any point. The head is secured by a webbing strap passing about the forehead from the extremities of the posterior uprights. In those cases before spoken of, where lateral deviation complicates the disease, it will be necessary to bend the brace up on one side and down on the other, and by enforced recumbency await the subsidence of the muscular spasm—or, removing the apparatus, apply a well-fitting plaster jacket with suspension, and subsequently resume the use of the spine brace. In cervical cases a modification

accredited to Dr. Goldthwaite[1] has proven efficient in my hands.
(Fig. 35.)   The brace and head-piece are all in one piece, the latter
being an extension upward of the spinal uprights.   (Fig. 36.)

The two most frequent complications—abscess and paraplegia—
require special treatment.   Iliac, lumbar, and psoas abscesses are
formidable complications.   It has been the habit of surgeons to treat
these expectantly—to allow them to open spontaneously, to aspirate

FIG. 35.   FIG. 36.

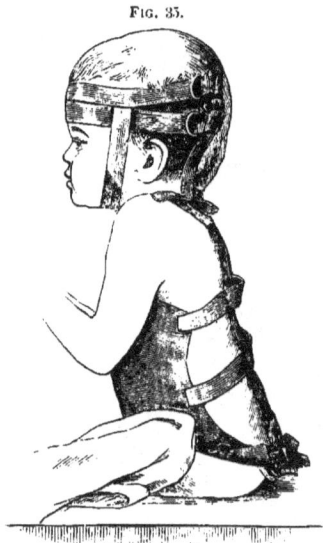
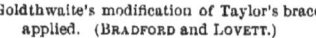
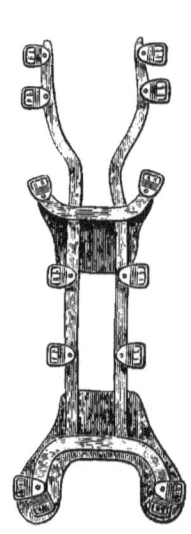

Goldthwaite's modification of Taylor's brace   Goldthwaite's modification of Taylor's
applied.   (BRADFORD and LOVETT.)   brace.   (BRADFORD and LOVETT.)

or incise, and then to close.   These plans are sanctioned by authority
and supported by excellent results, but there has been of late a
growing tendency to do something in an operative way.   Evacuation
and hyperdistention of the cavity[2] with various antiseptic fluids has
been practised by Wood, Israel, Agnew, Treves, Dollinger, and
others; but the operation did not fulfil all that was anticipated, and
Demours and Demoulin,[3] and Bradford and Lovett, have each recorded a death from it—two in all.

[1] Bradford and Lovett: Orthop. Surg., 1890, p. 82.
[2] Method of Mr. Collender.
[3] Progrès. Méd., 1886, 20, iv., pp. 1029, 1031.

Under strict antiseptic precautions the dangers of septicæmia are greatly lessened, and in suitable cases—where the pressure effects and distention of the abscess demand urgent interference—free incision, openings, and counter-openings, with insertion of drainage-tubes, should be advocated; particularly is this best where from the location and size it appears possible to remove by curetting the entire pyogenic membrane. Rupprecht, of Dresden, who four years before was an ardent advocate of the heroic plan of treatment, informed me during the summer of 1838 that about one-half of the patients subjected to the treatment died of tuberculosis, the other half recovered, and that he had abandoned it. This is not as favorable as the experience at Volkmann's klinik,[1] where 20 patients died of tuberculosis subsequently, and in 23 the abscesses healed.

At present there is a desire on the part of surgeons to treat abscesses and sequestra of spinal caries just as in other parts.[2] Dr. H. L. Taylor,[3] of Cincinnati, has shown anatomically that "the bodies of all the vertebræ are accessible to the surgeon," and "that there is little danger of opening the pleural cavity;" and Dr. Thomas Lafflin suggested about the same time before the Royal Academy of Medicine in Ireland, that this operation (sequestrotomy) should be extended to all the dorsal vertebræ. Podres, Reclus, Boeckel, Ashhurst, and others, have exposed the diseased vertebræ and removed, with the Volkmann's spoon or gouge, all the accessible diseased portion of bone. The results have not, as a rule, been successful. Doubtless these advances are all in the right direction and promise something, but it has occurred to the writer that the rule should be observed to operate in hospital cases where the disease is progressing in spite of appropriate treatment, but to avoid operation as long as possible in private practice, because in the latter patients are more apt to recover without operation. In psoas and lumbar abscess through-and-through drainage should be established where possible, and in iliac abscess a portion of the rim of the ilium may with advantage be removed by bone-forceps or trephine to permit thorough drainage. In all abscesses, after thorough cleansing with antiseptic solutions, iodoform emulsion in oil, or a 10 per cent. solution in equal parts of glycerin and water, after the method of P. Brun, should be injected, and a full antiseptic dressing employed. A proper fixation apparatus will

---

[1] Schreiber: Loc. cit., p. 112.
[2] Pradrez: Russk. Med., St. Petersburg, 1886, iv. 333. Treves: Brit. Med. Journ, 1884, i. p. 58.
[3] Medical and Surgical Reporter, 1888, lix. 136.

add much to the efficiency of any form of local treatment. In other cases the writer strongly favors the expectant treatment, and complete antiseptic dressings after the spontaneous evacuation of the abscess. I have myself seen a case of cervical caries, under the care of another surgeon, in which the fatal result could have been directly attributed to septic infection.

In retro-pharyngeal abscess and deep cervical abscesses burrowing toward the chest, prompt surgical measures must be adopted. The former can be relieved by a guarded bistoury, or in some cases by the administration of an emetic; the incision should be made in the median line of the posterior wall of the pharynx, and the head quickly thrown forward to avoid suffocation, or, what is decidedly better, the "Roser position," as recommended by Burrell,[1] of Boston, with the head dependent over the end of the table, with a suitable mouth-gag. In this manner one does not have to make a plunge in the dark, but the apex of the abscess can be freely incised, and a free discharge of pus will take place through the mouth and nostrils. In deep cervical abscesses which are approaching the thorax along the deep cervical fascia, the plan of Mr. Hilton is best adopted. An incision should be made through the sterno-mastoid, an exploratory incision carried through the deep fascia, and if pus be found, the incision further extended. The results of the different forms of treatment are well shown in the 75 cases collected by W. R. Townsend,[2] and which are here appended :

ANALYSIS OF 75 CASES OF ABSCESSES OF POTT'S DISEASE.

No treatment but brace; abscess disappeared	.	.	.	. . .	3
" " "	abscess *in statu quo*	.	.	. . .	8
" " "	abscess increasing, child doing well		.	. . .	8
" " "	child not doing well	.	.	. . .	2
					21

Aspirations.

Abscess disappeared	. 11	
Abscess opened spontaneously after aspiration failed .	. 3	
Abscess incised after aspiration failed . . .	. 4	
Abscess *in statu quo* after aspiration failed . .	. 1	
		19

No. of aspirations in each case from 2 to 6, average 3.

Incisions—Scraping Sac.

With use of iodoform emulsion or peroxide of hydrogen .	. . .	14
Results—Good	. 11	
Bad	. 3	
Infected at time of operation or at subsequent dressings	. 11	
Not infected	. 3	

[1] Medical News, Dec. 12, 1891.   [2] Trans. Amer. Orth. Assoc., vol. v. p. 169.

Opened Spontaneously.

Results—Good	15	
Bad	. 6	
	—	21
		—
		75

Deaths.

Tuberculous meningitis .		2
Amyloid liver . .		. 2
Suppression of urine .		. 1
		—
		5

The management of paraplegia consists in the accurate application of a spinal support, the enforcement of recumbency, and in some cases the use of large doses of iodide of potassium, as suggested by Gibney; but the experience of the writer corresponds to that of Ridlon, who declares himself convinced that it produces no effect unless there is reason to believe hereditary syphilis is present. The paralysis being spastic in its nature would strongly contra-indicate electricity until after the paraplegia has disappeared, when a mild interrupted faradic current may with advantage be employed to improve the tonicity of the muscles.

The operation of laminectomy, or trephining the spine, was revived by Macewen,[1] and has been performed for the relief of paraplegia a number of times with variable success. Dr. DeForest Willard[2] and Dr. J. William White[3] have written excellent monographs upon it.

In the 12 recorded cases the results were as follows:[4]

Macewen, 5, 3 successful; Horsley, 1 reported, doubtful; Abbe, 1, slight improvement; Wright, 1, no permanent improvement; Lane, 1, successful; White, 1, fatal; Duncan, 1, fatal; Burrell, H. L., 1, fatal. Thus, out of 12 operations, 4 have been successful. The objections to the operation are well stated by Willard as follows:

"1. It endangers life, and a certain percentage of cases will die from shock that would otherwise live for years and might even recover.

"2. It is uncertain in its relief, since when the compression is anterior it may be impossible to remove the cause.

"3. It weakens the only support of the head and shoulders, in the portion of the column upon which alone dependence is to be placed, since the anterior support—*i. e.*, the bodies of the vertebrae—has been already disintegrated.

[1] Encyclop. of Surg., vol. iv.   [2] Trans. Coll. of Phys. of Philadelphia, March 6, 1889.
[3] Annals of Surgery, July, 1889.
[4] *Vide* Bradford and Lovett: Orthopedic Surgery, 1890, p. 99.

"4. This weakening process must throw an additional strain upon both muscles and diseased bone, and the operation, if done before decided consolidation has occurred, would leave the trunk without any support, thus increasing the risk of sharp flexion and deformity."

In Pott's disease the operation is seldom called for, and should be reserved until all conservative measures have been exhausted and complete sensory paralysis has resisted all methods of treatment.

CHAPTER III.

SACRO-ILIAC DISEASE.

SACRO-ILIAC disease is an acute or chronic tuberculous affection of the sacro-iliac articulation.

Synonyms. *Sacro-coxitis* (Hueter); *Sacrarthrocace.* French, *Sacro-coxalgie.*

This disease is fortunately uncommon, although probably more common than is generally supposed. The first accurate account of the affection was made by Boyer, in 1821, since which numerous theses and monographs have been written, the best being those by Hahn, Nélaton, Erichsen, and Van Hook.[1]

Etiology. It is an affection of early adult life, but, as Velpeau remarked, it is met "at all ages, in private and hospital practice, among the rich and poor." The negro race is not exempt, as the writer has observed a typical case in a young mulatto woman. It is rare in children. Thus, according to Van Hook, in thirty-two cases in which the age was recorded, "less than 22 per cent of the cases were below fifteen years of age; the same proportion were between fifteen and twenty years of age; while in the fifth lustrum of life we find twelve cases recorded, just $37\frac{1}{2}$ per cent. of the whole number. All the remaining years of life furnish only six cases." Added to a specific constitutional predisposition, the exciting cause is usually found in exhaustion, exposure, or traumatic violence. One case directly due to injury is recorded by Louis. Among the cases recorded were gunners, exposed to the sudden jars of jolting caissons; laundresses, and children addicted to violent sports. Sex is an element, but the greater exposure of males to traumatism is partially offset by the greater liability of parturient women.

Pathology. The lesions are identical with tuberculous joint disease elsewhere. The affection may begin in the synovial membrane or bone, and extend rapidly to the cartilaginous constituents of the articulation. The anatomical construction and pathological features

---

[1] Annals of Surgery, St. Louis, 1888-89.

most resemble Pott's disease of the vertebra, as pointed out by Delens. Tubercular foci and sequestra are formed (Mueller, Koenig), and masses of bone may be discharged per rectum (Joyeux). Granulation tissue may form in the bone, or, invading the soft tissues, give rise to abscess formation. The inflammation is more often of the caries necrotica than of the caries sicca type, two varieties distinguished by Koenig as a "moist form" and a "dry granulating form."

FIG. 37.

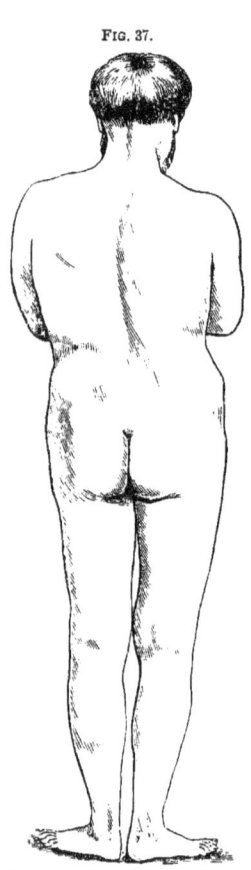

Position characteristic of sacro-iliac disease. (SAYRE.)

Symptoms. The cardinal symptoms of sacro-iliac disease are five, in the order of succession as follows: pain, lameness, changes in attitude and length, tumefaction, and abscess. The pain, at first intermittent or fugitive, becomes in time constant and severe. It is aggravated by coughing, laughing, urination, or defecation, and pressing the sides of the pelvis together produces pain in the joint, a symptom almost pathognomonic. The pain is often accompanied by the sensation as if the body was falling apart. After abscesses have formed, the pain may radiate to the anus, the leg, or extend to the knee. On account of the peculiar anatomical characters of the joint, muscular spasm and atrophy, such common and prominent symptoms in affections of other joints, are not conspicuous, but atrophy of the muscles of locomotion occurs. Lameness occurs early, walking is difficult, the diseased limb is favored as much as possible, the body is inclined to the sound side, and the pelvis tilted into the so-called "position hanchée" of Hattute, which Sayre regards as characteristic. Later, the patient becomes bedridden, lying upon the unaffected side. The elongation of the limb is apparent from downward rotation of the innominate bone, and not from actual increase in length, a fact easily demonstrated by measurement between the bony joints.

Swelling appears first over the head of the sacro-iliac joint, and later extends to and alters the shape of the buttocks. The local temperature is elevated, a fact utilized by Sayre for diagnostic purposes. Suppuration does not occur in all cases, and when present is most common during the late stages of the affection. In fifty-five collected cases, abscesses occurred in thirty-eight. When formed, abscesses find their exit in the direction of least resistance, becoming at once either intra- or extra-pelvic, in the proportion of 61.8 per cent. of the former to 38.2 per cent. of the latter. The direction and termination of these are well shown in the following table, modified from Van Hook :

TABLE OF ABSCESSES IN SACRO-ILIAC TUBERCULOSIS (AFTER VAN HOOK).

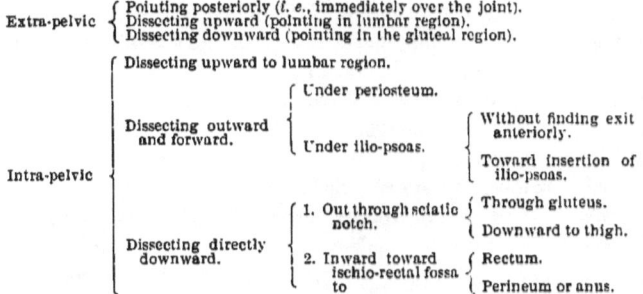

Diagnosis. Though the diagnosis of sacro-iliac disease can usually be readily made, there are several affections with which it may be confounded. These are : lumbo-abdominal neuralgia, sciatica, lumbago, psoitis, caries and necrosis of the iliac bone, lumbar Pott's disease, and hip disease, the diagnostic points of which have already been given. In addition to these, it may be remembered that this joint is liable to many forms of joint disease—acute and chronic suppurative inflammations, primary, osteomyelitic, and metastatic ; acute articular and gonorrhœal rheumatism and arthritis deformans, and is also the seat of tumors, fibro-plastic, hydatids, and enchondromas. Lumbo-abdominal neuralgia may be distinguished by the superficial and diffused character of the pain, its resemblance to other forms of neuralgia, and particularly by attention to the painful points of Valleix. Sciatica may be recognized by its occurrence in older persons, the situation of the pain below and not above the gluteal muscles and extending down the back of the thigh, slight flexion of the limb, deformity of the pelvis, or other signs of articular disease. In lumbago the tenderness is over the lumbar region, not over the joint; it is bilateral in distribution, increased by flexion and extension of the trunk, and unaccompanied by any symptoms of disease of the sacro-iliac joint. In psoitis the thigh, is flexed and the foot rotated inward; pain is absent from the posterior part of the joint, and pressure upon the sides of the pelvis does not elicit pain. Caries and necrosis of the ilium is to be distinguished by the absence of pain in standing and walking, elongation of the limb, and tilting of the pelvis on the affected side. The exploration of fistulæ will facilitate the diagnosis. Lumbar Pott's disease offers a condition which is very confusing, and one best distinguished by its occurrence during childhood, the presence of spinal deformity, of tenderness over the affected area, with general rigidity of the spinal column, and absence of elongation or other symptoms of disease of the sacro-iliac articulation. There is no lameness in walking.

Prognosis. The prognosis of advanced sacro-iliac disease has always been considered unfavorable, a fatal result usually occurring from long-continued suppuration and hectic fever. Among cases seen earlier, recoveries were recorded. Modern statistics indicate a strong tendency to an unfavorable prognosis in the moist variety, but a decidedly favorable tendency in the dry granulation form. Thus, in 16 out of 17 cases in which abscesses did not occur, the recoveries were 94 per cent., while in 38 cases in which abscesses occurred, the recoveries without operation were only 3, or 7.9 per cent.

Treatment. The proper treatment will depend upon the stage in which the disease is recognized and the variety of the affection. Whether of the dry granulation type, or the moist form with abscess formation, in either the hygienic conditions should be the best that circumstances permit, and the general health should be sustained and improved by stimulants, tonics, etc. In the early stage the joint should be immobilized with plaster-of-Paris, leather, or some other form of apparatus, and the patient, placed on crutches, wearing a high shoe upon the foot of the sound side, may enjoy the benefits of fresh air and sunshine. Extension by weight and pulley should be employed at night, and later may be used constantly with the patient in the recumbent position. When the pain is severe, counter-irritation with iodine, cantharidal collodion, or thermo-cautery is indicated. When abscesses form, they should be freely opened and thoroughly drained, and any sequestra found should be removed and the walls thoroughly curetted, the object being to remove, if possible, all tuberculous matter. Drainage is best effected with iodoform gauze, and full antiseptic precautions should be observed. If the abscesses be intra-pelvic alone, the disease may be reached without the extensive removal of healthy bone required in the operation of Tiling, by

FIG. 38.

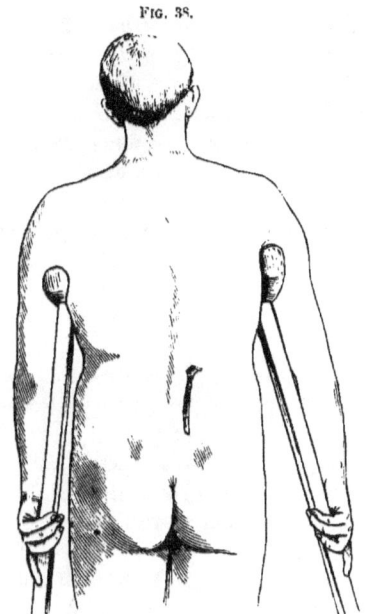

Line of incision for opening the lumbar abscess and operating on the sacro-iliac joint. (VAN HOOK.)

employing the method of Van Hook. A vertical incision, two or three inches in length, over the posterior spinous process of the ilium is first made, the bone is denuded of periosteum and connective tissue by scraping, and with a chisel small fragments are removed from the exposed bone until the anterior surface of the diseased joint can be exposed and thoroughly curetted with curved instruments. Iodoform gauze packing and an antiseptic dressing complete the operation.

CHAPTER IV.

HIP-JOINT DISEASE.

Hip-joint disease is a chronic tubercular lesion of the coxo-femoral articulation, beginning usually as an osteitis or synovitis and terminating in recovery, ankylosis, or complete destruction of the joint.

Synonyms. English, *Morbus Coxarius; Morbus Coxœ; Hip Disease; Tuberculous Disease of the Hip; Chronic Articular Osteitis of the Hip; Chronic Epiphysitis of the Hip; Medullo-Arthritis; Coxalgia; Coxitis; Morbo-coxario.* French, *Coxo-tuberculose; Cox-algie; Dislocatio Hanchœ* (Albucasis); *Goutte Sciatique* (A. Paré); *Arthritis Ischiatica* (Morgagni); *Luxation Symptomatique* (Dupuy-tren); *Luxation Spontanée* (Boyer); *Coxarthrocace* (Rust); *Femoro-coxalgie; Coxopathie.*

Many of these terms are misleading or at most unsatisfactory. The term here used, hip-joint disease, is now accepted by common usage, but for scientific purposes the terms chronic articular osteitis (Gibney), or coxo-tuberculose (Lannelongue), would best suit the purpose.

It affects usually but one side, but instances of double hip disease occur, one writer, Ridlon, having himself reported fourteen such cases. The hip is subject to a number of pathological lesions which have no association with true hip disease, such as inflammatory arthritis, specific in nature; arthritis from rheumatism and gout, arthritis from diseases of the central nervous system (Charcot's disease, etc.), and septic arthritis from numerous causes. These will be considered in their proper places and under differential diagnosis in this section, but need not be dwelt upon here, and have no connection with the subject under consideration.

The frequency of hip disease in surgical practice is illustrated in the fact that in 2292 patients treated in the Orthopedic Dispensary of the Hospital of the University of Pennsylvania, 298 were cases of this affection, and the relative number of cases of hip disease compared to diseases of the other articulations is shown in the following list of cases from the same source: Hip-joint disease, 298; knee-joint disease, 67; ankle-joint disease, 19; shoulder-joint disease, 2; elbow-joint disease, 6; wrist-joint disease, 2. Of the two sides, the left limb appears to be more frequently affected than the right; thus, Lansdale out of 112 collected cases found 65 on the left, and 47 on the right; and Wright out of 72 cases found the left hip affected in 42 instances, and the right in 27 cases.

Etiology. Until recently the exact nature of this affection was the subject of much discussion, and there are still a few, whose position entitles their opinions to respect, who doubt the tuberculous nature of this affection.

Two distinct lesions have been usually included under hip disease, simple traumatic synovitis and tubercular arthritis. The former is identical with synovitis elsewhere, and constitutes a large proportion, but not all, of the cases which recover promptly without suppuration under ordinary treatment.[1] These will be referred to under differential diagnosis, but will not be otherwise considered in dealing with hip disease proper, or chronic tubercular arthritis of the hip.

The causes of hip disease are both predisposing and exciting. The age, sex, heredity, hygienic surroundings and social condition, and the peculiar anatomy of the joint are all predisposing factors; the exanthemata of childhood, traumatism, and diseases of the neigh-neighboring organs being exciting causes.

Age. Though occurring later in life in exceptional instances, hip disease is essentially a disease of infancy and childhood, and age may be considered one of the most important predisposing factors. Congenital cases have been observed and reported by Broca, Verneuil, Morel-Lavallée, and Padieu. Brodie, Depuis, and Crocq have reported cases at one month, three months, and nine months respectively; but, as a rule, hip disease attacks children between the ages of two and thirteen years. Thus, of 365 cases collected by Sayre 221 were under fifteen years, and 121 of these were under five years; of 360 cases collected by Bryant 309 were under twenty years, and of these 126 cases were under four years; of 619 cases collected by Wright 480 were under fifteen years, and of these 130 were under six years; of 860 cases recorded by Gibney 84½ per cent. of all cases occurred before fourteen.

The relative frequency of hip disease at this early period is

---

[1] Trans. Amer. Orthop. Assoc., vol. i. p. 276.

probably due to the following causes : the frequency of tuberculosis
in childhood ; the active growth and immature nature of the
epiphyses and joint constituents generally ; the greater liability of
children to fall ; the greater bodily activity of youth, which favors
the development of grave disease from slight injury.

Instances of primary hip disease after twenty-five years are rare
and have never been observed by Barwell, though Bryant and Wright
have each recorded a number of cases. The latter has recorded one
occurring at fifty-four years, and Paget has pointed out the fact that
people over sixty are more often " scrofulous " than people between
thirty and fifty, and referred to the frequency of hip-joint disease in
the aged. Relapses, particularly between thirty and forty years of
age, are common.

Sex. Though sex is considered by most modern writers to be of
no importance as an etiological factor, the greater number of males
over females in all statistics remains. Thus, according to Ashhurst,[1]
of 100 cases admitted into the Children's Hospital of Philadelphia,
61 were boys and 43 were girls ; and Lannelongue[2] found in 100 cases
57 boys and 43 girls. Likewise, of 619 cases recorded by Wright,
371 were males ; and in 2307 cases reported by Holt,[3] 1178 were
males and 1129 were females. This preponderance of males over
females among patients suffering from hip disease has been assigned
to two causes : the greater liability of males to injury from their more
boisterous habits, and the existence of phimosis as an exciting cause.
The former will be considered under the exciting causes ; but the
latter, since it concerns the sexes directly, may be given here.

Barwell many years ago observed a singular coincidence between
hip disease and phimosis, and reported in 100 cases examined in 6
only was there no phimosis ; 66 had the affection severely, and 28
slightly.

Since this report, Sayre, Wright, and others have found it as one
of the exciting causes of hip disease.

More recently Roswell Park has shown in a series of observations
made upon 150 males of all conditions in hospital and private practice, that nearly 80 per cent. had phimosis in some degree, from
slight and partial adhesions with little or no retained smegma, to
cases where retraction was impossible. Other statistics are not
wanting to show the prevalence of this condition among both the

---

[1] Penna. Hosp. Rep., 1869, p. 148.   [2] Coxo-tuberculose. Paris, 1886, p. 52.
[3] *Vide* Gibney : Hip Disease, p. 206.

healthy and diseased, and while the author has observed the coincidence in many instances, and fully appreciates the deleterious influence of frequent and long-continued priapism upon the infantile spinal cord, he does not consider the relationship by any means established.

Barwell has likewise described a corresponding irritative condition in girls—vaginitis. Furthermore, according to Gibney, Bryant, and others, the sexes are about equally liable to the disease.

Heredity. The difficulty of correctly ascertaining the influence of inherited disease is at once evident when we realize both the inclination of parents and patients to deny the existence of tuberculous disease in their ancestors or near relatives, and their desire to establish, if possible, a traumatic origin for the affection.

For this reason the greater proportion of statistics are inaccurate and only approximate the truth, the error being, however, on the side against inheritance. Notwithstanding this, both experimental research and clinical investigation tend to establish the tubercular or strumous tendency as a predisposing cause. The direct transmission of tuberculosis has been of late established.[1] Clinical analyses of hip-joint cases have variously estimated the percentage of patients who have either an hereditary or acquired diathesis from 33 per cent. to 99 per cent. Volkmann claims that "individuals with fungous joint disease spring, practically without exception, from families in which scrofula and tubercle are hereditary," and Gibney, after a careful analysis of 596 cases of tubercular joint disease of the different joints, of which 265 were disease of the hip-joint, could only find one case which did not present either an inherited or acquired diathesis. Of these 265 cases, phthisis occurred in the father's family 53 times, in the mother's 56; and diseases unquestionably strumous in the fathers 10, and in the mothers 18 times, besides the syphilitic, rheumatic, and alcoholic diatheses a number of times.

Pathology. Though hip disease is fatal in a certain number of cases, the opportunity for anatomical study at an early period is rarely offered, hence the contradictory statements which exist in regard to the seat of the initial lesion. The pathological lesion in the advanced stage is a destructive tubercular osteitis, resulting in interstitial absorption or caries of a portion or the entire constituents of the joint, and as a tubercular lesion it does not differ from

---

[1] Landouzy and Martin: Faits clin. et expér. pour servir à l'Hist. de l'Hérédité de la Tuberculose. Sonnenberg: Arch. f. klin. Chir., 1881, xxvi. 789.

tubercular osteitis alsewhere (vide Pott's Disease). A sufficient number of early autopsies have fortunately been observed to definitely settle the fact that the initial lesion may originate in any one of the structures which constitute the articulation, occurring more frequently in certain localities than in others. Thus in a case reported by Agnew, of a lad suffering from incipient coxalgia, who died of tubercular meningitis, the inflammatory redness occupied the cartilage a short distance round the acetabular and femoral attachment of the ligamentum teres. Holmes also reports two autopsies in incipient coxalgia, in which he noted inflammatory lesions of the synovial membrane and round ligament, and one of erosion of the ligament. Marjolin and Gosselin, Martin and Collineau, report similar cases. Guéniot observed two cases, in the first of which there was, in addition to these early signs, an erosion of the posterior surface of the neck; in the second a loss of translucency in the articular cartilages. Barwell has reported an early case in which there was an excavation on the inferior face of the femoral extremity. In four specimens examined by Lannelongue, in two there were small cavities upon the femoral head, and in two there was caseous infiltration of the cancellous substance. Autopsies at a later period of the disease show the head of the femur and acetabulum frequently involved, but as far as known there are no specimens recorded of initial lesion of the acetabulum, it being usually secondarily affected.

In 61 specimens of hip excision analyzed by Müller,[1] it was found that the disease began in the bone in 47 cases, in the synovial membrane in 8, but it was impossible to state where it originated in 3.

According to Habern,[2] from an analysis of 132 hip resections in Volkmann's clinic, primary acetabular infection is more frequent. Thus, a caseous focus of the acetabulum was found in 50, with a sequestrum in 31; a focus was found in the femoral head, neck, or trochanter in 23; foci in both acetabulum and femur in 7; and the disease was so far advanced in 29 that it was impossible to locate the primary lesion.

The majority of surgical authorities at the present time believe in the osseous origin of chronic tubercular osteitis, while a few still cling to a purely synovial origin, and a few others to a purely ligamentous origin.

The origin of hip disease from a psoas abscess is primarily synovial .

[1] Koenig: Die Tuberc. der Knoch. u. Gelenke. Berlin, 1884.
[2] Centralbl. f. Chir., April 2, 1881.

by contiguity of structure. The recorded cases of primary infection of the ligamentum teres are sufficient to establish the occasional primary ligamentous origin of the affection. In adults a synovitis may terminate in a tuberculous arthritis. With these exceptions personal observation confirms the statement of Lannelongue, that in the majority of instances "la coxo-tuberculose est primitivement osseuse."

The microscopical appearance resembles tubercular osteitis elsewhere. Its extent and destructiveness will depend somewhat upon

Fig. 39.

Tuberculosis of head of femur. (KRAUSE.)

the initial lesion and the virulence of the process. In the head and neck of the femur the lesion may be limited to a circumscribed area; the epiphysis may be completely destroyed, separated, and lying loose within the joint—in the latter the sequestra being cut off by granulation tissue, the process being a *caries necrotica*. When the inflammation commences in the sound ligament microscopic section shows[1] active proliferation of the cartilage cells at the insertions of the ligament to the acetabulum and to the head of the bone, gradually extending to the cartilage and to the bone. In the acetabular variety

[1] *Vide* Agnew, vol. ii. p. 169.

the cartilage is frequently infected secondarily from contiguity of structure from the diseased head of the bone in contact with it.

In the severer forms all the components of the joint are infected and destroyed. The synovial membrane, at first thickened, ultimately disappears, the head and neck becomes carious and has a "worm-eaten" appearance, or the head and neck may entirely disappear; the acetabulum is excavated or entirely perforated by the ulcerative process. True dislocation seldom occurs. The absorption of the cavity by ulceration and new bone formation around the acetabulum—the so-called "travelling acetabulum" so characteristic of the disease, have given rise to this impression. A true dislocation is, however, possible, as pointed out by Erichsen, by the head of the bone being pushed out of the acetabulum by a "fungous fibro-plastic mass" within its cavity.

In those cases of osteitis in which interstitial absorption occurs—caseation without suppuration—a residual abscess may result. In this form of inflammation—the so-called *caries sicca*—the affection may exist for years with extensive destruction of the cartilages and bones, with consecutive dislocation, but without a drop of pus. In such the granulations are firmer, almost cartilaginous in consistence, tending to atrophy and cicatrization—a process analogous to cirrhosis. In a large proportion of cases, where the disease runs its course unchecked by treatment, suppuration occurs, the abscesses finding exit in the direction of least resistance, their course and termination being comparatively uniform.

Profuse suppuration from the joint, according to Billroth, is always a sign that part of the synovial membrane has not yet been destroyed, or that there are large abscesses near the joint; the secretion from fungous granulations is less abundant, serous or mucous. Periarticular abscesses, suppuration of the cellular tissue about the joint, whether associated with the joint disease or independent of it, increase the amount of discharge.

The disease may halt and recover at any stage. If the disease terminate before the articulation has suffered much mutilation, it may return by a process of repair almost to its original condition. If ankylosis occur, however, the femur and ilium become consolidated and fixed by firm fibrous or bony union, the articulation being protected against dislocation by an osteoplastic osteitis.

In this connection the association of hip disease with tubercular lesions elsewhere may be referred to. Patients with hip disease are exceptionally attacked by phthisis pulmonalis, though osteitis of other parts is common, as of the vertebræ, tarsus, carpus, elbow-joint, shoulder-joint, etc. The antagonism of external and internal tubercular processes, as pointed out by Treves, is tolerably well established.

This is why phthisis is not more frequent in subjects of hip disease; why subjects with external cervical gland disease are not more often attacked with pulmonary tuberculosis, and why subjects with angular curvature (caries of the vertebræ) are not more often attacked by it.

It is because of this apparent antagonism that exists between the tubercular diseases that it is, moreover, unusual for two grave manifestations of tuberculosis to be active at the same time—one will become manifest, while the other is observed to improve or subside.

When the localized tubercular process in the hip-joint infects the general system, tubercular meningitis or general miliary tuberculosis results.

Symptoms. The symptoms of well-established hip disease, taken together, are so characteristic that the affection is evident at a glance, and yet the symptoms indicating the commencement of this disease are so obscure that it is frequently mistaken for rheumatism, and still more frequently for knee-joint disease.

It is most convenient, for descriptive purposes, to divide hip disease into three stages, after the method employed by Mr. Adams and indorsed by Mr. Wright: *First stage,* from the beginning to the development of pus in the joint—the period of flexion; *second stage,* from the end of the first until the development of pus outside of the joint—the period of abduction and apparent elongation; *third stage,* from the end of the second to the termination of the disease—the period of adduction and shortening. Each of these stages is characterized by a group of symptoms peculiar to itself, and corresponding to the pathological changes occurring in the joint: (1) stage of localized bone disease, (2) stage of joint involvement, (3) stage of destruction of the capsule and external suppuration.

First Stage. The symptoms indicating the onset of the disease are very insidious, and consist of lameness, pain, induration about the joint, limitation of motion, muscular atrophy, glandular engorgement in the neighborhood of the joint, together with slight constitutional derangement.

Lameness. Among the earliest signs which denote this disease in its incipiency, is a slight limp observed in the gait of the child. This is most noticeable in the morning on rising from bed, and generally passes away in a few hours. It is more marked after a previous day of great activity. This limp is partially due to a stiffness about the joint, and to the pain which is also present at this time, and is aggravated by motion, though it may be absent in any or all of the stages of the disease.

Pain. Accompanying this limp pain is an early and frequent symptom, and is usually referred to the knee. It is usually periodical, evanescent, appearing suddenly while the child is at play, or coming on in the latter part of the day, and disappearing during the night. This pain in the knee has been variously explained. It has been ascribed by Bonnet to the pressure of the internal lateral ligament against the condyle of the femur, induced by the malposition of the limb. It is now, however, generally considered to be reflex, induced by pressure on the obturator nerve, transmitted to the short saphenous nerve, through the communicating branch, which passes between the femoral and profunda femoris vessels. Other explanations have been offered to explain this reflected pain, and Wright has ascribed it to three causes : First, the supply of both knee and hip by the obturator, sciatic, and anterior crural nerves ; second, sympathy between the ends of the bones or direct extension of inflammation ; third, muscular spasm ; but the one already given will be found most satisfactory, and is the one in which Wright places most importance. The pain is experienced upon the anterior and internal lateral surface of the joint over a considerable area, differing in this respect from the localized pain in knee-joint disease.

Induration. The marked swelling about the trochanter which is so characteristic of the later stages, is but slightly marked at this period. There is, however, if the trochanter be deeply and firmly grasped, some enlargement and hardening of the joint apparent ; particularly is this marked if the disease is pelvic in its origin.

Limitation of motion. A state of spasm or muscular rigidity of the joint is among the first and most important symptoms of this disease. This is one of the earliest signs always present ; very persistent, and is the result of reflex muscular spasm, accompanied by an unconscious automatic contraction of the muscles to fix the joint and diminish the jar in walking. This muscular rigidity, together with the atrophy which it induces, are the two most positive symptoms of hip disease at this stage. The limb is held in a position of slight flexion, and in some cases slight abduction. This constant tetanic contraction not only produces the malposition of the limb referred to, but by forcing the head of the bone against the acetabulum produces a destruction of the cartilage and an increase of pain. Associated with this rigidity about the hip-joint there is, often, also a muscular irritability of the lower erector spinæ muscles and the muscles adjacent to the joint, as pointed out by Bradford and Lovett. This muscular rigidity about the joint, when slight, may only be noticeable when extreme flexion and extension are made, the range of motion in the middle of flexion being slightly or not at all impaired. Anæsthesia reveals the true nature of this rigidity, the muscular spasm entirely disappearing if no adhesion or muscular contracture exist.

Atrophy. Wasting of the muscles of the thighs and buttocks is characteristic of the disease. It occurs early and persists throughout the course of the affection. That this is not the mere atrophy of disuse is shown by the fact that it occurs so early and advances so rapidly. Paget designated it reflex atrophy, but Brown-Séquard's experiments led him to believe that the wasting is due to an irritation of the nerves, independent of the trophic centres. That the trophic centres are, however, affected would seem to be proven by the fact that not only do the soft parts waste, but the bone also becomes diminished, both in diameter and length. The amount of atrophy is best estimated by circumferential measurements taken about the thigh, at points equidistant from the internal condyle, upon the affected and sound limb. In this manner the slightest degree of atrophy can be estimated, and will be found present in some cases even before the advent of pain. Similar measurements of the circumference of the calf, on either side, will demonstrate the absence of atrophy in them at this early period. This atrophy differs from the wasting of paralysis, in that the muscles retain their firmness or are unusually hard.

Glandular engorgement. Though the inguinal glands are frequently enlarged in patients of a strumous diathesis, the enlargement of the deep glands above Poupart's ligament are a frequent and early sign of this affection. Deep pressure in this situation reveals their presence, and palpation is painful. Attention has been called to this by Lannelongue, who states that even in the early months, and the first weeks, these enlarged ganglions may be recognized. The swell-

ing of these inguinal glands is stated by Mr. Barwell to indicate
osteitis, but Wright believes they often indicate disease of the pelvis
rather than of the femur.   Their suppuration usually indicates pelvic
disease.

General condition.   The constitution at this early stage suffers but
little, and in many cases the physical appearance but little indicates
the grave pathological process going on within.   The appetite may
be diminished, the digestion enfeebled, and the disposition irritable,
but in the majority of cases the premonitory stage, if any, passes unobserved.   In many of these children a large but unnatural accumulation of adipose tissue produces an impression of health which
really does not exist.

Second Stage.   The second stage is marked by an increase of all
these symptoms, and the advent of three other signs : " night-cries,"
suppuration, and grating, or joint crepitation.

Lameness.   The slight limp referred to in the first stage is in the
second stage a decided lameness, being due to the altered position of
the limb in flexion, tilting of the pelvis and apparent elongation of
the limb, or to actual shortening.   In standing at rest the body-
weight is transferred to the sound limb, and the affected limb is
advanced and rested as much as possible.

Pain.   The pain referred to the knee increases in intensity and
duration, and to this is added pain in walking and upon motion.   So
great may the sensitiveness of the joint become that the slightest
motion or jarring causes excruciating agony.   The pain is increased
by forcing the joint surfaces together, but this is cruel and entirely
unwarranted, since it adds nothing of diagnostic value.   The location
of the pain remains the same, and attempts to differentiate the locality
of the bone lesion are unsatisfactory.   Erichsen has suggested that a
pain in the knee indicated "femoral coxalgia;" pain in the joint,
"arthritic coxalgia," and pain in the iliac fossa, or side of the pelvis,
"acetabular coxalgia."   The pain increases with the distention of the
capsule, and during the exacerbation of the disease, and usually
ceases abruptly with the rupture of the capsule and the extravasation of the pus into the surrounding tissue.   Though an early and
persistent sign of the disease, pain may be entirely absent in any
or all of its stages, so that as an individual diagnostic sign it is of
little value.

Induration.   Thickening and hardening about the trochanter is
indicative of suppuration within the joint, and Wright considers it

pathognomonic of it. It is therefore particularly characteristic of this stage. It is best recognized by grasping the trochanters with each hand, the thumbs in front and the fingers applied well down into the post-trochanteric fossa. In this manner the two sides can be carefully compared.

Limitation of motion. The position of the thigh is changed by the muscular spasm and the formation and accumulation of pus within the joint. The thigh is flexed, abducted, and rotated outward. The coxo-femoral joint is more or less fixed, and if the pelvis retain its normal position the flexed, abducted, and rotated position of the limb is not only unsightly, but unfavorable to progression. To overcome this flexion the pelvis is tilted downward; to overcome the abduction, which is essentially an unfavorable position for walking, the pelvis is tilted laterally; to overcome the rotation of the head of the bone, the pelvis is tilted on its axis toward the sound side, so as to render the anterior superior spinous process prominent and removed far from the axis of the trunk.

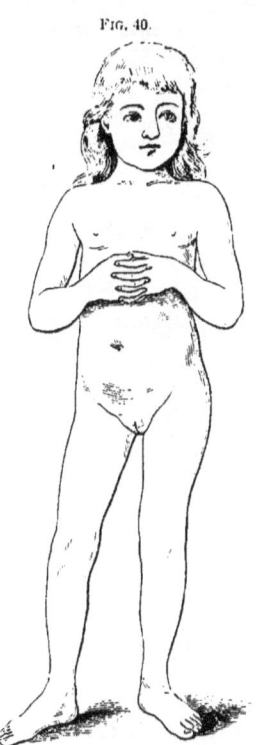

Fig. 40.

This alteration in the position of the pelvis is reflex in nature, and occurs consecutively with the osseous changes. That it is muscular and reflex is demonstrated by the fact that under profound anæsthesia the fixation disappears and the limb may be placed and examined in a normal position. Later, however, after the contraction has existed for some time, this simple muscular contraction is succeeded by structural alterations of the muscles, capsule, bones, etc., which render the position of the limb more or less permanent.

Position assumed in standing with slight abduction of right leg. (BRADFORD and LOVETT.)

The amount of deformity present can be ascertained by the employment of special instruments, as the compass of Martin and Collineau, the ingenious goniometer of Roberts, or they may be deduced with mathematic precision by the geometrical procedures of Giraud-Teulon, and the elaborate tables of Bradford and Lovett. The simplest method and the one most generally employed consists in placing the recumbent body in such a position that the anterior-superior spinous processes are on a horizontal line, and measuring from these on either side to the internal malleolus on the corresponding side. By this means the real or bony shortening may be ascertained. If the amount of practical shortening be desired, measurement must be made from the umbilicus to each malleolus, and by comparing these—the real or bony shortening and the practical shortening—the degree of adduction and abduction may be obtained. This will be referred to again in speaking of diagnosis.

Atrophy. The wasting of the muscles progresses during the second stage, and to the atrophy from the tetanoid spasm, always present, is added an amount of atrophy from disuse commensurate with the severity of the disease and the inability to use the leg. In this manner atrophy of the calf and other muscles not affected during the first stage are added and contribute a new symptom to distinguish this stage from the first.

Flattening of the buttock. The abduction and flexion of the thigh, and the atrophy of the gluteal muscles upon the affected side lead to flattening of the buttock and obliteration of the fold of the nates upon this side. This gluteal atrophy, like the adductor atrophy before referred to, is reflex and characteristic of the disease ; the muscles though wasted are hard and tense, and not soft and flabby. Peri-articular swelling also plays a part in the obliteration of the fold of the nates.

Night-cries. Though "night-cries," or night-shoutings, may occur early in the disease, they are significant of this stage. They occur early in the night, and some patients lie awake as long as possible fearful of their advent. As the child is losing consciousness, the muscular relaxation accompanying this stage allows pressure or friction of the tender surfaces within the joint, causes acute pain, a sudden awakening with a loud cry, and a violent spasm of the muscles to again fix the joint. After moaning and crying for some time sleep is again attempted, with, perhaps, a renewal of these disagreeable symptoms. This may be repeated several times during the night, and indicates extension or an exacerbation of the disease. They resemble somewhat, but should not be confounded with, nightmare, or "night terrors," from which the severe pain and the absence of disagreeable dreams serve to distinguish them. The condition is de-scribed by patients old enough to distinguish symptoms, as extremely sudden and severe pain, followed by an aching or bruised sensation in the thigh and hip.

Abscess.

Hip disease may run its entire course without suppurating, either extra- or intra-articular, the process being a *caries sicca*, a non-sup-purative osteitis, but, as a rule, abscesses are a frequent and serious complication. The presence of pus indicates the destructive char-acter of the osteitis, though extensive collections may spontaneously undergo caseation and absorption without much destruction of tissue, a "residual abscess" resulting. They occur but rarely during the onset, but may be the first symptom to attract the attention of the parent or the patient to the affected part. They are characteristic of the second stage, and occur, according to Lannelongue,[1] from the sixth to the fifteenth month, though they may be much later. Ac-cording to Lovett and Goldthwait,[2] of all the (sixty-three) cases examined with reference to this subject, abscess occurred within one year in 59 per cent., within two years in 13 per cent., in three to five years in nine cases, and in the seventh and ninth year one each. The frequency appears to be directly influenced by the efficiency of appro-priate treatment, and the stage during which it is begun. Thus, the percentage of abscesses occurring in the series of cases collected by Gibney, Marsh, the Clinical Society's Committee, and Lovett and Goldthwait, is from 23 per cent. to 69 per cent., of which from 41 per cent. to 50 per cent. developed before the patients came under treatment, and Mr. Marsh estimated that the formation of an abscess may be averted by early treatment in at least 80 per cent. of the total number of cases.

The origin and course of the abscess will depend upon the loca-tion and extent of the tubercular focus. From an epiphyseal osteitis the pus may extend in a course outside of the joint and become extra-articular; so, also, an extra-articular abscess may invade the joint. Suppuration in hip-joint disease coincides, in general features, with cold abscesses from osteitis elsewhere. The advent is usually without constitutional disturbance, but slight even-ing rise of temperature, slight rigors and perspiration may mark its

[1] Coxo-tuberculose, p. 96.   [2] Trans. Amer. Orthop. Assoc., vol. ii. p. 86.

progress. The complexion is pallid from the increased number of white blood-corpuscles in the blood—suppurative leucocythæmia—the appetite is capricious, but otherwise the general condition may be but little affected. Locally, the abscesses follow the fascias, and seek exit in the direction of least resistance, opening at some distance from the seat of disease, accumulating, in some instances, until they assume enormous proportions, or burrowing to great distances. The abscesses may open in many places, as in the case of Mr. Lund's,[1] in which there were no less than twenty-one different openings. The course and exit depend upon the location of the original disease, and attempts have been made to classify the different routes followed by the suppuration, and utilize the knowledge for diagnostic purposes. While the pointing is significant, it is not a positive indication.

Three varieties of abscesses may be distinguised, based upon the primary lesion : Arthritic, femoral, and acetabular. Suppuration, according to Ashhurst, occurs earlier in the acetabular than in the femoral variety.

In the arthritic variety, according to Bonnet, if the head of the femur does not press specially upon any part of the capsule, the pus will find exit at the inner side, the weakest spot at the capsule, and point at the inner side of the thigh among the adductors. If the limb be adducted and rotated outward, the head of the femur presses against the anterior and inner part of the capsule, and gives way at that point. In this event the pus enters the sheath of the psoas and iliacus, simulating a psoas abscess, or burrows downward toward the inner side of the thigh. If the limb be adducted and rotated inward, the pus escapes at the posterior part of the capsule to enter the pelvis along the course of the external rotators, or what is more common, points below the gluteus maximus.

In the femoral variety they open directly into the joint, to terminate eventually as the arthritic abscess before given, or escaping from the bone open upon the outer or anterior aspect of the thigh. Ashhurst says an abscess opening upon the outer part of the thigh, below the trochanter, indicates disease of the caput femoris, and Erichsen describes abscesses which burst below and in front of the great trochanter as indicative of disease of the femur.

In the acetabular variety, owing to the peculiar anatomical relations, purulent collections vary much in their course and exit. They

---

[1] Case 4, Wright: loc. cit., Append. ii.

may burst through the capsule and point in the inguinal region, or perforating the pelvis may pursue a circuitous route before being eventually discharged. Ashhurst says that abscesses opening in the pubic region denote disease of the acetabulum—the abscess being intra- or extra-pelvic, according as it opens above or below Poupart's ligament. Erichsen also considers pubic abscess pointing above Poupart's ligament due to pelvic disease on its inner aspect. Wright concludes that

FIG. 11.

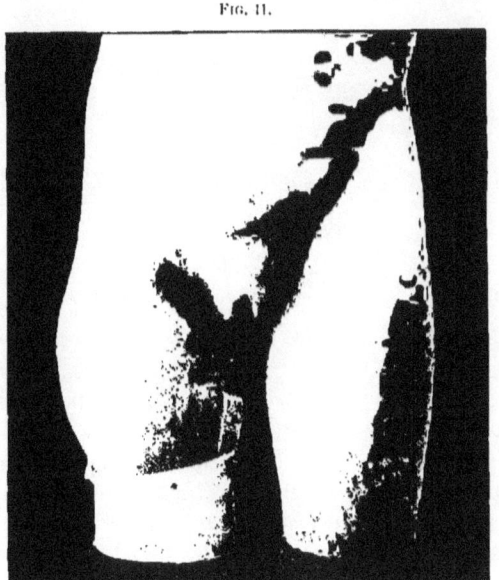

A hip abscess (LOVETT).
(By permission of the Trustees of the Fiske Prize Fund.)

when an abscess points on the front of the limb above a line drawn through the upper border of the great trochanter, there is disease of the pelvis, and this is the more certain the higher and the more internal the opening.

In the acetabular variety, in pelvic accumulations from perforations of the acetabulum, the pus (1) traverses the internal obturator muscle to become a gluteal abscess ; (2) fills the internal iliac fossa, and becomes a pubic abscess ; (3) burrows through the external obturator muscle to become an internal crural abscess, or, (4) opens internally, traversing the obturator internal muscle to open into the peritoneal cavity,[1] the rectum, urethra,[2] bladder, vagina, and upon the skin at the margin of the anus. According to Wright, abscess pointing between the scrotum or labium and the thigh is always of serious import, indicating pelvic disease. Gluteal abscesses, as pointed out by both Ashhurst and Erichsen, may be due to disease of either the femur or acetabulum.

Variety.	Course.	Exit.
1. Arthritic	*a.* Through inner side of capsule.	Inner side of thigh among adductors.
	b. Through anterior and inner side of capsule.	Enters sheath of psoas and iliacus, and burrows to inner side of thigh.
	c. Through posterior part of capsule.	Along course of external rotators or below gluteus maximus.
2. Femoral	*a.* Directly into joint.	As in arthritic variety, upon outer or anterior aspect of thigh below the trochanter.
	b. Outer or anterior course.	
3. Acetabular . . .	*a.* Through anterior part of capsule.	In inguinal region below Poupart's ligament.
	b. Perforates acetabulum :	
	1. Through internal obturator muscle.	As gluteal abscess.
	2. Fills internal iliac fossa.	As pubic abscess.
	3. Perforates external obturator muscle.	As internal crural abscess.
	4. Into peritoneal cavity.	Into rectum, urethra, bladder, vagina, and at verge of anus.

Though possible, it is rare for purulent accumulations to be absorbed after they have acquired certain volume. When the pus reaches the surface, it appears as a tense fluctuating swelling, of uniform color, with prominent superficial veins. When pointing occurs, the skin becomes red, thin, and ulcerates in one or more places. These, if small, readily heal after the exit of the pus ; they break down into large tuberculous ulcers, or remain indefinitely as orifices of sinuses or fistulæ filled with exuberant granulations. These fistulæ afford some clue to the seat of the disease, but cannot be readily explored on account of the tortuosity of their course. Fistulæ, originally pelvic, are distinguished, as pointed out by Barwell, by bearing-down efforts of the patient, causing the escape of purulent liquid.

Grating, or joint crepitation. From erosion of cartilage and exposure of the cancellous structure of the joint surfaces, grating, or joint crepitation from friction, may be elicited. Great care must be exercised lest injury result from rough manipulation. This may be obscured, even where considerable destruction exists, by the presence of abundant granulations, or where but one point of the bone is

---

[1] Martin et Collineau.　　　　　[2] Marjolin.

denuded, or where two or more roughened surfaces exist but do not approximate.

General condition. In the second stage, children suffering from this affection are often apparently robust, especially in the intervals between the exacerbations which characterize the affection. When, however, abscess occurs, and especially when suppuration is profuse from the sinuses, anorexia, pallor of the skin, fluctuations of temperature, irritability, and diarrhœa mark the progress of the disease.

Third Stage. The third stage is characterized by the adduction, shortening, dislocation, or ankylosis of the joint, ending in recovery; or suppuration, destruction of the joint, adynamic symptoms, and death from some visceral lesion.

Pain. The "starting" pains and night cries are, in the third stage, at times gradually, sometimes suddenly, relieved. A feeling of tension and tenderness remains, and may increase from the accumulation of pus.

Dislocation is often attended with considerable abatement of pain—sometimes complete and permanent, sometimes transitory. Dislocation, particularly subluxation, is often attended with sudden and severe pain. The character of the ordinary pain in this stage is not available to distinguish the form of the original disease.

FIG. 42.

Showing marked atrophy and deformity in advanced hip-joint disease.

Limitation of motion. Adduction, flexion, and shortening characterize the position of the thigh in this stage. This change from abduction and apparent elongation of the second stage into adduction and shortening of the third stage may occur rapidly, the position of mouths or years being reversed in a day.

Capsular contraction and muscular spasm are probably the cause of the adduction, while the elevation and backward thrusting of the pelvis and muscular spasm account for the early shortening. Later, actual osseous destruction and dislocation of the head of the femur are the true explanation of the real shortening observed.

6

Though frequently described in systematic treatises, it is probable that true dislocation from hip-joint disease very rarely occurs.

In destruction of the head of the femur, erosion of the brim of the acetabulum, in the so-called "travelling acetabulum" or "dislocation of the acetabulum" of Sayre, perforation of the acetabulum by the head of the femur, or where the trochanter is pushed upward, the head remaining in the acetabulum, dislocation cannot be said to have occurred. True instances of dislocation are chiefly confined to the femoral variety, and if, as sometimes happens, it occurs without suppuration, from the formation of a "fungous fibro-plastic mass" in the acetabulum (Erichsen), or where, by mere distention of the capsule, with rupture of the ligamentum teres (Holmes), a new socket may be developed upon the dorsum ilii or within the obturator foramen.

With the adduction and shortening there is associated undue prominence of the buttock on the affected side, marked upward and backward obliquity of the pelvis, lordosis, and a compensatory double lateral curvature of the spine. The lordosis is the result of fixation or ankylosis of the thigh in a fixed position, the arching of the lumbar region forward and the dorsal region backward being necessary to maintain the equilibrium. Flexion and abduction combined produce a compensatory lordosis and lateral curvature.

Recovery. Amelioration of all the symptoms, both local and constitutional, may occur at any period, but if suppuration has occurred, recovery with a shortened, somewhat ankylosed, and atrophied limb is all that can be attained by conservative measures. If the tendency is toward recovery, this will be indicated by the general improvement of all the constitutional and local symptoms. Gradually the sinuses cease discharging, and finally close, the swelling and induration diminishes, and the limb becomes more or less ankylosed in the position which it has last assumed, either from position of recumbency or the use of apparatus. In mild cases without suppuration, and some cases even with suppuration, perfect motion may be secured as the reward for efficient treatment.

Destruction. When the disease progresses toward a fatal termination the adynamia from excessive suppuration leads to death from exhaustion—the patient dies of amyloid degeneration of the internal viscera, tubercular disease of some distant organ, particularly meningitis, or succumbs from some intercurrent affection which otherwise would have been successfully resisted.

Double Hip Disease.

The course of the disease differs somewhat when both hip-joints are affected. According to Ridlon the disease seldom begins in both hip-joints at the same time, and the second joint may become diseased while the patient is resting in bed under treatment for the first joint, showing that traumatism may be excluded as a cause of the disease in the second joint in very many cases. The joint first affected is often the last to recover; the duration of the disease in the first hip is usually somewhat less than in an average case of hip disease, while the duration in the second hip is usually much less than that in the first. The amount of pain experienced in the second hip is usually less than that of the first.

Ankylosis, with more or less adduction in one or both legs, may result with or without treatment. The result in severe cases is a "scissor-legged" deformity; locomotion is possible, progression taking place entirely by movement of the knee-joint. In females impregnation and safe delivery are likewise possible, though in severe cases impregnation can only be accomplished *in positu animalis*, and parturition may become impossible from the marked adduction of the thighs, as in an instance some time since at the Philadelphia Hospital which terminated fatally. The kyphosis and other pelvic deformities will be much influenced by the age at which the hip disease occurs, and particularly upon the disturbance of the normal anatomical forces as pointed out by Schroeder.

Diagnosis. The general diagnostic signs of hip-joint disease are (1) limitation of motion, (2) atrophy, (3) lameness, (4) attitude of limb, (5) pain, and (6) swelling. These should all be carefully and thoroughly considered in arriving at a proper estimate of the condition of the joint in any given case. The symptom of "grating" is fallacious except in advanced cases, in which the associated symptoms will be so positive that its discovery will add but little of value. It can only be obtained in advanced cases where two opposed eroded bony surfaces can be rubbed together. The first two symptoms are peculiarly significant of hip disease, and upon them in the earlier stages the most reliance can be placed. Later, the attitude of the limb, the pain, and swelling are of great diagnostic importance.

Limitation of motion. Being a reflex tetanoid spasm, the first to appear, the most prominent while the disease exists, it is the diagnostic sign *par excellence* of hip disease. Its estimation in very young children, who through fright are apt to resist thorough examination, requires tact and patience. The examination should always be begun on the sound side to secure, if possible, the confidence of the patient, and flexion, extension, adduction, abduction, and rotation of the thigh, flexed at a right angle to the body, should each be separately investigated.

The tests in young and frightened children may be made with the child lying on the mother's lap, but are best made with all unnecessary clothing removed, upon a hard table or firm mattress.

In the earlier stages of disease forced flexion, forced extension, and forced rotation alone may give limitation, motion being perfect throughout a large portion of the arc in each direction and being limited only at the extremity of the arc of normal motion. Anæsthesia overcomes the spasm, and hence should not, for diagnostic purposes, be employed.

The patient upon the back, and the pelvis being fixed with one hand, the other should grasp the ankle of the sound limb first, and

Fig. 43.

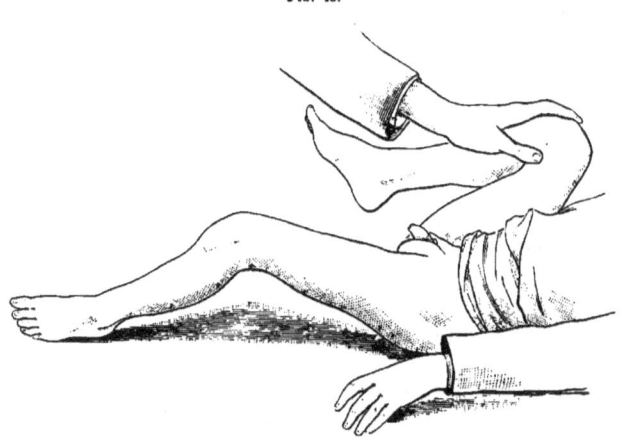

Estimation of flexion in hip disease. (BRADFORD and LOVETT.)

firmly but gently flex it until it touches the abdomen or meets with resistance. The suspected limb should then be flexed in a similar manner and the resistance, if any, be compared. The pelvis being again fixed, abduction can be estimated by separating first the sound limb and then the suspected limb as widely as possible from its fellow. Adduction can be estimated by crossing first the sound limb and then the suspected limb as much as possible, the pelvis being fixed. Extension is best estimated with the patient in a prone position. Fixing the pelvis with one hand, the sound limb is flexed at

FIG. 44.

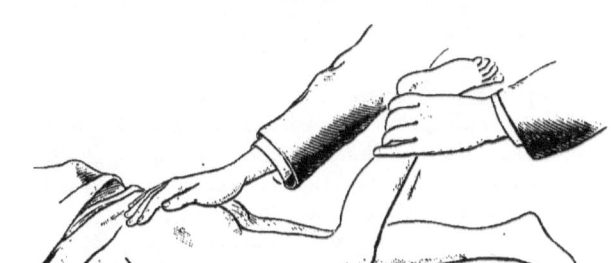

Examination for psoas contraction or limitation of extension in hip disease. (HOFFA.)

the knee to a right angle, and the thigh extended (backward) upward until it meets with resistance. The suspected limb is likewise placed in forced extension and the results compared.

The degree of rotation present is best ascertained by rotation with the hip flexed at a right angle to the body.

In the more advanced stages the estimation of limitation is easier, since the pelvis readily moves with the affected limb before the limit of motion is reached. Thus, in flexion from contraction of the psoas and iliacus muscles the popliteal space cannot (as on the sound side) be placed upon the hard surface upon which the patient lies without the arching (lordosis) of the lumbar region. Likewise the amount of abduction and adduction may be estimated by placing one hand upon the anterior superior spinous process of the ilium of the sound side, and gently but firmly abducting and adducting the suspected limb, the pelvis moving with the limb as soon as the limit of motion is reached.

In these estimations the greatest reliance should be placed upon limitation of motion at the extremity of the arc of normal motion.

Atrophy. Atrophy of the muscles being the result of the reflex muscular spasm, appears very early in some instances while the disease is still confined to the epiphysis, and is one of the most important symptoms in arriving at a correct early diagnosis. The atrophy is not only greater than that in simple disease of the muscles, but being unilateral (in a single hip disease) and being confined to certain groups of muscles, is very characteristic. In the earliest stages the adductors only are atrophied, but later the obliteration of the fold of the nates occurs through atrophy of the glutei, and the calf muscles and all the thigh muscles share in the general atrophy.

Measurement for atrophy is made by taking the circumference at the same level of both the thighs and calves with an ordinary tape-measure.

For the purpose of record, or where greater accuracy is desired, the levels at which the circumferences are taken should be measured from some bony point. Thus, in measuring the thigh the writer prefers to take two points measured in the direction of the sartorius muscle from the internal condyle, three and five inches or five and seven inches, according to the size of the individual. Such measurements are of great value for future reference. The calf measurement should be taken through its thickest part. Atrophy is seldom absent where hip disease exists.

Lameness. A limp in the gait is one of the earliest objective signs of hip disease, but since it is intermittent in the earliest stage, and may be exactly simulated by other conditions, as a single positive sign it is of little value.

Later, the attitude and fixation of the limb, and shortening, induce an attitude and gait that are very characteristic.

Attitude of Limb. The abnormal position of the thigh to the pelvis and to the opposite thigh are due in the early stages to muscular contraction; in the later stages to fibrous or bony ankylosis. Adduction and abduction are recognized by the patient, not as lateral deviations of the thigh, but as apparent shortening or lengthening of the affected limb.

In adduction the pelvis is tilted upward on the affected side, giving rise to apparent shortening of the limb; and in abduction the pelvis is tilted downward on the affected side, giving rise to apparent lengthening. This lateral tilting of the pelvis may be recognized by drawing an imaginary line between the anterior superior spinous processes, which should intersect at right angles a line drawn from the umbilicus to the pubis, if the pelvis is in its normal position. This apparent or practical shortening and lengthening, the result of adduction and abduction respectively, have led to some confusion which will require explanation. The accompanying diagrams will illustrate this.

This obliquity of the pelvis depends upon the fact that in walking or standing the limbs must be made parallel.

If, however (as in Fig. 45), the thigh is fixed by muscular spasm or ankylosis in an adducted position, progression is impossible while

FIG. 45.

FIG. 46.

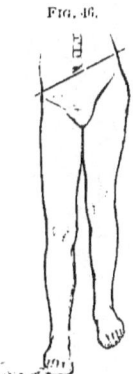

Position of the adducted limb when the pelvis is brought straight. (BRYANT.)

Position of patient when standing, with disease of the left hip-joint and an adducted limb. The pelvis is tilted up on the affected side, and the limb thereby apparently shortened.

the affected limb crosses the sound limb. Elevation of the pelvis (as in Fig. 46) brings the thighs parallel and permits of walking, but the affected limb is apparently shortened in proportion to the degree of adduction present.

In the same manner, if (as shown in Fig. 47) the thigh be abducted, progression is impossible while the limbs are so widely separated. Lowering of the pelvis (as shown in Fig 48) brings the thighs parallel, but the affected limb is apparently lengthened in proportion to the degree of abduction present.

The position recommended as best suited for examination is that in which the sound and affected limbs are parallel; the nature and amount of the obliquity of the pelvis as it affects the diseased side is then more evident than when the pelvis and sound side are perfectly straight. For measurement of the apparent and practical shortening or lengthening, the position best suited is that in which the trunk and the sound side are perfectly straight. The practical measurements are then made with an ordinary tape-measure from the umbilicus to each malleolus. Thus, in adduction the tilting of the pelvis has caused a practical shortening of the affected limb, and in abduction the opposite tilting of the pelvis has caused a practical lengthening of the diseased member.

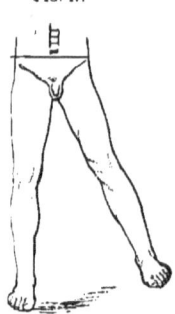

FIG. 47.

Abducted position of the diseased left limb when the pelvis has been brought into its natural position at right angles to the spine. (BRYANT.)

FIG. 48.

Apparent elongation of the left lower extremity in hip disease on the left side, due to abduction of the limb and the necessary tilting upward of the pelvis on the sound side to allow of the abducted limb being brought into a line with the body. (BRYANT.)

Real or bone shortening results from destruction of bone from disease or retarded growth, and is a permanent condition. It is best estimated by measuring from the anterior-superior spinous process to the malleolus on each side. Real shortening is sometimes due to spontaneous dislocation from disease. In such cases the degree of subluxation, or the amount of change in the acetabulum, may be estimated by ascertaining the relation of the great trochanter to the Roser-Nélaton line. Thus, with the affected thigh slightly flexed, if a line be drawn from the anterior-superior spinous process to the most prominent part of the tuberosity of the ischium, it should pass immediately above the upper border of the trochanter. If there is any alteration of this, luxation is evident, and its degree may be estimated.

The degree of adduction or abduction may be estimated with a goniometer, or joint measure. The horizontal arm is placed upon the anterior superior spinous process, the vertical arm is then placed parallel with the diseased limb, and the degree of deformity estimated. To estimate flexion or extension with the goniometer the instrument rests with the graduated arc against the body over the affected joint, with the horizontal arm in the line of the body. The movable arm rests against the limb parallel with its popliteal axis, while the index records the degree of deformity.

A simpler and more accurate method is that introduced and employed by R. W. Lovett,[1] of Boston. It is based upon the mathematical relationship existing between real and practical shortening. The measurements with the patient lying straight are taken in the usual way, with an ordinary tape measure—*i. e.*, the real or bone shortening is measured from the anterior superior spinous processes to each malleolus, and the apparent or practical shortening is measured from the umbilicus to each malleolus. An additional measurement between the spinous processes is necessary to complete the data. Upon these Dr. Lovett has constructed an elaborate table, and deduced the following rule: "If the practical shortening is greater than the real shortening, the diseased leg is adducted; if less than real shortening, it is abducted." For example: Length (from anterior superior spinous process) of right leg, 23 inches; left leg, 22½ inches; length (from umbilicus) of right leg, 25 inches; left leg, 23 inches; real shortening 1½ inches, apparent shortening 2 inches; difference between real and practical shortening, 1½ inches; pelvic measurement, between spines, 7 inches. By following the line 1½ inches until it intersects the line for pelvic measurement of 7 inches, 22° is found to be the angle of deformity, and as the practical shortening is greater than the real, it is 12° of adduction of the left thigh.

Distance in inches between anterior superior spines.

	3	3½	4	4½	5	5½	6	6½	7	7½	8½	9	9½	10	11	12	13
¼	5°	4°	4°	3°	3°	3°	2°	2°	2°	2°	2°	2°	1°	1°	1°	1°	1°
½	10	8	7	6	5	5	4	4	4	4	4	4	3	3	3	3	2
¾	14	12	11	10	8	8	7	7	6	6	5	5	4	4	4	3	3
1	19	17	14	13	11	10	9	9	8	7	7	6	6	6	5	5	4
1¼	25	21	18	16	14	13	12	11	10	9	8	8	7	7	7	6	6
1½	30	25	22	19	17	15	14	13	12	12	10	10	9	9	8	7	7
1¾	36	30	26	23	20	18	17	15	14	13	12	11	10	10	9	8	8
2	42	35	30	26	23	21	19	18	16	15	14	13	12	12	10	10	9
2¼		40	34	30	26	24	21	20	19	17	16	15	14	14	13	12	11	10
2½			39	34	29	27	24	22	21	19	18	17	16	15	14	13	12	11
2¾				38	32	29	27	25	23	21	20	19	18	17	16	14	13	12
3				42	35	32	29	27	25	23	22	21	19	18	18	16	14	13
3¼					39	36	32	30	27	26	25	22	21	20	19	17	15	14
3½						40	35	33	30	28	26	24	23	22	21	19	17	16
3¾							38	35	32	30	28	26	25	23	22	20	18	17
4							42	38	35	32	30	28	26	25	23	21	19	18

(Left margin label: Difference in inches between real and apparent shortening.)

[1] Boston Medical and Surgical Journal, March 8, 1888.

The amount of flexion in degrees may be estimated in a similar manner, after the method of G. L. Kingsley,[1] of Boston. (Fig. 49.) The patient lies on his back upon a table or other hard flat surface, and the surgeon flexes the diseased leg by the heel until the lumbar vertebræ touch the table, showing that the pelvis is in the normal position, and the angle (A B C) which the leg makes with the table

Fig. 49.

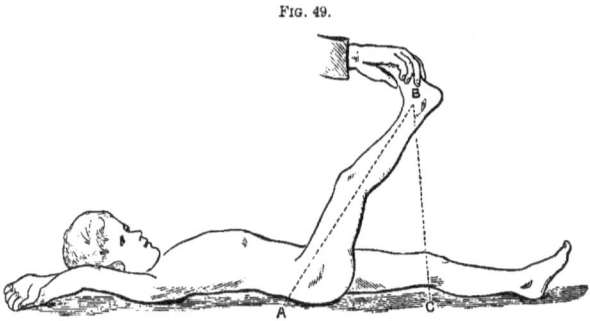

Estimation of flexion. (BRADFORD and LOVETT.)

is the angle of flexion of the thigh from the normal position. In this position the surgeon measures off two feet on the external aspect of the leg with a tape measure, one (the zero) end of which is held on the table, in the direction of the line of the leg (A B). From this point (B) one measures the perpendicular distance in inches to the table (C), and from the number of inches in this line (B C) can be ascertained in the table the degrees of flexion of the thigh from the normal position. For example, if the distance between the point on the leg and the table is 10 inches, 25° flexion of the thigh are present.

Inches.	Degrees.	Inches.	Degrees.	Inches.	Degrees.	Inches.	Degrees.
0.5	1	6.5	16	12.5	31	18.5	50
1.0	2	7.0	17	13.0	33	19.0	52
1.5	3	7.5	19	13.5	34	19.5	54
2.0	4	8.0	20	14.0	36	20.0	56
2.5	6	8.5	21	14.5	37	20.5	58
3.0	7	9.0	22	15.0	39	21.0	60
3.5	9	9.5	24	15.5	40	21.5	63
4.0	10	10.0	25	16.0	42	22.0	67
4.5	11	10.5	27	16.5	43	22.5	70
5.0	12	11.0	28	17.0	45	23.0	75
5.5	14	11.5	29	17.5	47	23.5	80
6.0	15	12.0	30	18.0	48	24.0	90

[1] Ibid., July 5, 1888.

If the leg is so short that it is impracticable to measure off 24 inches one may measure 12 inches; from this point ascertain the distance to the table in a perpendicular line just as before, double this distance, and look as before for the amount of flexion present.

Pain. Pain may be absent in any or all the stages of hip disease, but the characteristic pain is usually experienced in the affected joint or in the corresponding knee. The latter is often intermittent, and usually signifies the femoral form of disease. Pain in the arthritic form is constant, acute, and accompanied with a feeling of tension and tenderness above the great trochanter. Tenderness and pain on jarring the hip or on motion is more marked in the acetabular variety. Pain elicited by striking on the knee or heel, and thus pressing the joint surfaces together, is untrustworthy, and as a diagnostic sign should be abandoned.

Swelling. Local swelling is considered by Wright as one of the most important symptoms. It is most marked in the arthritic variety, which may be considered the acute form of the disease. Confined to the front and back of the joint, it indicates effusion into the synovial sac; confined to the great trochanter, it has been considered pathognomonic of suppuration within the joint. Thickening of the great trochanter as a late symptom is of some value as a confirmatory sign, but swelling and thickening alone are of but little importance.

In conclusion, the general diagnostic signs of hip-joint disease are limitation of motion, atrophy, lameness, attitude of limb, pain, and swelling, and of these the first two are peculiarly significant of early disease, while later are added the attitude of the limb and swelling.

Differential Diagnosis. Many diseases have been mistaken for hip disease through ignorance of its characteristic symptoms, or at a very early period of the disease. The diagnosis of contusions and sprains can readily be ascertained by exclusion, and of muscular rheumatism of the hip by the inherited or acquired history of the affection, the lameness preceding the pain, the absence of reflex muscular spasm, and local hyperæsthesia, perinephritis, perityphlitis, rhachitis, and pseudo-hypertrophy are entirely dissimilar, and do not require consideration, although the writer has seen a rhachitic elongation of the internal condyle diagnosticated as hip disease through ignorance and carelessness.

Hip disease could only be mistaken for abscess external to the joint, for disease of the knee, or for caries of the great trochanter, by neglect of careful examination.

The diagnosis from separation of the upper epiphysis of the femur with abscess is difficult, if not impossible—a matter which, fortunately, as pointed out by Ashhurst, is of no practical moment, as excision would be equally indicated in either affection. The same may be said of the differential diagnosis of traumatic, rheumatic, or blennorrhagic arthritis from the arthritic variety of morbus coxarius, since conservative treatment would be indicated in either affection.

There, are however, several affections commonly mistaken for hip disease which deserve thorough consideration. They are as follows:

1. Synovitis of the hip-joint.
2. Lumbar Pott's disease.
3. Periarthritis.
4. Infantile spinal paralysis.
5. Congenital dislocation.
6. Sacro-iliac disease.
7. Hysterical affections.

Synovitis of the Hip-joint.

The greatest difficulty will be experienced in distinguishing acute primary synovitis from the arthritic form of chronic articular osteitis. Acute synovitis from trauma is a disease of adult life, but it also undoubtedly occurs in children, particularly of from eight to twelve years. The acute nature of the affection, the short course, the absence of muscular spasm and atrophy, with eversion and outward rotation of the limb, are all of value in distinguishing synovitis. The acute form may, however, become chronic, and the consensus of medical opinion favors the possibility of a synovitis eventually becoming an osteitis. Of this nature undoubtedly is the so-called arthritic variety of hip disease. Of such nature are the cases with marked characteristic symptoms which recover promptly and permanently within a few months.

From synovitis the following considerations will be sufficient to establish the distinction:

Synovitis.	*Chronic articular osteitis.*
1. Pain coincidental with lameness, and invasion sharp and clear.	1. Pain preceded by lameness, and invasion seldom, if ever, sharply defined.
2. Locomotion speedily impossible.	2. Locomotion continues possible.
3. Occurs after eighth year.	3. Occurs before eighth year.
4. Joint tenderness present.	4. Joint tenderness absent.
5. No bone tenderness nor periarticular or articular infiltration.	5. Bone tenderness early; infiltration beginning of second stage.
6. Muscular atrophy absent.	6. Muscular atrophy present.
7. Rotation, eversion, and apparent elongation.	7. Limbs parallel in early stage.

Lumbar Pott's Disease.

In the early stage of lumbar Pott's disease it is difficult and in some instances almost impossible to arrive at a positive diagnosis from hip disease until the symptoms become more thoroughly established. Among the first symptoms may be a limp and limitation of motion in one limb from irritation or the presence of pus in the psoas muscle. This limitation of motion is confined usually to forced extension, but later, from extension of the pus as a psoas abscess, limitation of all motion at the hip will still further complicate the diagnosis. In either event the presence of muscular spinal rigidity from reflex spasm will determine the true nature of the affection, due allowance being made for the spasmodic contraction upon manipulation of these muscles in hip disease. The amount of abduction present in the hip-joint is also important, since abduction is very early lost in hip-joint disease, while it often remains free in advanced cases of psoas irritation and contraction. In advanced cases of lumbar Pott's disease, where psoas abscess is in contact with and irritates the hip-joint, it is extremely difficult to ascertain whether or not hip-joint disease coexists. The amount of abduction and rotation present, and the effect of rest and quiet to the joint, will best determine the exact condition. In conclusion, in lumbar Pott's disease the following peculiarities will be of service in arriving at a correct diagnosis: The lameness is subject to complete remissions, depends upon psoas contraction, and has more lôrdosis associated; the patient can stand as well upon the lame limb as upon the other; and reflex muscular spasm is never excited by passive motion of the lame hip.

Periarthritis.

It is only in the very early stage and only in phlegmonous inflammations about the joint that much difficulty is experienced, since the fibrous form does not, as a rule, occur in children. Under this head may be included inflammation of bursal and lymphatic glands and psoas abscess. The acuteness of the attack, the rapid development of the physical signs of acute abscess, the constitutional disturbance, the absence of reflex muscular spasm and atrophy, and the osteitic "night cry," would serve to distinguish periarthritis.

The movements of the muscles, whose mechanical execution is interfered with by inflammation, is limited, but the reflex element is absent, as likewise it is in cases of sarcoma of the hip, in which the greater hardness of the swelling, as well as the absence of the typical symptoms, would distinguish it.

Infantile Spinal Paralysis.

In the initial stage of infantile spinal paralysis, the age, tottering walk, history of fall, accentuation of pains and hyperæsthesia of the joint, closely correspond to early hip disease. Later, the absence of pain and swelling, the abnormal mobility, the extreme atrophy of the whole limb and coldness, render the diagnosis patent. Acute hip disease may, however, occur in a leg affected by infantile paralysis, of which Mr. Savory[1] records an instance. The differential points which establish the diagnosis in infantile spinal paralysis are: the character of the walk, which is a tottering, not a stiff gait; the absence of reflex muscular spasm; the degeneration reactions given with the galvanic current, the normal formula of $CCC > ACC$ being reversed (*vide* Infantile Spinal Paralysis); and the loss of the faradic reaction within the first week, an important and easy diagnostic point for the general practitioner.

Congenital Dislocation.

The persistent limp and the pain of sprains from repeated falls are the only symptoms in congenital dislocation which have any resemblance to hip-joint disease, but the congenital nature of the limp, the excessive mobility of the joint, the entire absence of reflex muscular spasm or limitation of motion, and in fact, of all the important symptoms of the latter disease, lead to its easy recognition. Moreover, congenital luxation is more frequently bilateral than is hip disease.

Sacro-iliac Disease.

This affection imitates hip-joint disease in the lameness, pain when the joint surfaces are pressed together, elongation of the limb, and often in the presence of abscesses. It differs from hip disease in that the seat of greatest tenderness is different, the limb is not abducted, there is no shortening, and no pain on moving the hip if the pelvis be fixed, nor from pressure on the trochanter and counter pressure

[1] Brit. Med. Journ., 1887.

on the tuber ischii; the patient inclines the body to the opposite side and not forward in walking, and the pelvic distortion is permanent and absolute, and not, as in hip disease, temporary and relative. Tubercular osteitis of this articulation, though exceedingly rare, does undoubtedly exist.

Hysterical Affections.

Functional joint disease of the hip occurring in young anæmic women, particularly of the upper class, coincides closely with the *bona-fide* organic disease. As *coxalgie hystérique* it is exceptionally rare in men, Charcot having observed but one case in the Salpêtrière. Pain, which is so exquisite as to resent the slightest contact, is localized in the region of Poupart's ligament, over the sacrum, or in the lower part of the thorax. The limb, rigidly fixed during any manipulation, relaxes under gentle pressure when the attention is withdrawn or the examination completed; the nutrition of the limb is intact; there is no trace of muscular atrophy. The greatest difficulty is experienced where the neurotic element exaggerates a true arthropathy. In such, a true estimate can only be made by the discovery that the symptoms are out of all proportion to the objective signs. A correct diagnosis may be made by observing the presence of an inherited or acquired neurotic diathesis; the simultaneous appearance of the pain and lameness; the correspondence of the areas of hyperæsthesia and paræsthesia to the distribution of certain nerve-branches; the yielding readily to forced movements of the muscular spasm about the joint, and of the effect of anti-neurotic medication upon the disease. In the later stage anæsthesia will be an important aid, since the muscular rigidity of the neuro-mimetic coxalgia will disappear under its administration, while the permanent contraction of hip-joint disease will remain.

Prognosis. Hip disease tends to recovery, with more or less deformity in the majority of cases, the prognosis in individual cases being directly influenced by the age of the patient, the variety of the disease, the associations of complications, and the efficiency of treatment. The prognosis is likewise affected by the hygienic surroundings and social status of the patient, the mortality being much higher in dispensary and hospital than in private cases. When the disease makes its appearance after puberty it is less manageable, more extensive, and more fatal than when it occurs at an earlier period.

In the arthritic variety of the affection the prognosis is most favorable, being more serious in the femoral, and still more grave in the acetabular variety. The coexistence of some grave organic disease of an internal organ, as Bright's disease, phthisis, tubercular meningitis, or caries of some other portion of the skeleton, as the vertebræ, renders the prognosis most unfavorable.

In a like manner does efficient treatment, early commenced, favorably influence the subsequent course and termination of the disease. Spontaneous recovery of extremely severe cases sometimes occurs,[1] but the early employment of proper and efficient treatment will prevent complications and otherwise directly and favorably influence the prognosis in all cases.

Under prognosis the results of efficient treatment, relapses, the causes of death, the time required for treatment, the prognosis in double cases, and the effect of operation upon the prognosis, must all be considered.

Efficient treatment. The mortality in all cases which have been under treatment varies from 7 per cent. to 73.2 per cent., the increased mortality in all cases being directly influenced by the absence or presence of suppuration and complications. Thus, in the Alexandra Hospital, London, there were 100 deaths in 384 cases, a mortality of 26 per cent.; of these 260 were suppurating cases, among which the mortality was 33.5 per cent.; whereas, in the Clinical Society's Report, in 1880, the mortality reached 30.4 per cent. in the suppurating, and about 7 per cent. in the non-suppurating.

Cazin reports in 80 cases of suppurative cases treated at the hospital in Berck, 12 per cent. died.

Gibney reports in 80 cases, 48 of which had abscesses, 12½ per cent. died.

Jacobson records 73.2 per cent. of deaths out of 63 suppurating cases that were treated without operation (*vide* Hueter).

Wright found in 100 cases treated without operation, only 35 of which could be traced as to sequel, 9 were unrelieved or relapsed, and 9 were dead, or 25.2 per cent.

Hueter reports a mortality of 27 per cent.; Billroth, of 31 per cent., and Yale concludes, after carefully considering all the statistics bearing on the subject, that the mortality was not above 30 per cent., so that the mortality in all cases of chronic arthritis of the hip joint may be said to range between 7 per cent. and 30 per cent.

---

[1] *Vide* Bradford and Lovett, Orthop. Surg., p. 295.

The frequency of relapse is also shown in these statistics, there being 9 unrelieved or relapsed in 35 cases treated by Wright, and 6 relapsed in the 51 cured cases investigated by Shaffer and Lovett. These result from the infliction of fresh injury, the too early use of the limb, or are induced by the failure of the health from intercurrent disease, as scarlet fever, measles, etc., and consequently influence the prognosis. In such cases, as pointed out by Wright, it is important to distinguish between true *relapse* and a *residual abscess*, the latter being the result of irritation of some local product of former disease, with little tendency to spread, the former demonstrating a lack of sound repair in the original lesion, and tending to progress, as in the first instance.

The reports of the results of conservative treatment would tend to emphasize the importance of the influence of treatment upon the prognosis in this disease.

In the 80 cases treated by Dr. Gibney[1] by internal medication and counter-irritation alone, at the end of the disease—which ran its course in 33 cases in three years, and in 28 cases in from three to six years, and in 19 cases from six to ten years—61 of the patients could walk well, and run without discomfort; 12 walked only fairly, requiring a support at times, and 7 could not walk without crutches. In these 80 cases, 12 had, at least, an arc of 15° motion in the diseased joint, the shortening amounting, in the majority of cases, from one to three inches. Abscesses had existed in 48 cases.

Of the 41 cases who remained well of the 51 cases recorded by Shaffer and Lovett, none were incapacitated from doing a full day's work at his or her trade or occupation, and only one, a boy who had suffered from associated Pott's disease, used a cane.

In the 76 cases more recently reported by Mr. Howard Marsh,[2] the favorable results of conservative mechanical (fixation) treatment are well shown. Of 37 suppurative cases one year after discharge, recoveries were perfect in 1, excellent in 6, good in 17, and moderate in 13. In 39 non-suppurative cases recoveries were perfect in 9, excellent in 9, good in 12, and moderate in 9. The average amount of shortening was two-thirds of an inch, while 50 per cent. had a degree of movement in the affected joint classed as "free movement." In bilateral hip disease the duration of the disease in the first hip, according to Ridlon, is usually somewhat less than that of the

[1] N. Y. Med. Record, March 2, 1878.
[2] British Medical Journal, August 3, 1889.

average case of hip disease, while the duration in the second hip is usually much less than that of the first. The joint that is first attacked is often the last to recover.

Causes of Death. Death in hip disease may occur from the general dissemination of tuberculosis, as in tubercular meningitis, phthisis pulmonalis, or general miliary tuberculosis; from lardaceous disease of internal viscera; from pyæmia and septicæmia; from exhaustion from suppuration; from intercurrent disease, as measles, scarlatina, etc.; and from operation. Of the 614 cases recently analyzed at the Alexandra Hospital there were 85 deaths, as follows: meningitis, 12; disease of the lungs, 5; amyloid disease, 9; following amputation, 3; exhaustion, 2; uncertain, 4. Of the 96 deaths after suppurative hip disease at the same hospital there were from meningitis, 16.7 per cent.; albuminuria and dropsy, 20.8 per cent.; phthisis, 5.2 per cent.; exhaustion, 9.4 per cent.; intercurrent disease, 7.8 per cent.; and after operation, 9.4 per cent. In the Clinical Society's report, 1881, in 260 cases with suppuration treated without excision, 30.4 per cent. died from causes connected with the disease, of which 9.2 per cent. died from tubercular disease; and in 124 cases without suppuration treated without excision, the total mortality was 10.5 per cent., of which 7 per cent. died of tuberculosis. In the same report of Mr. Croft's 45 cases of excision, 7 cases died from results of operation, and in Mr. Bryant's and Mr. Baker's 203 cases of excision, 13.7 per cent. died directly from the operation. Likewise Mr. Wright,[1] after summarizing the 15 deaths which occurred in his 100 cases of excision, says an examination of these cases will show that in only two instances did death result directly from the operation.

Time Required for Treatment. Cases of hip disease under thorough and efficient conservative treatment will require from two to four years to permanently establish a cure, and without mechanical treatment the disease will run its course in from three to ten years.

Thus, in the 39 cured cases reported by Shaffer and Lovett, 31 required from two to four years to effect a cure, and the remaining 8 from four and a half to eight years.

In the 80 cured cases reported by Gibney, 33 ran their course in three years, 28 in from three to six years, and 19 in from six to ten

---

[1] Hip Disease in Childhood, p. 128.

years. Even after all the signs have disappeared, it is better for a time to continue the use of the splint, or to substitute a convalescent splint in order to avoid relapse.

Amount of Deformity. Mild cases may recover with perfectly free motion and without either deformity or shortening (Fig. 50);

FIG. 50.

A.   B.   C.

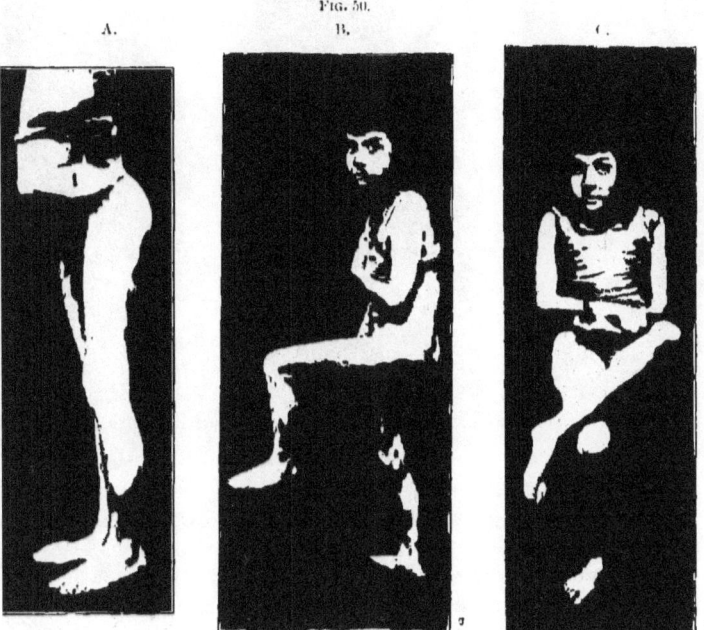

Author's cured case of disease of the left hip-joint, one-eighth inch shortening, showing, A, extension ; B, flexion ; and C, rotation of joint.

but where suppuration has occurred, and particularly in the femoral and arthritic varieties, unless persistent precautions have been taken with regard to position, ankylosis with great deformity will ensue. If ankylosis occur, the position of the limb as regards locomotion is most important. Obviously the less the flexion and abduction the more favorable the position. Even severe grades of distortion may, however, subsequently be entirely removed by osteotomy.

In all cases where suppuration has occurred, shortening to a greater or less degree is the rule. (Fig. 51.) The degree of ulti-mate shortening will depend upon the position of the limb, dislocation true or false, actual destruction of osseous tissue by disease or operation, and arrest of growth. The amount of shortening increases slightly subsequently from permanent retardation of growth.

FIG. 51.

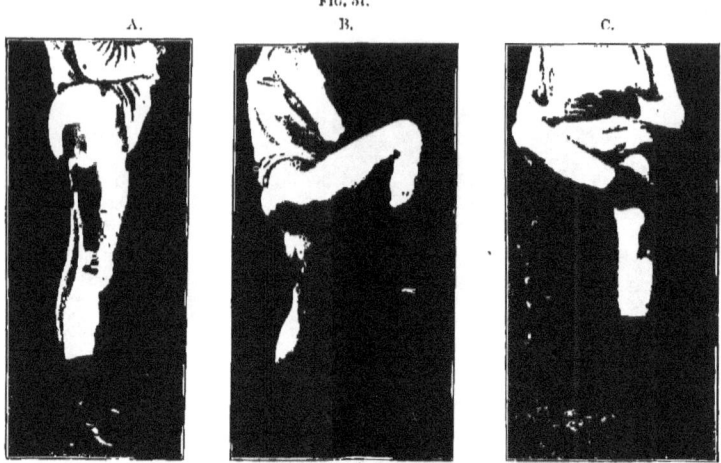

Author's cured case of disease of the right hip-joint, one-fourth inch shortening, showing, A, extension; B, flexion; and C, rotation of joint.

By massage and exercise muscular atrophy may be diminished but never entirely disappears. The appearance of abscess is significant either of inefficient treatment or of the advancement of the disease in spite of thorough treatment. In the former event it demands greater care and vigilance, in the latter it betokens a serious prognosis and a high mortality.

Treatment. The treatment of hip-joint disease has in later years received much attention and been greatly modified. From the time when the remedies employed were entirely of a constitutional nature, little attention being given to local measures, to the present time, when complicated mechanical appliances are employed to the complete exclusion of local remedies, many decided advances have been accomplished and many important principles have been established.

In conjunction with the employment of mechanical means the writer would urge the importance, as adjuvants to a speedy cure, of improved hygiene, generous diet, sufficient sleep, and such constitutional medication as the general condition of the afflicted individual may indicate. In some instances counter-irritation and local applications may be required to meet some indication, but the expectant treatment as a plan of treatment is happily a thing of the past, as is also the Hutchinson physiological method of the use of crutches and high shoe without a splint.

Local treatment in hip disease may be divided into

1. Conservative or mechanical, and
2. Operative or surgical.

The conservative or mechanical method aims to put the inflamed joint at rest by recumbency with traction, by fixation, or by traction, the latter two being especially employed as portative appliances.

Recumbency with extension is a well-recognized plan of treatment, and one employed at times to meet certain conditions by those who most valiantly support locomotion with portative apparatus. This may be accomplished by means of the Buck extension, the so-called "stretcher splint," or "extension tray," the gouttière de Bonnet or wire cuirass, and the "portable bed."

The "Buck extension" is employed as in the treatment of fractures of the femur and as described in all systematic works upon general

FIG. 52.

Author's plaster hip extension applied.

surgery. To secure counter-extension by means of the body-weight the foot of the bed should be elevated. From one-half to two bricks or an equivalent weight should be employed to make extension. A more elegant extension may be applied by cutting the plaster as for the long traction-splint. Two strips of adhesive plaster the length of the entire limb, about four or five inches wide at the upper end, and one-third the width at the lower, are prepared by cutting the plaster into five tails. From the upper end of the centre tail a piece four to six inches long is cut and added to the lower end to reinforce it. The two applied ends are folded upon themselves and buckles attached, and the whole thus prepared are applied to the lateral aspect of the leg, the buckles immediately above the malleoli and the centre tails extending the entire length of the limb. (Fig. 52.) The lower tails are wound spirally about the leg, overlapping each other, the

FIG. 53.

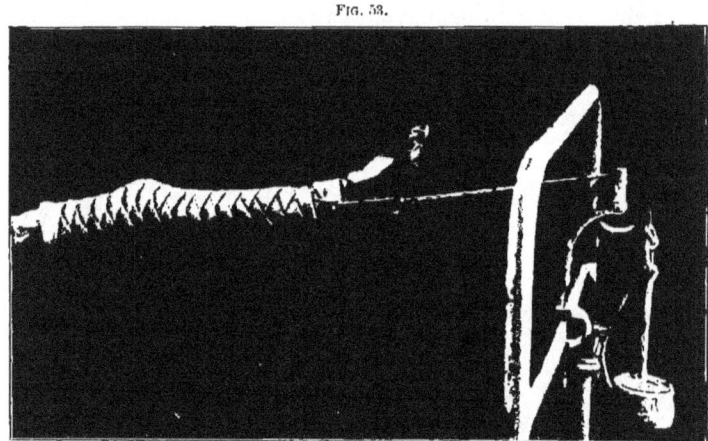

Author's weight and pulley apparatus for bed traction in hip-joint disease.

other two pairs are wound spirally about the thigh, and the whole is secured by a spiral-reversed roller bandage extending from the ankle to the perineum. Extension from the buckles is made with a steel bar about six inches in length, perforated in the middle for the extension cord, and having the leather straps extending from either extremity. (Fig. 53.) The advantage of such an apparatus is the ease with which a long traction splint may be applied at one time and extension at another.

The stretcher splint or extension tray consists of an oblong frame made of bar iron one-fourth by one inch for small children, one-fourth by one and one-fourth inches for older children, and shellacked, varnished or galvanized to prevent rusting. The frame is then covered with canvas, excepting a two- or three-inch space in the centre corresponding to the anal opening, tightly laced on the under side. Traction is made with adhesive strips in the usual way, the cord from the centre of the stirrup passing over a wheel attached to an upright steel piece slid on the lower end of the frame, so arranged that the wheel may be elevated or lowered to make extension in the line of the deformity. Counter-extension is made by two perineal straps attached to the frame or to an arm arching over the hips from the affected side and firmly secured to the side bar of the stretcher.

FIG. 54.

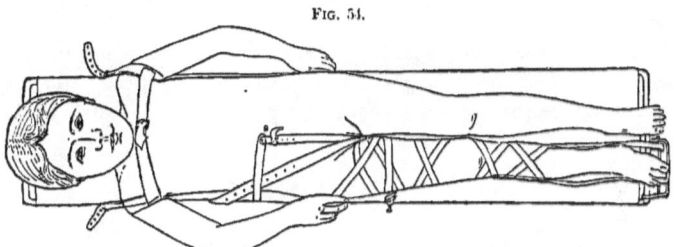

Bed frame with traction apparatus added. (BRADFORD and LOVETT.)

(Fig. 54.) These straps pass through the central opening and upward, to be attached to the stretcher. Shoulder-straps to prevent rising may be added, but are seldom required. Upon this stretcher or extension bed the child may be carried from place to place.

The gouttière de Bonnet, or wire cuirass fulfils the same indications—recumbency and extension—but is more expensive. The

FIG. 55.

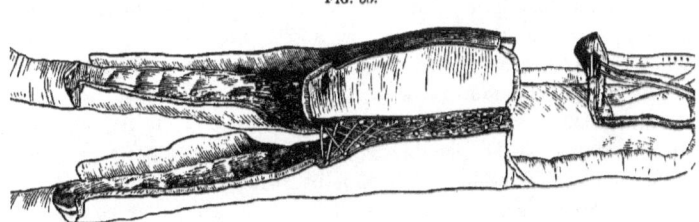

Phelps' plaster-of-Paris portable bed for hip-joint disease.

portable bed, as constructed by Dr. Phelps, of New York, possesses all the excellent qualities of the wire cuirass, and has the advantage of cheapness, being a plaster and wooden cuirass which any practitioner may readily construct. (Fig. 55.) A board of three-quarters of an inch spruce is cut to correspond to the shape of the child, four inches longer and three-quarters of an inch wider than patient. The child is then laid upon the board and enveloped with a plaster-of-Paris bandage from the feet to the axilla to a thickness of three-eighths of an inch. As the plaster bandages are rolled on they should be nailed to the edges of the board, thus making the board and plaster one. When the plaster is set the front is cut away. The bed is then lined, a front put on and lacings put in, or the child can be held in place by bandages. Extension is made to the foot-piece.

This furnishes a cheap and excellent mode of treatment, particularly in the more acute stages of the disease, and in it the child may be readily carried about without the possibility of injury to the affected part.

The English Method.

Fixation, as a principle in the management of hip-joint disease, enters into the modes of treatment before described, but as an element in the employment of portable splints it deserves further notice. Fixation without extension is the principle of the so-called English method, to distinguish it from splints constructed on the principle of extension, or the so-called American method.

They are all constructed upon a combination of the physiological and fixative methods, and aim to immobilize the hip-joint by plaster-of-Paris bandages, leather or metal splints applied to the hip, pelvis, and thigh, and the use of crutches and a patten. Plaster-of-Paris bandage may be applied from the ankle of the affected limb up to the axilla, encircling the limb, pelvis, and thorax. (Fig. 56.) As a temporary measure it has advantages, particularly if applied under ether for the correction of deformity, but as a permanent dressing it is clumsy and uncleanly. Moreover, it does not firmly fix the trunk above the pelvis, the possible motion of the lumbar vertebræ interfering with the fixation of the hip-joint through the movement of the pelvis. Moreover, it has no effect upon reflex muscular contraction and the coincident intra-articular pressure, since no traction is applied, and consequently will not prevent the destruction of the head of the femur and perforation of the acetabulum from absorption. The usual deformities peculiar to the disease may be prevented.

The same remarks apply to the fixation leather and metal splints, which, not extending so high in the thorax or so low in the limb, do

not fix the joint so well as the plaster dressing. They fulfil certain
indications, however, and in some cases accomplish good results, but
as a mode of treatment to the exclusion of others they are not to be
recommended. Dr. Vance's leather splint represents the best of this
class and affords excellent fixation. It is made of best saddle skirting, the pattern being taken of the sound limb in the position most
desirable for the diseased
hip. It has the advantages
of being easily constructed,
readily removed, a n d
cleansed without injury,
and the ease with which it

FIG. 56.

FIG. 57.

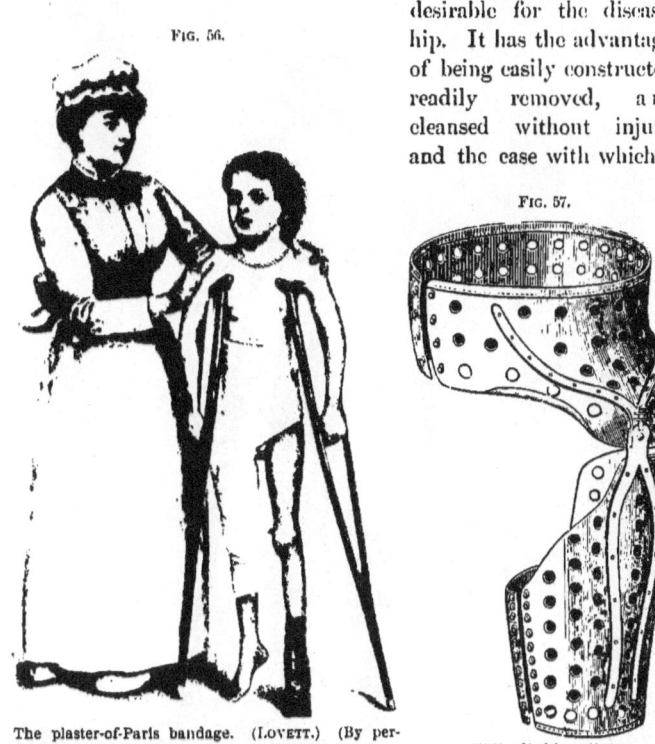

The plaster-of-Paris bandage. (LOVETT.) (By permission of the Trustees of the Fiske Prize Fund.)

Willard's hip splint.

can be changed to suit the position of the limb. A very ingenious
fixation splint is the one devised by Dr. Willard. (Fig. 57.) It is
made of enamelled leather over a cast, and has a simple joint over
the articulation so that the patient can sit down with comfort. It
is always used in connection with crutches and a high shoe, and, as
Dr. Willard says, it is applicable to a limited number of cases—
those in which the inflammatory symptoms are not acute.

The most perfect type of a pure fixation splint is that of the late
Mr. Thomas, of Liverpool, of which many of the leading English
surgeons speak highly at the present day, and which, in Mr.
Thomas' hands, was undoubtedly an efficient apparatus, as almost
any apparatus might become considering the skill, accuracy, and
thoroughness with which it was applied.   (Fig. 58.)

FIG. 58.

Thomas' splint applied.

It is simple in construction, and can be readily made by the practitioner.   It consists of a malleable iron bar extending from the lower
angle of the scapula to the lower third of the leg—just where the
calf begins.   This should be one inch by one-quarter inch for an
adult, and three-quarters by three-sixteenths for children.   Three
strips of hoop iron, one (the chest crescent) for the upper extremity,

four inches less than the circumference of the thorax, attached at right angles to the upright; one (the thigh crescent) for the thigh, two-thirds the circumference of the thigh at its upper third; another (the calf crescent) for the calf, one-half the circumference of the limb at this point. The splint, fitted to the posterior part of the trunk and affected limb, is held in position by a strap and buckle attached to the upper band; suspenders are used over the shoulders, and a roller bandage firmly applied to the lumbar portion and the limb. A patten is worn on the sound side, high enough to clear the foot of the diseased limb, and crutches are employed.

Its proper application requires skill, and Mr. Thomas deserves credit for the great attention to detail exhibited in describing the splint. While the method of treatment has much to recommend it to the profession in general, the writer, after a considerable experience in his own practice and observation in that of others, has a decided preference for other methods.

As a fixation splint in recumbent cases *with traction* it is an excellent appliance. For the correction of deformity the upright is bent at the buttock, and the splint is applied in the deformed position, the curve of the upright being lessened as the deformity yields. While the fixation of the Thomas splint is not perfect, the principal objection is that there is no traction to prevent intra-articular pressure from reflex muscular contraction. My own observation leads me to believe that abscesses are more frequent, and this impression is confirmed by comparison of the report of 62 cases of hip disease observed in the practice of Mr. Thomas, of Liverpool, by John Ridlon, M.D.,[1] of 58 of which 23 had one or more abscesses, or 39.6 per cent., and the results of a series of 63 cases from the Boston Children's Hospital,[2] in which abscesses occurred in only 23 per cent. of all cases of hip disease under out-patient treatment. Since these reports represent the best results of both the fixation and traction methods of treatment in the hands of surgeons of equal skill, under similar conditions, the comparison may be considered just and the result obvious.

Fixation versus Ankylosis.

The dread of ankylosis, according to Verneuil,[3] has led to much bad surgical practice, but recent experiments and discussions have

[1] Trans. Am. Orthop. Assoc., 1890, vol. iii., 163.
[2] Boston Med. and Surg. Journ., Nov. 24, 1889, p. 503.
[3] Society of Surgery in Paris, 1879.

clearly established "that whatever ankylosis occurs in a joint subjected to immobilization, occurs by reason not of the immobilization, but of the nature and intensity of the inflammations and of the inefficiency of the apparatus employed."[1]

Perfect immobilization, or absolute fixation, is almost an impossibility by fixation apparatus alone. False ankylosis is due to contraction of the periarticular muscles from reflex irritative spasm. Fixation by means of traction applied early will relieve muscular spasm, prevent trauma of the articular surfaces, diminish intra-articular pressure, and permit a subsidence of the inflammation, after which either motion or fixation may occur in the joint, since ankylosis, if it occur, depends directly upon the amount of disease, and not upon either fixation or traction.

FIG. 59.

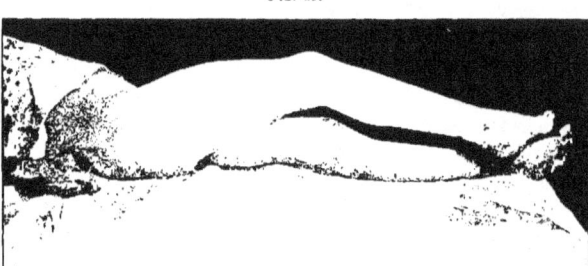

Instinctive effort at traction. (LOVETT.) (By permission of the Fiske Prize Fund.)

Phelps,[2] in a series of carefully conducted experiments upon dogs, concludes: 1. That a normal joint will not become ankylosed by simply immobilizing it for five months. 2. That motion is not necessary to preserve the normal histological character of a joint. 3. That when a healthy joint becomes ankylosed or its natural histological character changed, it is not due to prolonged rest, but to pathological causes. 4. That immobilization of a joint in such a manner as to produce and continue intra-articular pressure will result in destruction of the head of the bone and the socket against which it presses. 5. That atrophy of the muscles of the limb will follow prolonged immobilization of a joint.

If, as he remarks, these experiments prove that prolonged fixation

---

[1] Gibney : "Immobilization in Articular Disease," Trans. Orthop. Assoc., vol. i., p. 232.
[2] N. Y. Med. Journ., May 17, 1890.

will not produce ankylosis of a natural joint, that motion is not essential for the preservation of its normal function, then the causes of ankylosis must depend upon pathological conditions, and not upon prolonged fixation.

These experiments are valuable as illustrating the effect of prolonged fixation in healthy joints. It does not solve the problem of the control of reflex muscular spasm, the most destructive element in the disease. Traction, with or without fixation, will best control this muscular spasm, diminish intra-articular pressure, secure rest, and diminish inflammation. (Fig. 59.)

Traction by extension and counter-extension is the keynote of correct surgical treatment in hip-joint disease.

The American Method.

It is upon this principle of traction in the line of deformity that the long extension splint, or the so-called American method of treatment for hip-joint disease, is established. To relax the muscles by overcoming the muscular spasm, and to induce fixation and prevent concussion, are the objects sought. It would appear from recent experiments that a distracting force sufficient to separate the head of the bone from the acetabulum is neither necessary nor desirable. Experiments upon the cadaver are unreliable, since reflex muscular spasm, the most important factor, is absent. Koenig and Paschen, upon the cadaver, found slight separation, using eight pounds or more; while Morosoff was unable to separate the surfaces with less than sixty pounds. Lannelongue, however, by frozen sections found ten pounds sufficient to produce separation in a well-marked case of hip disease.[1] Upon anatomical grounds a much less weight is known to be required to produce distraction if the limb be abducted, since when the limb is in the line of the body, or adducted, the cotyloid ligament surrounding the head renders separation almost impossible.

Lovett[2] estimated from an experiment upon a healthy boy of ten years that the thigh muscles are capable of exerting a force of thirty-six pounds. It is highly probable that separation of joint surfaces in hip disease is seldom or never attained, but that the traction force employed for a long time overcomes the muscular spasm, secures fixa-

---

[1] For a thorough *résumé* of the subject of distraction, *vide* "An Experimental Study of Distraction of the Hip-joint," by E. G. Brackett, Trans. Amer. Orthop. Assoc., vol. ii. p. 207.

[2] Trans. Amer. Orthop. Assoc., vol. i. 196. *Vide* also Trans. Amer. Med. Assoc.: Sayre, 1860, p. 467; Davis, 1863, p. 141.

tion and rest to the affected part. The amount of fixation at the hip-joint secured by the long traction splint has been variously estimated, but the only experiments of any practical value are those of Lovett, before referred to. These experiments illustrate the practical

Fig. 60.

point that traction in itself furnishes very incomplete fixation, but that with the original Taylor hip splint, with a rigid pelvic band and two perineal straps, better fixation to the joint was obtained than with the newer form of splint, with only one perineal strap, although it still lacked a very little of being complete. Traction will therefore best fix the joint and fulfil the indications in the acute forms of disease when the joint between the upright and pelvic band is fixed by a screw which will permit or arrest motion, and where two perineal straps are employed.

Straight two-band hip splint applied. (LOVETT.)
(By permission of the Trustees of the Fiske Prize Fund.)

Traction Splints.

All long traction splints at present in use are patterned more or less after Taylor's modification of Davis' splint. (Fig. 60.) The principles of the so-called mechanical treatment are founded upon the two following aphorisms:[1] "1. All organs while in a state of disease require rest from the performance of their functions in the direct ratio of the amount, quality, and intensity of the abnormal movements. 2. What is rest for an organ in one condition is not necessarily rest for it in another condition; that is to say, an organ in a certain degree of progressive inflammation presents conditions essentially different from the same organ in the same relative degree of inflammation in the retrogressive stage."

The first object sought is to overcome the muscular contraction by extension and counter-extension applied to the line of the deformity. This is accomplished by means of a long steel bar extending from the

---

[1] Taylor: Boston Med. and Surg. Journ., March 6, 1879.

trochanter to below the foot, to which above is rigidly attached a sheet-steel pelvic girdle and one or two perineal bands, and which below is attached to the limb by adhesive plaster straps or bandages. To more readily adjust the appliance to varying lengths of legs, as well as to apply extension, the long steel bar is provided with a tube and sliding ratcheted bar moved by a key (Fig. 61), and secured by a spring and sliding catch. The lower part of such a splint, bent at right angles to the long upright bar, is covered with a flat leather shoe, and has attached a leather strap for attachment to the buckles

<table>
<tr><td>FIG. 61.</td><td>FIG. 62.</td></tr>
</table>

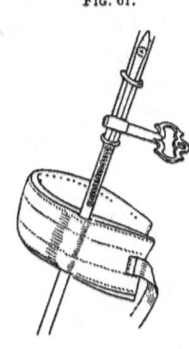 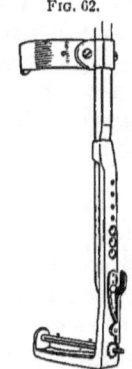

Ratchet and key extension. (LOVETT.)
(By permission of the Trustees of the Fiske Prize Fund.)

Windlass and extension as used n the long traction splint. (LOVETT.)
(By permission of the Trustees of the Fiske Prize Fund.)

upon the adhesive appliance on the leg. A cheaper extension appliance of the windlass pattern can be attached directly to this foot-piece, dispensing with the costly tube and ratcheted rod. (Fig. 62.) The pelvic band may be rigidly attached to the long bar, but is best secured by a bolted screw, which will permit motion or fixation. The perineal straps may be made of military webbing covered with canton flannel, silk, or chamois. Pads are objectionable, straps being more comfortable and less likely to chafe or excoriate. With ordinary care to cleanliness and the local application of alum and alcohol (3ij to Oj), followed by dusting with talcum or ordinary toilet powder, little difficulty will be experienced, and any reasonable amount of pressure may be made upon the perineum. Should the parts chafe or excoriate, simple cerate or vaseline will relieve, or failing in that, the splint may be removed for a few days and the patient be confined to bed.

The long bar may be variously modified to meet indications. The upper portion may be attached by two portions for convenience in dressing sinuses; it may have attachments to secure the knee laterally, or antero-posteriorly, as the U-shaped attachment of Judson, for the purpose of better fixation of the thigh; or it may be made stronger or less flexible, as in the Judson brace, and so twisted upon itself and tapered that the weight is centred near the upper part. The traction straps applied for the extending force should extend beyond the knee and well up the thigh to avoid undue traction upon the lateral ligaments of the knee. Strong surgeon's adhesive plaster is efficient. Maw's moleskin plaster, made in England, is the best, but Shivers' swan's-down plaster, made in Philadelphia, answer equally well. They should be changed about every four weeks, but, no irritation or displacement ensuing, they may remain on from three to four months. If chafing occurs beneath the dressing the parts should be dusted lightly with powdered boric acid when being renewed. Substitutes for plaster where the skin is extremely sensitive, such as cloth or leather legging, a stocking extension, may be employed, but are all inferior.

The plaster should be cut as before described under Buck's extension, and the roller bandage applied to retain them should be overseamed its entire length, or better, be secured by a narrow strip of adhesive plaster wound spirally up the leg and thigh. In applying the splint the pelvic band and perineal straps should first be secured, next the foot-piece be attached by the straps to the buckles; extension is then to be made, and finally the knee is to be secured. Such a splint may be at once employed for ambulatory purposes, a shoe with a sole sufficiently high to equalize the length of the limbs being worn on the opposite limb, but in acute cases it is better for the patient to maintain the recumbent position until the acute symptoms have subsided, or to resume this position should increased pain or deformity indicate an exacerbation of the disease. Taylor's method of reducing deformity by placing the patient upon an inclined plane, with conveniences for adapting the angle to the amount of relaxation gained, is practical and efficient.

So applied in the line of deformity, the splint has necessitated certain modifications. If simply flexion exist, the long bar may be set at an angle to the pelvic girdle, but for adduction or abduction,

Shaffer's modification of the pelvic attachment may be employed. This consists of two parts, joined by a lateral hinge. The first part is fastened to the pelvic band, and the second part is attached to the shaft of the splint. Through the everted lip there passes a screw, which operates through a button (which revolves on a horizontal axis), and which is fastened into another button (also revolving on a horizontal pivot) in the first part. By turning the screw, we can either approximate the lip toward the the first part (producing *abduction*), or, by reversing the screw, we can separate the lip from the first part and *adduct*. An efficient and more simple device is figured in Fig. 63.

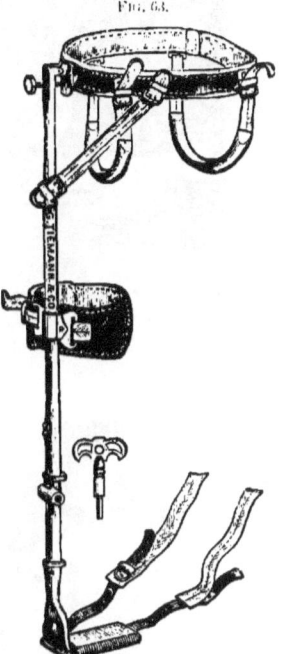

Fig. 63.

Traction splint for the left leg, seen from the back, provided with two perineal bands and an abduction screw. (BRADFORD and LOVETT.)

The writer prefers to overcome the lateral deformity by extension and counter-extension in bed (Fig. 64) before applying the splint, or should deformity become marked after the use of the splint, to remove it for a time and place the patient in bed until the exacerbation, which this signifies, has subsided. If, however, such a course is inconvenient, the splint may be worn and the patient be placed upon crutches, with a high patten upon the sound limb. This was Taylor's combination method, and the one for which Wyeth[1] claimed advantages superior to all others. Upon this principle also are the combination splints of Lovett[2] (Fig. 65) and Phelps,[3] (Fig. 66) modified combinations of the English Thomas splint with the American Taylor splint, both excellent splints for use in cases in which the fixation afforded by the Taylor traction splint is not sufficient. The latter possesses many other excellent qualities, since

[1] Medical Gazette, April 17, 1880.
[2] Trans. Amer. Orthop. Assoc., vol. i. p. 196.
[3] New York State Med. Soc., February, 1889.

8

FIG. 64.

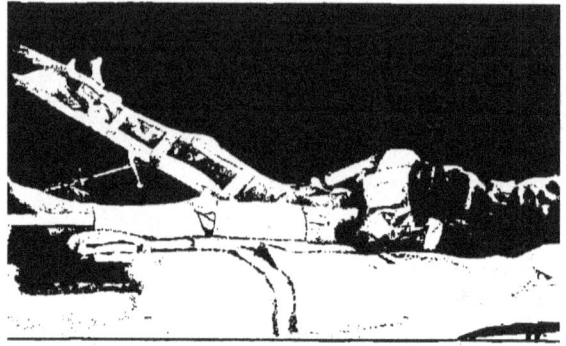

Bed traction in a case of flexion and abduction. (LOVETT.)
(By permission of the Trustees of the Fiske Prize Fund.)

FIG. 65.

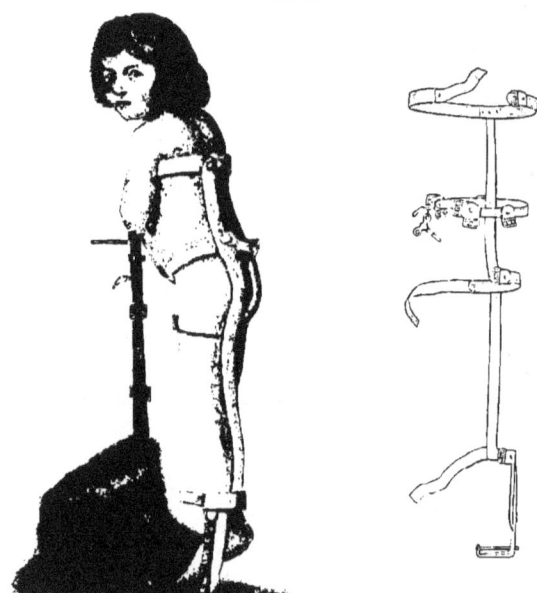

Combined fixation and traction splint. (LOVETT.)
(By permission of the Trustees of the Fiske Prize Fund.)

it is intended to prevent every motion at the hip-joint, and at the same time apply extension in a line with the neck of the femur, but is a little heavier and more unwieldy than the ordinary Taylor splint.

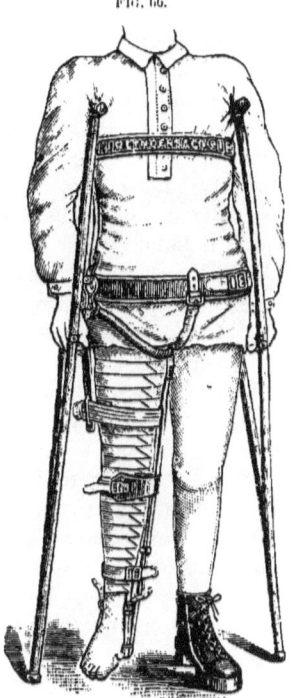

Fig. 66.

Convalescent Protective Splints.

After all the characteristic signs have disappeared for a considerable period under traction treatment, it will still be necessary to protect the limb from the jar in walking, and as a precaution against injury. For this purpose the ordinary extension may be employed by simply diminishing the extension and counter-extension until the shoe rests nearly or completely upon the ground. If it is desirable to have a joint at the knee, the convalescent splint of Taylor or one of its many modifications may be employed. In the former, "the lower steel plate is riveted to the upright, but the upper one is fastened by three 'keepers' which enable it to be raised or lowered in adapting the instrument to the length of the leg."

Phelps' combination traction hip splints.

In children, and where expense is considered, the lower portion of the long rod may have substituted for it a slotted piece, to be inserted into a slot attached to the shoe, or a piece with a flat steel sole-plate screwed to the shank of the shoe. The use of crutches as a means of protection, the sound limb being raised upon a patten, will be found useful in convalescent cases before abandoning the splint altogether. Protection will be necessary from eighteen months to two years after the complete disappearance of all the active symptoms, and in severe cases even longer, it being better to wear the splint too long than to suffer relapse from its too early removal.

Treatment of Complications. The complications which will demand special treatment are "night-cries," abscesses, and malpositions of the limb.

Night-cries usually at once subside under the use of traction in the line of deformity and fixation. Should these measures fail, a large blister over the trochanter, followed by hot poultices, with the internal use of bromide of potassium and morphia, should be employed. Salicylate of soda in full doses, as for acute articular rheumatism, has been highly recommended. If they continue unabated, operative means must be resorted to. Deep puncture of the joint or trephining of the head of the femur offers the greatest chance of relief, if induration or great fulness of the capsule can be recognized. Where these cannot be discovered a formal incision into the joint, as an exploratory operation, should be undertaken. In a recent case of this kind in the practice of Dr. Willard, where excruciating pain was the principal symptom, excision revealed the entire separation of the epiphyseal cap of the head by granulation, without a drop of pus being within the joint. The sudden subsidence of night-cries indicates the efficiency of treatment or the rupture of the capsule of the joint and the extravasation of pus.

Abscesses in many cases will rupture spontaneously, but should they assume enormous size, or exhibit a tendency to burrow to great distances, they should be incised. Small recent abscesses will sometimes disappear under the use of compound iodine ointment and pressure. Aspiration is usually unsatisfactory, and the subsequent injection of antiseptic solutions is not unattended with danger, since a death has been recorded[1] from the washing out of a small cavity with a few ounces of a 1 to 40 solution of carbolic acid.

Some orthopedic surgeons believe in the spontaneous absorption or rupture of abscesses, urging the danger of septic or tubercular infection for their fear of interfering. Judson[2] ably defends this ultra-conservative view in a recent paper, in which he says: "I should not omit to give my reasons for failing to see the importance of incision, scraping, and antiseptic closure of the abscesses in question. Incision is a tardy and fruitless procedure. The most painful stage in the history of the abscess is long past. It was present when the pus was collecting under the periosteum and in the cells of bone. If we could interfere early with the bistoury and knew where to direct

---

[1] Bradford and Lovett : Orthop. Surgery, 1890, p. 326.
[2] N. Y. Med. Journ., March 2, 1889.

its point, we might relieve the pain, and perhaps, in favorable circumstances, shorten the case and save bony tissue by dividing the thickened periosteum or breaking the shell of compact bone. But when the pus is in the cellular structures or the cavity of the joint, I do not see that the progress of the case can be materially affected by interference. If the abscess is cold, there is no painful tension to be relieved. If it is phlegmonous, tension is the result of inflammatory infiltration, and can be relieved only by extensive and multiple incisions. If we operate in either case, we substitute artificial for natural closure, and with the best antisepsis we gain nothing by operating unless we reach and scrape out the purulent depot or the interior of the joint; and then nothing unless we remove the eroded cartilage and exfoliating bone, and excavate the focus; and then nothing, in many cases, unless we remove large quantities of bone or excise the joint. And if we operate in the manner described, we do not avoid the necessity of bringing to bear the best mechanical treatment and hygienic control—which, if they are supplied, will bring about a recovery, whether we operate or not, by the slow but sure process of natural repair, with the better result the less we interfere with the soft parts, as a general rule."

My own opinion is that if they are enlarging rapidly, or tending to spread, or if pressure is being made upon important organs, they should be evacuated at once under antiseptic precautions by incision. As Gross[1] suggests, great care should always be taken to avoid pressure upon the part. The cavity should be entirely washed out by, first, a continuous flow of bichloride solution and then boiled water, and one or two drachms of iodoform emulsion be injected, and the whole sealed by a full antiseptic dressing.

In this manner the pyogenic membrane remains intact, and the dread of both septicæmia and tubercular infection may be guarded against; and a few subsequent dressings will entirely remove even abscesses of considerable size.

The rule regarding the early evacuation of pus wherever formed is somewhat modified in these cold abscesses, since the encapsulation wall differs in thickness, strength, and development from that formed in acute abscess. Hence, if the development has been gradual, and the abscess remains stationary and gives the patient no inconvenience or elevation of temperature, it can with perfect propriety remain

---

[1] On the Bones, 1831, p. 370.

unopened, in the hope that absorption, even at a late date, may occur.

In opening an abscess an exploration of the joint cavity, if accessible, may, under antiseptic precautions, be made.

In all instances in operating upon an abscess the surgeon should be prepared to remove sequestra or to proceed with an erasion or excision should the exigencies of the case demand it.

Deformity.   The correction of the malpositions of the thigh incidental to the disease has already been given.   Recumbency, with continued extension, will often accomplish correction, even in severe grades of fibrous ankylosis.   In bony ankylosis, as in fibrous ankylosis, where mechanical means have failed operative measures should be employed.

These include multiple myotomy and tenotomy, *brisement forcé,* osteoclasis, and osteotomy.

Multiple Myotomy and Tenotomy.

The contracted structures which will require division are the tensor vaginæ femoris muscle, the fascia lata, and the intermuscular ligament between the rectus femoris and tensor vaginæ femoris.   These can be divided subcutaneously in the majority of cases; but, if the muscular contraction be more extensive, an open incision under antiseptic precautions should be performed after the V-shaped method of Billroth.   This would prevent the injury of large bloodvessels or nerves, but would not otherwise facilitate the correction of the deformity, since, as pointed out by Volkmann, the skin is unimportant in maintaining the contraction.   In rare cases the psoas and iliacus may require division.   This should always be performed through an open incision on account of the immediate proximity of the femoral artery and its branches, although a successful case of subcutaneous section has been recorded.[1]

After multiple tenotomy the limb should be brought into a corrected position, manual force being employed if necessary, and retained either by plaster-of-Paris dressing or preferably by weight-extension.   If fracture of the surgical neck occur before the deformity yields to the manual force it does not complicate the case, since the deformity may then be corrected and the fracture be dressed subsequently as after mechanical osteoclasis.

[1] Case of Dr. Abbe, Trans. Amer. Orthop. Assoc., vol. II. p. 4.

In forcible manual correction there is always some risk of the manipulations lighting up an active process in an old quiescent tubercular deposit, and osteoclasis or osteotomy is therefore preferred. Mechanical osteoclasis for the correction of hip deformity lacks precision, and has of late been abandoned for the more exact and, under the present antiseptic measures, equally safe operation of osteotomy.

Osteotomy.

Under this head are included several operations, all of which aim at a correction of the bony deformity, with or without a movable joint.

To Dr. J. Rhea Barton,[1] of this city, in 1826, belongs the credit of first successfully correcting osseous hip ankylosis by osteotomy. This operation did not consist of the excision of a wedge-shaped piece of bone, as has been stated, but of a linear section through the femur between the trochanters.

Dr. J. Kearney Rodgers, of New York, in 1830 modified the operation by removing from between the trochanters a disk-shaped piece of bone. Dr. Lewis A. Sayre still further modified the operation by removing a segment of bone from the same situation, the upper section being semicircular, with its concavity downward, and the upper end of the lower fragment being rounded off in imitation of a ball-and-socket joint. Mr. Adams, in 1869, divided the neck of the femur through a small wound. Mr. Gant, in 1872, introduced his infra-trochanteric operation, dividing with a saw the femur below the lesser trochanter. Volkmann, in 1873, removed a wedge-shaped piece from the outer side of the greater trochanter of the femur, breaking the rest; subsequently substituting an excision of the joint with a chisel and a gouge, first performing a regular linear osteotomy.

More recently Barwell has modified Sayre's operation, using a chain-saw to divide the femur above the great trochanter, and Reeves modified Barton's operation by performing an incomplete infra-trochanteric section of the femur, just below the greater trochanter. Gant's operation has been modified by Maunder by dividing the femur below the lesser trochanter with chisels instead of the saw, and Macewen's osteotomes have also been successfully employed by a

[1] North. Amer. Med. and Surg. Journ., 1827, iii. p. 279.

number of surgeons. Of the various methods it may be said that false joints are of doubtful utility if obtained, besides increasing the danger of the operations—the mobility of the pelvis compensating for the loss of a movable articulation. The operations, therefore, of Adams and Gant (modified) are to be preferred in suitable cases, the former being employed where the ankylosis has resulted from acute traumatic inflammation and the neck remains long and intact, and the latter in all other cases of bony ankylosis of the hip-joint. " When," says Mr. Gant,[1] " in consequence of continued disease of the hip-joint, the head of the femur has disappeared, leaving only a stunted nodule of bone representing the neck above the trochanter, in such a case the operation of section in the femoral neck cannot be performed, there being no neck to divide. Even when supra-trochanteric section is practicable, the state of the neck may render this operation abortive. The seat of the operation will be in an almost carious portion, which is unfit to yield a fibrous union."

The application of the operation of division of the neck of the bone is necessarily limited, and Mr. Adams considers most favorable for operation : 1. Cases of rheumatic ankylosis, because in rheumatism no destruction of the bone exists, and the head and neck of the bone always remain of their full natural size. 2. Cases of ankylosis after pyæmic inflammation, most especially in its subacute form, from which the patient often recovers ; in these cases destruction of the bone rarely, if ever, exists, the cartilages only being more or less destroyed. 3. Cases of ankylosis after traumatic inflammation of the joints, in which little or no destruction of the bone occurs. 4. The most unfavorable cases for operation are those which occur in strumous subjects, where destruction of the head and neck of the femur has taken place.

The object sought in either operation is the correction of the deformity by more or less firm fibrous union. Adams' operation, from the destructive character of hip-joint disease, is inapplicable in most cases of this nature.

The fact that section of the femur below the lesser trochanter, in addition to the advantages previously mentioned, gives better results in correcting the deformity and in lessening the chances of recurrence, is explained by Wharton[2] by the contraction of the psoas magnus and iliacus internus muscles inserted into the lesser trochanter.

[1] Brit. Med. Journ., Oct. 18, 1879.
[2] Amer. Journ. Med. Sci., July, 1883, p. 105.

When the section is made below this insertion these muscles remain attached to the upper fragment, and do not, therefore, interfere with the straightening of the thigh. The section should be made as close as possible to the lesser trochanter, for the shorter the upper fragment the less perceptible will be the resulting angle of union, and the more natural the appearance of the limb.

Fig. 67.   Fig. 68.

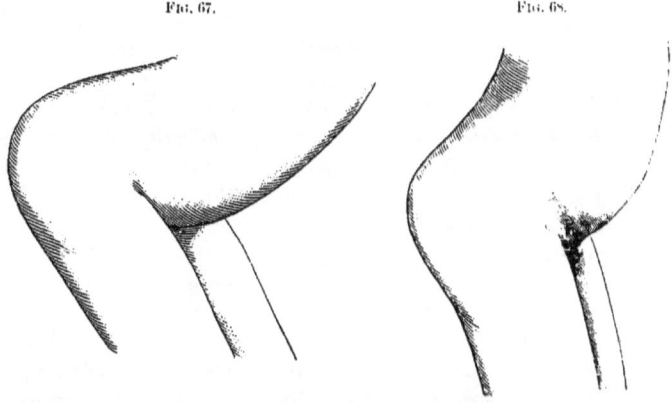

Aukylosis of both hips; subcutaneous osteotomy below the trochanters. (ASHHURST.)   Ankylosis of both hips; osteotomy below the trochanters. (ASHHURST.)

Adams' operation is thus performed : The parts having been sterilized, a narrow, straight bistoury, or a special tenotome with a very long rounded or blunt portion and a cutting surface a little over an inch long, is entered a little above and in front of the great trochanter, and carried straight downward and backward to the neck of the femur, dividing the muscles and incising the capsule freely. The knife is then withdrawn, the narrow, firm, pistol-handled saw of Mr. Adams is introduced and the neck divided. If, after the neck be severed, the rectus, tensor vaginæ femoris, adductor longus, or sartorius tendons resist correction they should be divided with the tenotome. Antiseptic precautions are essential, and recumbent extension should be maintained for five or six weeks.

Gant's operation, as modified by Maunder, of London, is thus performed : The parts having been sterilized, and a marble slab or large sand-bag having been placed between the thighs as high up as the perineum will allow, a sharp-pointed, narrow, straight bistoury is entered just below the great trochanter upon the outer aspect of the thigh, or the posterior aspect, as appears to be Mr. Gant's practice, and carried directly to the bone, completely incising the soft part. It is withdrawn, and the osteotome introduced with its blade in the direction of the long axis of the femur, and turned at right angles as it reaches the bone. The osteotome is then driven with sharp blows until the spongy structure of the bone is divided, when the instrument is to be withdrawn, and fracture completed with but little force. The limb is then brought straight, tenotomy being performed, if necessary, upon the resisting structures. An antiseptic pad and a weight-extension complete the dressing. Should the limb be shorter than its fellow, provision may be made for this by abducting the limb sufficiently to compensate for the subsequent obliquity of the pelvis after recovery. The permanent shortening may be relieved by the use of a high shoe or patten, or one of the special apparatuses designed for this purpose.

The statistics of these two operations are at present unreliable, but under strict antiseptic methods these operations are the most satisfactory in surgery, and the risks as regards suppuration, limb, or life are no greater than in simple fracture of these portions of the femur.

Operative Treatment. So efficient is conservative mechanical treatment in the management of this disease that operative treatment is seldom or never required in private practice where the patient has been under complete control of the surgeon from the inception of the disease. In hospital practice, and in neglected and badly treated cases, operative measures become absolutely necessary for the salvation of life or limb. The proper selection of cases for operation is most difficult, and will depend much upon the diagnostic skill of the surgeon. Operative treatment should always be preceded by thorough mechanical treatment. There are conditions in which the propriety of operative interference cannot be questioned, as in albuminuric cases from prolonged suppuration, or where from great destruction of tissue or exquisite prolonged pain, as in acute cases of caries sicca without suppuration, dissolution is imminent. Other cases will depend almost entirely upon the surroundings and social condition of the individual, since in private cases in whom mechanical treatment is thoroughly and intelligently carried out, the necessity for operation is exceedingly rare.

The operative measures employed in the treatment of hip disease include :

1. Aspiration of joint fluid.

2. Ignipuncture.
3. Trephining and drilling.
4. Incision of joint.
5. Exploratory incision, erasion, and drainage (Willard).
6. Excision.
7. Amputation.

Aspiration of Joint Fluid. The removal by aspiration of joint fluid in cases of induration and great distention of the capsule is of great utility, especially in acute cases. In traumatic synovitis, which may speedily degenerate into a tuberculous arthritis, aspiration may at once arrest the disease. If the parts, hands, and instrument be sterilized, no harm can possibly result from aspiration. The puncture may be made anteriorly a little above and in front of the great trochanter, but is preferably made posteriorly behind the great trochanter. In removing the aspirator care should be taken to prevent the ingress of air, and the wound should be sealed with iodoform collodion and cotton or an antiseptic dressing. Rest and extension in the line of deformity should be subsequently continued until all active symptoms have subsided.

Ignipuncture. As advocated by Kolownin,[1] and practised by Park, Bradford, and others, ignipuncture consists of electrical tunnelling or cautery of the bone, the object being to destroy the localized tuberculous area. The trephine may be first employed in the direction of the longitudinal axis of the neck of the bone, and either the Paquelin cautery or the electric cautery is then introduced and the spongy tissue in the region of the epiphyseal juncture is thoroughly cauterized. The antiseptic technique must be employed. It is especially advocated in femoral coxalgia, but as yet it is not an established surgical procedure.

Trephining and Drilling. As proposed and performed by Fitzpatrick, in 1867, the operation consists in trephining into the great trochanter a short distance and attempting to destroy the diseased area by inserting a stick of potassa cum calce. As revived by Stoker and performed by Lovett and others, it consists in trephining with a small trephine from the outer surface of the great trochanter in the direction of the axis of the neck as deeply as possible without injury to the joint, and subsequently with a curette evacuating the diseased focus; a drainage-tube and antiseptic pad complete the dressing.

---

[1] Boston Medical and Surgical Journal, April 26, 1885, p. 392.

Greig Smith and MacNamara employ a drill or gouge in the same manner. The operation is a recognized and valuable one where the disease is confined to the epiphyseal juncture of the upper extremity of the femur, but even these would appear to the writer to be better treated by exploratory incision and erasion, presently to be described.

Incision of the Joint. A straight posterior incision behind the great trochanter is of great service where in acute cases aspiration in the same locality has failed to relieve the capsular distention and acute pain. It has also the advantage of permitting the removal of diseased and softened bone by means of a curette. (Fig. 69.) A

Fig. 69.

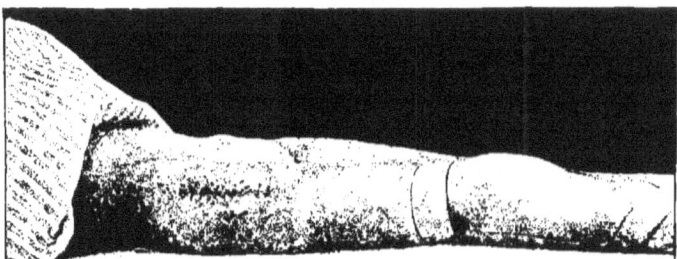

Hip abscess incised and sewed up; firm cicatrix. (LOVETT.)
(By permission of the Trustees of the Fiske Prize Fund.)

narrow straight bistoury is entered behind the great trochanter and thrust directly to the joint, incising the capsule. The benefit, if any, is at once experienced. As a simple incision it has no particular advantage over aspiration, except cases in which the joint contents are too thick to pass through the aspirator.

Exploratory Incision, Erasion, and Drainage. As performed by DeForest Willard[1] the operation consists in making an incision over the most dependent portion of the abscess, avoiding the large vessels and nerves, and also paying attention to the integrity of the muscular fibres as far as possible, or following sinuses if they exist. By irrigation with hot sublimate solution (1 : 1000) a protective layer of coagulated albumen is formed upon the freshly-cut surface to pre-

[1] Trans. Amer. Orthop. Assoc., vol. ii. p. 145.

vent tuberculous infection. The opening into the joint is enlarged and with a hollow-handled sharp spoon "every particle of diseased tissue that can be reached is cut away." The spoon Dr. Willard employs is a sharp Volkmann spoon with a bulky handle tunelled longitudinally

FIG. 70.

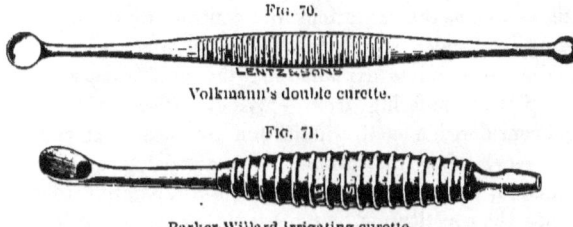

Volkmann's double curette.

FIG. 71.

Barker-Willard irrigating curette.

with a large bore so as to allow a free flow of water from the rubber supply-tube. The roughened ends of bone are sawed off or gnawed by strong rongeurs, and diseased fragments of soft tissue are removed with knife and scissors. The wound is then allowed to fill with fresh blood-clot, and a rubber drainage-tube inserted and closed with

FIG. 72.

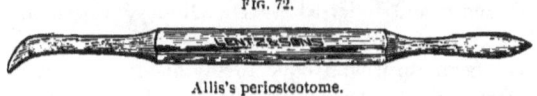

Allis's periosteotome.

catgut. A large antiseptic dressing and compress cover the wound. The operation as thus performed substitutes in many cases the graver operation of excision, and where the bone destruction has been small it offers great advantages over the latter operation.

Rest, weight-extension, and fixation are essential to a cure, and should be continued until every sign of the disease has disappeared.

Excision. In cases requiring operative interference where exploratory incision reveals areas of diseased bone too extensive to be removed by erasion, excision will be required. In well-selected cases it is an operation of exceeding great value, and one which, since its introduction forty years ago by Sir William Fergusson and his school, has had an important effect in diminishing the number of amputations for joint disease. It must not, however, be forgotten, as pointed out by Dr. John Ashhurst, Jr.,[1] that the operation of

[1] Principles and Practice of Surgery, ed. 1878, p. 605.

excision is, in every region of the body, at least as fatal as the corresponding amputation. It should not become a routine practice in bad cases, nor should it in selected cases be considered as a *dernier ressort*.

For a description of the operation of excision and its various methods, as well as the indications and contra-indications for excision in general, the reader is referred to works upon general surgery. One exception requires explanation : the performance of excision upon a patient suffering from amyloid disease of the kidneys. Though considered a contra-indication by most systematic writers upon surgery, when indications of commencing amyloid disease make their appearance, excision is almost imperative to check the drain upon the constitution, notwithstanding the immediate risks of the operation, and may be undertaken with strong hope of arresting the kidney disease.[1]

The views of accepted authorities upon the indications for excision are contradictory, some considering it a last resort, others employing it as an accepted method of treatment in selected cases, while still others resort to the operation in all cases once suppuration has occurred. Thus the remarks of Mr. Holmes, quoted by Mr. MacNamara and indorsed by Dr. Gibney,[2] represent the first class, when he says :

" I would sum up what I have to say about excision of the hip in a very few words by the simple statement that it ought to be very rarely indeed required if the disease were treated properly at its commencement. In cases seen in an advanced stage of the disease, it is chiefly when sequestra exist that the operation is *necessary*, though it may be *advisable* as a means of shortening the treatment in other cases ; also, when the patient cannot obtain the prolonged surgical care which is essential to natural recovery."

The indications for the selection of cases are well given in the report of the Committee of the Clinical Society of London appointed to investigate the subject in 1880 :

"As to the indications for resorting to the operation of excision there are certain conditions which, when developed in the course of hip disease, either preclude recovery or make recovery improbable unless some operative interference is adopted."

---

[1] *Vide* Gibney: Diseases of the Hip, p. 393. Also Willard : Trans. Amer. Orthop. Assoc., vol. ii. p. 149.

[2] Gibney : Loc. cit., p. 397.

These, in the opinion of the committee, are :

" 1. Necrosis and separation of the entire head of the femur, and its conversion into a loose sequestrum.

" 2. The presence of firm sequestra, either in the head or neck of the femur, or in the acetabulum.

" 3. Extensive caries of the femur, or of the pelvis, leading to prolonged suppuration and the formation of sinuses.

" 4. Intra-pelvic abscesses following disease of the acetabulum.

" 5. Extensive and old-standing synovial disease and ulceration of the articular cartilages, with persistent suppuration.

" 6. Displacement of the head of the femur on the dorsum ilii, with chronic sinuses and deformity.

" With respect to the general question of operative interference, the committee are of opinion that the effect of complete rest and weight or other modes of extension, and the withdrawal of matter, should always be patiently tried in the first instance, and that operative interference should be resorted to only when these means have failed to secure the favorable progress of the case."

Mr. G. A. Wright,[1] in contrast to these conservative views, declares that "treatment short of excision, when once suppuration occurs, is useful only as a palliative or a means of temporizing."

These contradictory opinions can only be reconciled by considering the circumstances under which such opinions were formed, the character and surroundings of the

Fig. 73.

Photograph of case showing good result after excision of the hip-joint for hip-joint disease.

_____

[1] Wright: Loc. cit., p. 97.

patients, the facilities for thorough conservative treatment, etc. Such a comparison is obviously impracticable here. In general, it may be said that careful conservative treatment should always be carried out for a long time; that where required early excisions are preferable to late operations, particularly in hospital and dispensary patients, where continuous conservative treatment is impracticable; that whereas the time required for treatment by excision is shorter than by conservative methods, the mortality is higher and the resulting limb less useful. (Fig. 73.)

Ashhurst[1] says: "The operation is, indeed, such a grave one under any circumstances, that I do not consider a resort to it justifiable in any case in which it is not evident that life will be imperilled by longer persistence in expectant measures."

The writer firmly believes that where cases are treated by conservative methods from the incipiency of the disease, operative treatment will be only exceptionally required, and that exploratory incision, erasion, and drainage should have the precedence over excision in all cases where it is possible to employ the more conservative operation.

Amputation. Since the introduction of excision, amputation for hip disease has fallen into disuse. Since, however, it is occasionally performed, the question of its indications, mortality, etc., require consideration. It is indicated where the femur is so extensively necrosed that excision would not entirely remove all the disease, and in descending osteomyelitis. It is not indicated where caries of the pelvis is so extensive that removal of the limb would be of little service except in arresting the drain upon the constitution, nor does the writer believe it indicated in amyloid disease, where excision is all that is required to check suppuration and all the shock such patients can bear, nor is it justifiable where the patient is moribund. According to MacNamara, in adults excision is less satisfactory than in children, and according to Wright, "amputation should always take the place of excision after puberty."

The mortality of amputation at the hip-joint is not so great after hip disease as for injury or other disease. The death-rate for amputation after injury, according to Bradford and Lovett, is 70.9 per cent., and after other disease is 42.6 per cent. Ashhurst has collected 34 cases of *primary* and 31 of *consecutive* (i. e., after excision) amputations, with 19 deaths—a mortality, rejecting 5 cases in which the

[1] International Encyclopædia of Surgery, vol. iv. p. 500.

result was undetermined, of 32 per cent. In amputation for hip disease, then, according to Ashhurst's table, the mortality was 19 deaths in 60 cases, or 27 per cent., whereas in the more recent table collected by Bradford and Lovett,[1] it was but 3 deaths in 22 cases, or 14 per cent., thus confirming the statement of Wright that "amputation at the hip, performed with due precautions as to hemorrhage and shock, and special care during the first twenty-four hours, is not a very fatal operation in children." In 7 amputations performed by Wright, in 6 of which excision had been previously performed, 6 recovered well from the operation, and 1 died from hemorrhage.[2]

The absolute control of hemorrhage, or the so-called "bloodless amputations," are of the greatest service here.[3] For the various methods of operating and the best means of controlling hemorrhage the reader is referred to works upon general surgery.

In conclusion, amputation at the hip-joint for hip disease may be considered as the very last resort, to be employed only when the disease in the femur is too extensive to be removed by resection, contra-indicated by extensive amyloid disease or a fatal asthenic condition of the patient.

In conclusion, this section upon the operative treatment of hip disease may be appropriately terminated by quoting in full Dr. Willard's able conclusions to his paper upon the same subject before the American Orthopedic Association in 1889 :

"CONCLUSIONS : 1. Mechanical measures, which enforce strict and long-continued rest of the bone, are the best preventives of operative procedures, and when they can be properly employed, as in private practice, operation is seldom necessary.

"2. Antiseptic aspiration of the joint is desirable if it is distended with serum ; sometimes when it is filled with pus.

"3. Ignipuncture and trephining give great relief provided the focus of the disease is reached.

"4. If in doubt as to the presence of pus, employ rest and fixation, and wait ; or aspirate antiseptically. Repeat the aspirations if the fluid drawn is sero-purulent, or if the liquefaction of caseation is present, since organization may occur without the opening of the abscess. If there are no signs of caseation, wait. If caseation takes place, make early exploratory incision, remove, and drain.

---

[1] Orthop. Surg., loc. cit. p. 356.
[2] Case 28, loc. cit.
[3] *Vide* method of Mr. Jordan Lloyd : Lancet, May 20, 1883. Also, Lamphear : University Medical Magazine, July, 1891, p. 613. Also, Wyeth : Medical News, December 9, 1893.

"5. When certain that true pus has formed, provided drainage alone is decided upon, aspirate and inject with iodoform, in order to strengthen the walls of the abscess cavity, as a preliminary to further operation.

"6. If the abscess is to be opened and no further operation performed, incise freely as soon as the sac refills. Wash out the contents with a slow current of 1 : 2000 hot sublimate solution, but on no account employ pressure to expel the pus, lest fissures be produced in the internal wall and infection thus favored. Drain thoroughly, dress antiseptically, and fix the hip. Continue to keep it at rest until closure occurs, or until further steps are necessary.

"7. When extirpation has been decided upon, make an exploratory incision either along the line of a sinus or directly upon the bone and thoroughly examine the joint. If the destruction of tissue is small, try erasion and remove the carious bone and other tissues with scoop, scissors, and knife. Endeavor to remove every particle of tuberculous tissue of both hard and soft parts, including abscess and sinus walls. Irrigate thoroughly with strong antiseptics to prevent bacillary infection of the freshly cut tissues, and drain freely. With antisepsis the dangers of opening the joint are but slight, and the desirability of operation is now based upon the question of time and the ultimate results as regards usefulness of the limb.

"8. Perform a formal excision, not as a routine practice, but when the disease is found too extensive to be removed in any other way. In desperate cases excision becomes of exceeding value, but should never be adopted as a routine practice. In adults, especially when poor, a typical resection gives the most useful results. As erasion is less mutilatory than an excision, it should be performed provided all tubercular foci can be removed.

"9. Always follow operative procedures by most careful attention to rest, fixation, and protection of the joint for months or even years."

CHAPTER V.

KNEE-JOINT DISEASE.

KNEE-JOINT disease, or white swelling, is a chronic tubercular lesion of the knee-joint, beginning usually as an epiphyseal osteitis, or a synovitis, and terminating in recovery, ankylosis, or complete destruction of the articulation.

Synonyms. English, *Strumous Arthritis; Scrofulous Disease of the Knee; Chronic Purulent or Fungous Synovitis of the Knee; Chronic Tubercular Osteitis of the Knee.* German, *Scrofulöse Gelenkentzündung; Fungöse Arthritis; Scrofulöse Caries; Tuberculöse Caries.* French, *Tuberculose Articulaire; Ostéo-périostite; Tuberculose Chronique.* Latin, *Tumor Albus; Fungus Articuli; Caries Sicca; Caries Mollis sive Fungosa.*

The terms tumor albus and white swelling are the best established, but they do not indicate, as do the more modern terms, the pathological character of the affection. For general use the term knee-joint disease, here employed, is useful and inoffensive, since it involves no etiological or pathological theory.

Etiology. Tuberculous knee-joint disease occurs at all ages, but is most frequent in children and young adults. It is second in order of frequency, there being in 2292 cases of orthopedic diseases treated at the Orthopedic Department of the University Hospital, 67 cases of this affection. Three forms, as in hip disease, may be distinguished, based upon the primary origin of the disease—femoral, tibial, and arthritic. According to the experience of Koenig, the primary synovial and osseous varieties are about equally frequent in youth, but three times as many osseous as synovial in adults. The investigations of Willemer[1] show that under ten years of age the disease is primarily synovial in 39 per cent., and primarily osseous in one or both articular extremities in 61 per cent.; between ten and twenty, synovial in 49 per cent., osseous in 51 per cent.; and above twenty years of age synovial in 33 per cent., and osseous in 65. The extent and peculiar anatomical complexity of this articulation confer a

[1] Deutsche Zeitschr. f. Chir., B. xxii. p. 268.

peculiar chronicity upon all its diseases. The causes of knee-joint disease are both predisposing and exciting, and the remarks upon the etiology of hip-joint disease apply with equal force here. The predisposing causes are : age, sex, heredity, hygienic surroundings, and social condition. As exciting causes, traumatism and cold are more frequent than in hip-joint disease, and primary synovitis, particularly in adults, is much more frequent.

Pathology. The pathological lesion is usually a chronic epiphysitis, or chronic purulent synovitis. Kocher asserts that he has observed primary tuberculosis of the semilunar cartilages, and Kummer[1] has reported a case of extirpation of the patella for primary tuberculosis of this bone.

Symptoms. The symptoms of this disease, which is essentially chronic, can most suitably be discussed by classifying them under three stages : 1, an incipient stage; 2, an acute stage, and 3, a later stage.

This classification is practically identical with that of Mr. Adams, and attempts to give information as to the condition of the joint. First stage, from the onset to the formation of pus within the joint ; second stage, from the end of the first stage to the formation of abscess outside the joint ; third stage, from the formation of abscesses to complete ankylosis or the death of the patient. It must, however, be remembered that the disease may terminate at any period and retrograde changes supervene.

These divisions correspond to : 1, Localized bone destruction ; 2, suppuration of joint contents ; 3, reparative or destructive processes.

First Stage. Beginning usually as an epiphyseal osteitis, its onset, as a rule, is very insidious. A slight pain or stiffness, attributed to growth or rheumatism ; a slight limp or halt in walking, transient, recurrent, disappearing after a night's rest, mark the gradual onset of this grave malady. Associated with these there may be an indisposition, a languor, an unnatural inactivity, and an indescribable something included in the comprehensive word *malaise;* or the general health may remain unimpaired.

With the destructive advance of the local process the symptoms assume a more positive character, and a decided limp, intermittent pain, swelling of the joint, discoloration of the skin, defective movement and muscular wasting, with marked constitutional disturbance,

---

[1] Revue Suisse Romande, 1889, No. 11.

indicate the true character of the morbid affection and attract attention to the part.

Lameness. The slight limp observed in the early stage is due to the pain experienced on motion, and the efforts of the patient to diminish the shock of the impact of the foot upon the ground. Later the contraction of the knee, together with the impairment of motion and complete stiffness, produce a positive shortening and interference with free motion which are characteristic. In some instances an enlargement of one of the condyles, usually the internal, causing knock-knee, adds to the amount of disability and increases the limp.

Pain. The pain at first is paroxysmal and slight, of a dull aching and gnawing character, produced or increased by jarring. Later the suffering during the acute exacerbations is severe and excruciating. This is somewhat diminished by the muscular spasm by which the joint surfaces are held rigid, and motion is limited. Tenderness or sensitiveness on pressure is also present, the tender spot being usually on the internal condyle, about half an inch from the patellar edge.

Swelling. The tumefaction of the joint is characteristic—shapeless, uniform, obliterating the natural configuration of the part, filling the

FIG. 74.

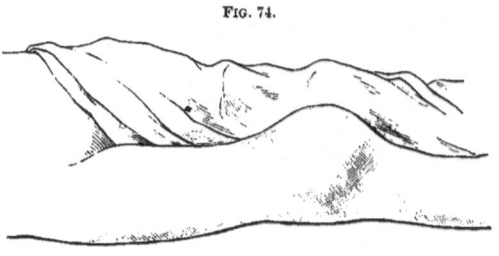

Articular osteitis (internal condyle). (BARWELL.)

depressions of the sides of the patella—giving an indistinctness of outline and roundness of appearance that cannot be mistaken. (Fig. 74.) Its consistence is soft, inelastic, and doughy, softer in some spots than in others. Unless the effusion be large, the patella is not floated up as in the synovitis, but appears fixed in a soft gelatinous mass, or may even appear depressed from prominences upon either side.

Discoloration. Instead of the bright-red color typical of inflammation, the skin retains its natural hue or loses its color and becomes white, from whence it has derived the name of "white swelling." The surface is marked by blue, tortuous, superficial veins.

Heat. Elevation of temperature is also absent, and in some cases the part may present to the hand the impression of cold. Pyrexia is likewise absent.

Limitation of motion. Among the earliest positive symptoms the writer would lay particular stress upon this symptom. Preceding even the symptoms of local joint mischief, a limitation of motion, particularly in forced flexion and extension, is characteristic of osteitis. This is due to a reflex tetanoid spasm, present in both the first and second stages of the disease, but increased during the exacerbations, and leading to the atrophy subsequently observed. It is most marked in the flexor muscles and produces the subluxations and complete luxations of the second and third stages.

General condition. Even in this early stage the expression is anxious, careworn, and apprehensive; sleep is restless and disturbed, but without the sharp and sudden cry and pain to be noted directly; the appetite is capricious, the digestion feeble, the disposition irritable, and in every way the general condition denotes a local annoyance and physical distress that is harassing and debilitating to the constitution.

SECOND STAGE. As the disease progresses, the second stage is ushered in by night-cries and starting pains. The local pain changes its character, and abnormal movement, with joint crepitation and subluxation, muscular atrophy and abscesses, indicate the destructive nature of the pathological process.

Night-cries. These starting pains are characteristic of bone mischief, and Barwell considers them to indicate commencing ulceration of cartilages. They are not so common as in chronic arthritis of the hip, and they occur usually when the child is first sinking to sleep. Their exact nature is uncertain, but they probably result from an unguarded twist, and the subsequent reflex muscular spasm.

Pain. In addition to the acute suffering experienced during the exacerbations, there is also present during this stage what Barwell has well described as "intra-articular tenderness," and which he believes indicates that the articular lamella has given way, and the bone cancelli are open to the joint. It is a soreness, increased upon the slightest motion, and rendering the patient very apprehensive.

Abnormal movement. From intra-articular destruction the joint surfaces admit of motion in abnormal directions, eliciting also a grating or joint crepitus from friction of the exposed osseous surfaces. This must not be confounded with the posterior displacement of the head of the tibia upon the condyles of the femur, which, commencing earlier, may at this stage be excessive and amount to a subluxation or even to a complete backward dislocation. These are often associated, and indicate softening or destruction of the ligaments, and great intra-articular disorganization. Joint crepitus may, however, not be elicited, even though great destruction exist, from the luxuriance of the granulations filling the joint cavity.

General condition. The constitution likewise suffers. Pyrexia, anorexia, sleepless nights, emaciation, and debility characterize the second stage of the affection.

THIRD STAGE. The third stage is marked by repair and ankylosis of the joint, or its total destruction, with hectic, exhaustion, and death from some visceral lesion.

Abscesses. Though abscesses may develop earlier, intra- and extra-articular suppuration is characteristic of the third stage. The pus does not find ready exit, and may open at some distance from the seat of the disease, or after separating the integument from the underlying connective tissue will gradually open and discharge a quantity of ill-formed, ichorous pus, filled with flocculi.

Upon microscopical examination the pus was found by Koch, in two out of four cases, to contain the bacillus tuberculosis, and Schuchardt and Krause, at Volkmann's clinic, and Kanzler, Müller, Castro Soffia, Roswell Park, and others elsewhere have found them in still greater relative proportions. The importance of the direct association of tubercular osteitis and the tubercle bacillus cannot be overestimated. The course of abscess formation about the knee-joint, from the anatomical peculiarity, is not so constant as in other localities. Femoral and tibial osteitis both tend to invade the joint. In femoral cases in which the pus does not enter the cavity of the joint, the abscesses find their exit in the inner side (Fig. 75), near the epiphyseal juncture, or on the outer and anterior aspect of the external condyle of the femur. In tibial cases in which the joint is not primarily involved, the abscesses usually open upon the inner side over the inner tuberosity of the bend of the tibia. In arthritic cases the abscesses usually open by one or more sinuses upon the anterior inner surface of the joint, but in rare instances may open posteriorly in the popliteal space, or burrow long distances up or down the posterior aspect of the limb before opening. The disease, however, may run its entire course without suppuration, or it may easily undergo cheesy degeneration and absorption, and caries sicca, identical with that found in the vertebræ and elsewhere, may be present.

Fig. 75.

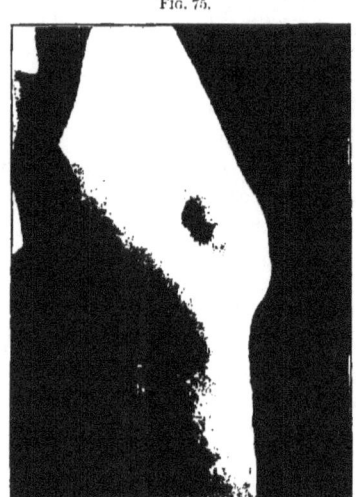

Photograph of cured case of knee-joint disease, showing cicatrix.

Recovery. If the tendency is to recovery with ankylosis, the gradual amelioration of all the constitutional and local symptoms—increased appetite, undisturbed sleep, reduction of temperature, improved circulation and better complexion, diminished suppuration, and the stationary condition of the local symptoms—indicate the change for the better. Later the sinuses close, the joint shrinks, hardens, and becomes stationary in its abnormal position, and ankylosis, with shortening, muscular atrophy, and fixation marks the cessation of the morbid process within the joint.

Destruction. As the disease advances, its destructive progress is marked by an increase of all the symptoms; the patient becomes apathetic, the hectic is marked by greater ranges of temperature, the evening exacerbations are higher, complete anorexia, greater restlessness; scanty, albuminous urine; profuse night-sweats, and possibly diarrhœa, lead to great exhaustion. Notwithstanding this formidable and melancholy array of symptoms, patients do not die of the joint affection alone, but of amyloid disease of the abdominal viscera (liver, kidneys, and spleen), from the excessive suppuration; or of other tubercular affections, particularly meningitis.

Pari passu with these several retrograde general symptoms, the local disease is making rapid progress toward the total destruction of the joint. The joint increases in size and softness, the muscles atrophy greatly, the bones become more movable or are completely luxated, the skin over the parts desquamates in large flakes, and the sinuses become filled to overflowing with luxuriant granulations.

Diagnosis. The general diagnostic signs of tubercular osteitis of the knee-joint are intermittent lameness, paroxysmal pain, general rounded swelling, heat over the joint, and a tender spot on pressure to the inner side of the patella, with limitation of motion and atrophy. The last two symptoms are peculiarly significant of osteitis, and upon them the writer places most reliance. Being a reflex tetanoid spasm, it appears very early, in some cases even before much intra-articular mischief has been set up, or, in other words, while the disease is still confined to the epiphysis or epiphyseal cartilage.

Later, the character of the swelling, the location of the tender spot, and abscess formation, with night startings and cries, render the diagnosis certain. But, for treatment to be efficient, it is essential that an early diagnosis be made, and as direct aids to this, great stress should be placed upon the two cardinal symptoms before referred to.

Differential Diagnosis. Tubercular osteitis must be distinguished from synovitis, rheumatitic and pyæmic arthritis, arthritis deformans, periarthritic disease, and neuromimesis.

Synovitis.

It is with this affection, particularly in the earlier stages, that tubercular osteitis is most frequently confounded, and if subacute and without effusion, it is difficult or impossible to distinguish between them. It is chiefly upon the character of the swelling (the effusion) and the reflex symptoms that the differential diagnosis rests. If the *swelling* in synovitis be large, the patella, by pressing upon it, can be depressed until it is felt to strike against the bone; and on relaxing the pressure it floats up again upon the fluid within the joint. Fluctuation, too, may frequently be detected. In osteitis the swelling is soft, inelastic, and doughy, filling the sides of the patella and fixing it immovably. The localized pain, the reflex spasm and atrophy, are all absent in synovitis, and motion is often in the latter but little impaired.

Acute Rheumatic Arthritis.

This affection, which partakes of the nature of an acute synovitis, is characterized by the sudden onset, the local manifestations being preceded by pyrexia, the association of other joints, localized œdema about the affected joints, and the acute, essentially pyrexial nature of the disease.

Pyæmic Arthritis.

Acute purulent arthritis, as a complication of traumatic infective fever, resembles the tubercular osteitis in its destructive nature and the presence of chills, fever, sweating, progressive emaciation, and loss of strength, as it also does in the local symptoms; but it is readily distinguished from the latter by the rapid course, the association with traumatic fever, and by the character of the suppuration, which contains the bacteria of the acute infective processes, and does not contain the bacillus of tuberculosis.

Rheumatoid Arthritis.

This affection, particularly that division known more accurately as osteo-arthritis, resembles the tubercular osteitis in the character and location of the pain, in the contraction and limitation of motion, and the enlargement of the joint and the muscular wasting. Moreover, it may occur at any period of life. It is, however, to be distinguished by the synovial character of the tumefaction, by the absence of fever, rigors, sweating, suppuration, and muscular spasm, and by the roughness of the crepitation, the participation of other joints, and its predilection for mature and advanced life. The muscular wasting is the atrophy of disuse.

Periarthritic Disease.

Periarthritic or peri-articular abscesses and inflamed bursæ are to be distinguished from the graver affection by the position and fluctuation of the swellings, their relation to the patella, the entire absence of muscular spasm, atrophy, joint stiffness, and constitutional disturbance.

Neuromimesis.

Functional or *hysterical* joint disease occurs most frequently in young females of the upper class, and usually in those of a pronounced brunette type. This pseudo-arthritis can, however, readily be distinguished from the genuine disease by the absence of swelling, redness, atrophy, or muscular spasm, by the superficial but exaggerated nature of the pain, the suddenness of the attack, and the fact that the appetite and general condition are not affected. Moreover, the position of the limb is unchanged, and there is no tendency to suppuration. The limitation of motion disappears under mental emotion, and its free motion is not then accompanied by increased pain.

Prognosis. The prognosis in knee-joint disease is, in general, more favorable than in the same affection in the larger joints, the termination depending upon the early recognition, the social condition and general health of the individual. According to Billroth, a good nutritive condition is the most important point for a favorable prognosis, which would not be very greatly affected even by early and extensive suppuration. Where treatment is instituted early and faithfully carried out, recovery may be anticipated in the milder cases. Where much destruction exists ankylosis may supervene, with flexion, subluxation, or permanent distortion of the joint in an unfavorable position for locomotion. In severe cases flexion is the rule, often even when treated with the greatest care. When not arrested in its early stage, the course is apt to be very chronic and suppuration profuse from the vast extent of the synovial envelope, and the presence of the semilunar cartilages, which likewise undergo liquefaction. In the severest grades of the affection complete destruction of the joint results without ankylosis and renders excision or amputation necessary. According to Billroth, slight swelling of the joint, with great pain and early muscular atrophy in anæmic children, but with little or no suppuration, indicates primary disease of the bone, and renders the prognosis very bad. In mild cases, even after recovery, protection will be necessary for a considerable period.

Treatment. *Conservative measures.* As in all chronic inflammations, the treatment must be both general and local, the general treatment being more important the more chronic and insidious the disease. The constitutional treatment will include all therapeutic, hygienic, and other measures calculated to invigorate the system, and these will not require enumeration here. Local measures are in general more effective the more acute the stage. They include, besides counter-irritation, the mechanical and operative treatment of the knee-joint. Painting with tincture of iodine, flying blisters, the actual cautery, compression with adhesive plaster, are often of great benefit in mild cases. Billroth has recommended a strong salve of nitrate of silver (ʒj to ʒj) rubbed into the joint. Mercurial ointment has also been been employed. The writer has experienced much satisfaction in the use of compound iodine ointment, and mercury and belladonna ointment (ung. bellad. ʒj, ung. hydrarg. ʒiij), employing the latter where pain was a prominent symptom. In severer forms, while these local measures are not to be abandoned, more radical measures must be employed. According to Bradford and Lovett, the introduction of the actual cautery into the bone tissues softened by osteitis seemed to have a beneficial effect in stimulating the development of a cicatricial granulating tissue, but only in connection with mechanical treatment.

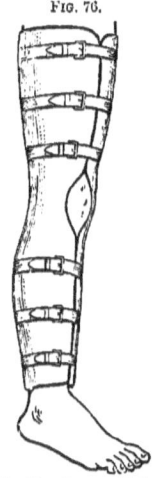

FIG. 76.

Leather knee-splint.
(MARSH.)

Mechanical Measures. In conjunction with local remedies, rest of the joint, not only articular but general, the patient being confined to bed for five or six weeks, should be insisted upon. If this is impossible, some portative apparatus which will combine protection and fixation to the affected joint will be needed. These may be used with even greater advantage while the patient is confined to bed. A sufficient amount of fixation for practical purposes may be secured by splints of plaster-of-Paris, silicate of soda, leather, hatter's felt, poroplastic felt, wood bandages, etc. (Fig. 76.) These should be as long and extend as high as possible, preferably from just above the ankle to the middle of the thigh, and, moreover, should be applied tightly and uniformly, so as to restrain the joint as much as possible from motion. Such splints are of great utility, especially where the more expensive appliances presently to be described cannot be employed. In all portative apparatus, a high shoe must be worn on the sound foot and the

patient walk on crutches. Plaster-of-Paris splints are applied by
crinoline bandages impregnated with fine, dry dental plaster. The
limb is incased in a thin sheet of cotton batting secured by a flannel
or plain gauze roller, and the bandages, previously saturated in luke-
warm water free from salt, are evenly and uniformly applied.
Silicate of soda splints are made with gauze bandages soaked in
silicate of soda, and applied similarly to the plaster-of-Paris bandages.
As silicate of soda hardens very slowly, it is an advantage to apply a
plaster-of-Paris bandage over the completed silicate of soda dressing,
to be removed on the next or the following day, when the latter has
thoroughly hardened. Leather may be softened and moulded to the
part, or a plaster cast may be taken, and over the counter-cast of
this an indurated leather splint may be made. Poroplastic felt, or
felt soaked with a saturated solution of shellac in alcohol, may be
moulded over a plaster counter-cast, or may be applied directly to the
limb and held in position until it hardens. The fixation splint preferred by the writer consists of hatter's felt strengthened with steel
bands (*vide* Fig. 77). The circumferences of the limb are taken above
the ankle, around the knee over the patella, and above the middle

FIG. 77.

Author's hatter's-felt knee splint.

third of the thigh. The angle of flexion is also taken with a strip
of lead, and from these measurements the splint may be constructed
by any harnessmaker or blacksmith. It is a firm, removable splint
that may be used for a long time as a fixation splint. In all cases
where a fixation apparatus is employed for locomotion, the limb must
be protected by a patten or high shoe upon the sound foot and the
use of axillary crutches. Of the more elaborate splints employed,
some of which combine traction with fixation and protection, are
Thomas', Sayre's, Shaffer's, Roberts', the author's, and Taylor's
knee-splints.

Thomas' knee-splint (Fig. 78) consists of an ovoid iron ring
encircling the thigh at the perineum, from which two iron rods pass
down the side of the leg to a metal plate several inches below the foot,
so that the toe of the diseased limb may be fully one inch short of
reaching the ground. In measuring for the splint, the length of

the outer bar is taken from midway between the crest of the ilium
and the top of the great trochanter, to three inches below the sole
of the foot. The circumference of the thigh at the groin is taken,
allowance being made for padding. A patten must be worn upon
the sound limb, and the apparatus must be suspended by a strap
over the shoulder of the sound side, attached posteriorly to the ovoid
ring and buckled anteriorly. Any blacksmith and saddler can construct the splint from the following directions: The upper crescent
is formed of an iron ring three-eighths of an inch in thickness, varying according to the age and weight of the patient. It is nearly
ovoid in shape, covered with boiler felt and basil leather, and attached
to the inner rod at an angle of 55 degrees, which, when correctly
padded, becomes reduced to 45 degrees. To sup-
port the limb, an apron of basil leather is stretched
across the two bars, and in the leather are two slits
for the insertion of the bandage by which it is applied. It is customary in this country to secure the
apparatus to the limb by two broad bands of leather,
either buckled or laced anteriorly. Traction may
also be applied by utilizing the metal sole-piece, but
the use of traction is not in accordance with the
views or practice of the author of the splint. The
side bars may also be made adjustable by a simple
hollow tube and bar arrangement, in case it should
be desirable to lengthen or shorten the splint. The
Thomas splint not only fixes and protects the joint,
but retains it during the progress of reduction of any
deformity.

FIG. 78.

Thomas' knee-
splint.

The Sayre knee-splint was designed by its distinguished inventor to keep the knee-joint in a state of
absolute rest and extend the parts so as to remove
all pressure from the articular surfaces. As a fixation splint it
serves an excellent purpose, but it can hardly be admitted at this
time that it is capable of separating the joint surfaces. What it
probably does do, however, is to overcome reflex muscular spasm,
which is such a destructive element in this affection. (Fig. 79.) It
consists of "two sheet-iron bands or collars, connected by two bars
so constructed that they can be made longer or shorter, as required.
The bands are about an inch in width, have a joint behind, and slots
and a pin for fastening in front. The hinge-joint at the posterior

portion of the band that is to surround the *leg* is made by cutting straight across the band, and then fastening the pieces in the proper manner for forming a joint. The hinge-joint at the posterior portion of the band that is to surround the *thigh* is made by cutting out a V-shaped piece, and then fastening the pieces in the proper manner for forming a joint. This V-shaped piece is removed for the purpose of securing a smaller circle at the lower edge of the band than at the upper, which will better adapt it to the natural tapering shape of the thigh. The band which surrounds the leg should be immovably attached to the side-bars.

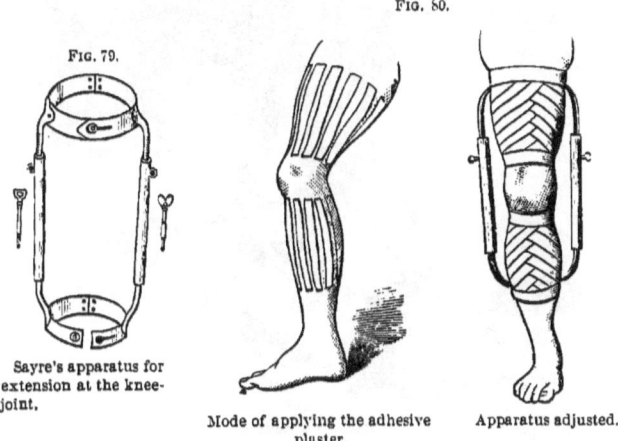

FIG. 80.

FIG. 79.

Sayre's apparatus for extension at the knee-joint.

Mode of applying the adhesive plaster.

Apparatus adjusted.

The band which surrounds the thigh should be attached to the side-bars in such a manner (by a single rivet or hinge) that it can be tilted about at pleasure, which permits the use of the instruments when the leg is flexed upon the thigh at a slight angle. The bars which connect these bands or collars are divided into two pieces, one of which carries the cog and the other the ratchet, by means of which extension is to be made. The ratchet is moved by means of a key, and in this manner any amount of extension desired can be readily obtained."[1]

The ingenious apparatus of Shaffer (Fig. 81) is designed to apply the desired force directly to the head of the tibia, throwing the same

[1] Sayre: Orthop. Surg., ed. 1883, p. 210.

forward and downward by a simple movement. The force applied
to overcome the muscular contraction is in a direct line with the
deformity, and the effect is to relieve joint-pressure and correct the
deformity simultaneously. The apparatus is thus described : [1]

"It consists of three parts—the thigh, leg, and intermediate. The
first two are secured to the limb by adhesive plasters, which are
attached at the points A A. Extension is made with a key at the
extension rod proper at B. The joints at C and D move upon

FIG. 81.

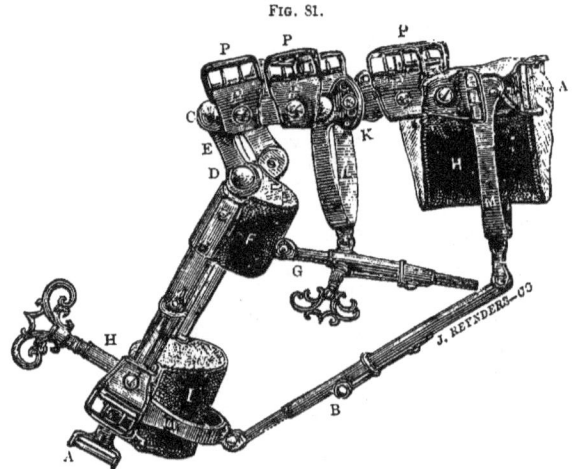

Shaffer's knee splint.

pivots, and as the extremities of the apparatus are secured by their
adhesive straps at A A, the joint D moves forward and downward,
describing a circle, the radius of which is the bar E. Pressure is
thus made directly upon the head of the tibia by the band F, and
this can be very greatly augmented by using the extension rod at G,
which further relieves the joint of pressure by additional extension
in the position already acquired by the preliminary extension of the
rod B. H is an accommodation—not properly an extension rod—
which glides forward as the extension is applied at B. As soon as
the leg is thrown sufficiently forward the accommodation rod is
secured by a slide, and an extra turn of the key at B and G leaves

[1] "On Reflex Muscular Contraction and Atrophy in Joint Disease." Archives of Clin. Surg.,
June, 1887.

the joint free from pressure, and with adequate extension applied directly to the contracted flexors. The thigh and leg bands at H and I move upon pivots, so that they adjust themselves readily to any position, and at K there is an arrangement by which the curved bar L may be adjusted to suit the requirements of the extension rod B. The bars M and O are secured to the thigh and leg parts by double rivets. Through the buckles at P P P webbing straps (padded) are passed, producing counter-extension in addition to that secured by means of the adhesive straps."

This apparatus gives most perfect fixation and extension to the joint, relieving intra-articular pressure, overcoming muscular contraction, and correcting deformity. It requires skill in its application and constant attention on the part of the surgeon or attendant, and is not, therefore, adapted for general extensive use. The same may be said of the Roberts[1] brace, which is the one the writer prefers and employs where the patient or institution can afford the more expensive apparatus. It consists of two light steel troughs padded or coated with rubber, that are firmly secured to the limb by encircling bands of surgical webbing, affording absolute fixation to the joint when the extending rods are locked after adjustment. Three ratchet extension bars arranged in the form of a triangle are placed posteriorly, corresponding to the long axis of the limb. This enables the surgeon to produce extension, to overcome flexion and the tendency to the production of deformity, the extension rod B acting directly upon the head of the tibia and parallel with the line of contraction of the flexor group of muscles, and not low down on the tibia, as in the popular Stromeyer's splint, and all splints constructed upon that faulty mechanical principle. Moreover, force applied below establishes a fulcrum at the surface of the diseased joint, thereby increasing the disease. In this splint the long extension rod C is used only as a compensating bar, adjusting the angle of the splint to the angle of the flexion of the limb. The power for restoring the head of the tibia and overcoming the spasmodic contraction of the flexor muscles is applied with moderation directly in the axis of their contraction. Through the extension bar B (Fig. 82) the head of the bone describes in its restoration the arc of the circle A B (Fig. 83). The compensating bar in correcting the angle of flexion carries the limb through the arc C to D, having a centre in the end of the femur. Mechanically this ar-

---

[1] Trans. Med. Soc. of Penna., 1884, p. 408.

10

rangement of force, as Dr. Roberts remarks, corrects the deformity and relieves, by extension, the reflex spasm of the flexor muscles without crowding together the diseased joint surfaces, or aiding in subluxating the head of the tibia, as would be the case should the limb proper be used as the long arm of our lever, with the insertion of the hamstring tendons, instead of the normal centre of motion of the joint, as the centre of motion of the splint. Sup-

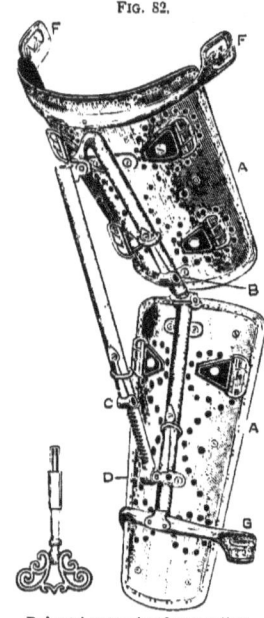

FIG. 82.

plementing the direct extension upon the contracted muscles, another bar D has been added to aid in steadying the joint and relieving intra-articular pressure. The splint acts upon the limb through adhesive plaster applied above and below the joint, to which surgical webbing has been attached. This webbing is firmly secured to the counter-

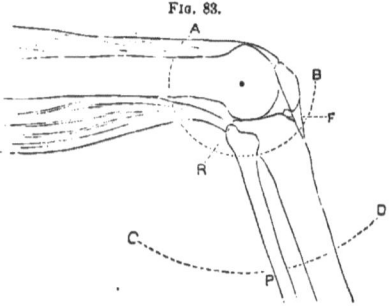

FIG. 83.

Roberts' extension knee splint.

Diagrammatic illustration of mechanical principle of Roberts' extension brace.

extending band F, and to the extension rod G, affording a means of direct extension always corresponding to the angle of flexion. The extension bars are controlled by a key and ratchet movement, held in place, after adjustment, by a small ring and pin.

This splint meets all the indications better than any apparatus with which the writer is familiar, and enables the patient, by means of crutches, to enjoy all advantages of exercise in the open air. It overcomes reflex muscular spasm, relieves joint pressure by making extension in the line of deformity, and fixes the articulation in the most favorable position for recovery. The author's knee splint is similar to Roberts', but simpler in construction. (Fig. 84.)

A very useful walking splint, and one highly recommended by Bradford and Lovett, is one similar to that described under hip disease as Taylor's protection splint. It consists of a long outside bar provided with a perineal band above, a lock-joint at the knee that can be set at any angle, and an extension bar below the foot. Application

FIG. 84.

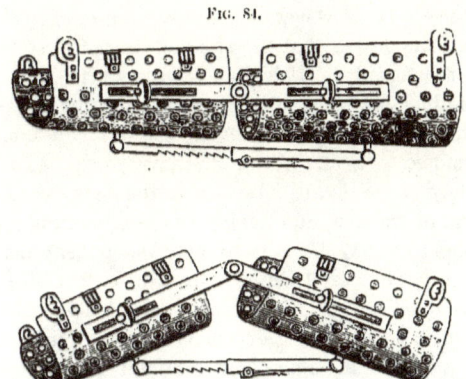

Author's extension knee brace.

of the splint is made with adhesive plasters and surgical webbing, attached above to the splint and below to the extension bar. The knee-joint on the splint is set at an angle corresponding to the angle of deformity, extension is made, and the splint secured to the limb by leather lacings.

Treatment of Complications. The complications requiring special treatment are deformity and abscess.

Deformity. Flexion of the knee is an early and persistent complication of knee-joint disease. It occurs at two separate periods of the affection, as a painful symptom of the early acute stage, and as an insidious, painless complication of the later stage. The means required for straightening the knee will differ at these two periods, and may be classified into

 1. Traction.
 2. Gradual forcible extension.
 3. Rapid forcible extension.

Traction should always be applied in the line of the deformity, and may be made while the patient is recumbent in bed, or by means of portative traction apparatus. In the earliest stages, and when much sensitiveness exists, it is best to confine the patient to bed. The traction by weight and pulley can be made by adhesive plaster applied below the knee, as in the Buck extension for fracture, or, better, by means of the plasters applied as for hip-joint extension, but applied only as far as the knee. By means of a small foot-piece fastened by a leather strap, the weight-extension is applied. The foot of the bed must be elevated to secure counter-extension, and the limb can be arranged so that the traction is applied in the direction of the line of deformity. This may be accomplished by placing the limb over an inclined plane of pillows, or swinging it from a cradle. Extension must be conducted in a very gentle manner, and the limb must be restored to a straight position as quickly as is consistent with the safety of the joint. As soon as the sensitiveness subsides, under the use of traction, and flexion has been overcome, a portative traction apparatus may be applied and the patient may go about upon crutches. The best traction apparatus is that of Dr. Roberts, already fully described, or the brace of Dr. Shaffer, which is upon the same principle, or the long extension traction splint before recommended as a protection splint for hip-joint disease. With these splints traction is applied in the direction of the deformity, overcoming the flexion and subluxation without establishing a fulcrum at the surface of the diseased joint, which is always the case when the force is applied at a distance below the insertion of the resisting flexor tendons.

Gradual forcible extension may be applied by means of the Thomas knee splint, the Billroth splint, or Stillman's sector splint. In employing the Thomas knee splint to correct deformity, the roller bandage is firmly applied in front of the thigh and knee and behind the calf. The more firmly these are applied the greater the pressure that is exerted, and great care must be taken lest an acute exacerbation be excited from improperly applied force.

The Billroth splint is a simple and very efficient method of applying plaster-of-Paris extension. It consists of a plaster bandage applied to the limb, in which at the knee are incorporated two hinged iron strips. The bandage is made thicker under the knee, is allowed to harden, and a window is cut above over the knee, and below, beneath the knee, a transverse division of the plaster is made. Into this slit wedged-shaped pieces of cork of increasing size are daily inserted, until the knee is straight. Extension must not be begun before twenty-four hours. The writer is personally familiar with its application in Prof. Billroth's clinic in Vienna, where its excellence has been fully tested in many cases. The force exerted is great, and the splint must be carefully watched. There is no better appliance for public charity use. In both the Thomas splint used for extension and the Billroth splint, the power is applied upon the erroneous mechanical principle of the popular Stromeyer splint, *i. e.*, the power is applied low down on the tibia, establishing a fulcrum at the surface of the diseased joint and exerting injurious pressure upon the diseased surfaces. Stillman's sector splint is open to the same objection. For retention after extension, the sector splint of the author will be found useful and inexpensive. Instead of the simple free, or locked joint, a sector is added to one arm of the joint, which enables it to be refixed at any angle by a set-screw. A large leather knee-pad secured by straps and pins extends and fixes the knee-joint. A considerable amount of pressure may likewise be exerted. The cut (Fig. 85) sufficiently explains the details.

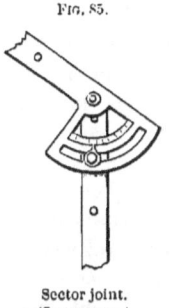

FIG. 85.

Sector joint.
(SCHREIBER.)

Rapid forcible extension—*redressement forcé* —should not be employed in the early stages or in acute cases, but where firm fibrous ankylosis exists, this will be found to be the most efficient method of overcoming deformity. Manual force will be sufficient in the majority of cases, and powerful mechanical appliances should be avoided if possible. The knee should first be forcibly flexed and then forcibly extended. All contracted tendons and fascias should be freely divided, open incision being the best method in this situation. A free incision across the popliteal space allows ample inspection, and in this manner rupture and division of the popliteal artery may be avoided. Fracture of the femur and separation of the epiphyses may also be avoided with proper care, but should they occur, they do not necessarily increase the gravity of the prognosis, and their cure will be effected by the subsequent fixation requisite to treatment. If a mechanical appliance is required, the best will be found to be the one recommended and illustrated by Bradford and Lovett.[1] Its action is thus described: Pressure forward on the head of the tibia is exerted by turning the handle; this, by means of a screw-force pushes the plate forward,

[1] Bradford and Lovett: Orthopedic Surgery, p. 385.

working through the band. The calf muscles protect the artery and nerve from injurious pressure. Counter-pressure is secured by means of leather straps, which are passed respectively over the knee and leg, protected by a thick layer of saddler's felt. Several straps will be needed at the knee to prevent loss of counter-pressure as the limb is made straighter. Another strap under the leg, secures the lower part of the leg. The side-bars, bands, and plate of the apparatus should be of strong steel.

Where firm bony ankylosis has existed for a considerable period, resection of the knee or osteotomy will be necessary. Authorities differ in regard to the relative value of these two operative procedures. Poinset believes resection preferable, for the reason that the removal of all diseased tissue obviates the possibility of subsequent inflammation, and has recorded seventy-seven collected cases, with a mortality of 8 per cent. He regards the operation as entirely free from danger in patients under fifteen years of age. Jacobson, on the other hand, thinks excision should be abandoned for the better operation of osteotomy, strongly urging caution in rapidly and completely straightening a knee joint which has long been the seat of a bony ankylosis in a bad position, the dangers being pressure upon the popliteal vein and tetanus from stretching of the contracted popliteal fascia and the popliteal nerves. The best methods would appear to be forcible manual extension, forcible mechanical extension, osteotomy, and resection, in the order named. Resection should be reserved for bony ankylosis in adults, and should be abandoned for osteotomy in all cases where possible. The operative procedures, surgical technique, etc., will be found described in works upon general surgery.

Where operative measures are declined, an apparatus to overcome the shortening will assist the patient in walking.

Abscess.

Under efficient mechanical treatment suppuration and fistulæ are less frequent, but when abscesses occur they may be treated on the same general plan as hip abscesses. Small localized abscesses may be allowed to remain with the hope of reabsorption, or to open spontaneously while the limb is firmly fixed in some retaining apparatus, after the method recommended by Volkmann and Billroth. Large burrowing abscesses should be freely opened under full antiseptic precautions, but small circumscribed ones may be allowed to burst spontaneously. If the accumulation of fluid be great, it should be removed by aspiration under the most rigid antiseptic precautions. The general management of abscesses will be the same as described under abscess in hip disease. Even after numerous abscesses have formed and discharged, spontaneous recovery may follow.

Expectancy. As long as the general condition of the patient remains good, an attempt may be made to secure recovery by rest and fixation and the use of antiseptic injections. The sinuses should be irrigated daily with a 1 : 2000 bichloride solution, to be followed, if the sinus be free, with injections of pure peroxide of hydrogen, washed again with the bichloride solution, and injected finally with from 1 to 4 drachms of emulsion of iodoform in sterilized sweet oil in the strength of 10 per cent., after which a full antiseptic dressing should be applied, and the limb secured in a plaster-of-Paris cast or other apparatus. Under this plan of treatment the discharge may lessen, the sinuses close, and ultimate recovery quickly ensue. Some of the injections formerly used may still be prescribed with benefit, as solutions of nitrate of lead, potassium permanganate, or carbolated water.

The disintegration and discharge of necrotic tissue may be hastened by the use of solutions of nitric acid (3 per cent.), with or without the addition of active solutions of pepsin. After these preparations have been allowed to remain in the sinuses for twenty or thirty minutes, the cavities must be irrigated with bichloride solution to remove the digested tissue.

Operative Measures. When disorganization of the joint has resulted, operative measures are to be resorted to. These may be considered under three heads: arthrectomy, resection, and amputation.

Arthrectomy, or erasion, consists in laying open the joint and by means of the bone curette removing all of the diseased tissue, irrigating the joint freely with bichloride solution during the operation, passing drainage-tubes through the joint, dressing the wound antiseptically, and immobilizing the joint. As a substitute for excision, this operation has been employed by many surgeons, both at home and abroad, and offers advantages over the latter operation in children and where the disease is not extensive, where thorough removal can be accomplished, and especially where the synovial membrane alone is affected.

Excision and amputation are only to be resorted to where the general health of the patient is gradually failing and the disease is extensive and steadily progressing. According to Agnew, " excision is always to be preferred for children, and amputation for adults," and this agrees, in the main, with the opinion of most surgeons of large experience, the question being largely one of individual judgment. In children, in rare cases, where the resection must remove a very large portion of the shafts of both bones, amputation may be considered. In adults, excisions, under strict antiseptic precautions, yield better results than formerly, and may be resorted to where they offer a prospect of cure, amputation being reserved as a life-saving measure. In many cases a positive decision cannot be formed until the joint has been freely exposed. For a description of the different methods of excision and amputation, the reader is referred to works upon general operative surgery.

CHAPTER VI.

ANKLE-JOINT DISEASE.

THIS is an acute or chronic tuberculous affection of the ankle-joint. It occurs both in children and in adults, and is next in order of frequency to that of the knee. Thus, in 2292 patients treated in the Orthopedic Department of the Hospital of the University of Pennsylvania, there were 19 cases of this affection.

Etiology. Added to a constitutional predisposition, the disease is usually excited by traumatism and by exposure to dampness or cold. When we consider that the ankle-joint is more frequently the seat of injury than any other joint in the body, it is remarkable that it is not more frequently affected by tuberculous osteitis. The disease may begin in the synovial membrane (Münch, Cheyne), in the ligaments (as suggested by Agnew), or in the bone (Erasmus). Thus, in 28 cases reported by Münch, where the ankle-joint alone was affected, the disease was primarily synovial in 23, in the lower end of the tibia in 1, and in the astragalus in 4. Erasmus, on the other hand, found only 2 primarily synovial in 11 cases.

Pathology. The pathological changes met with in ankle-joint disease are identical with tuberculous joint disease elsewhere. Tuberculous foci may be deposited in the lower end of the tibia, the astragalus, os calcis, or in any of the tarsal bones, and invade the joint secondarily, or the synovial membrane or ligaments may be primarily involved. In the latter, fungous granulations luxuriate and the joint cavity is early filled with fungosities.

Symptoms. The onset of this disease is very insidious, but the cardinal symptoms of chronic joint disease may be early recognized. These are muscular spasm, atrophy, lameness, the peculiar attitude of the limb in walking and standing, pain, swelling, and the occurrence of abscesses. As the result of muscular spasm, the foot is in a position of equinus with the toes pointed downward, and later, wasting of the calf and thigh muscles occurs. Lameness is an early symptom, at first a little soreness or stiffness in the joint after exercise, later increasing in severity, and aggravated by walking or motion, and accompanied by tumefaction and heat. (Fig. 86.) The swelling, at first in front of the malleoli, appears later behind the malleoli, obliterating the normal outlines of the joint. Pain is not a constant symptom, but the entire joint may be exquisitely sensitive to pressure and accompanied by "night cries." When suppuration occurs, the abscesses point in the direction of least resistance, either anterior or posterior to the malleoli.

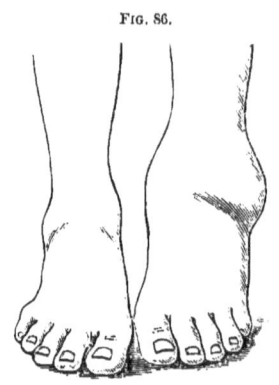

FIG. 86.

Ankle-joint disease. (BARWELL.)

Diagnosis. The disease can be easily recognized by attention to the foregoing symptoms, but the affection must be distinguished from teno-synovitis and caries of the astragalus and os calcis. In teno-synovitis the swelling and tenderness are in the course of the tendons posterior to the malleoli, motion is more painful, and creaking over the tendons may sometimes be elicited. In caries of the astragalus the swelling is over the dorsal aspect of the foot, in which locality also abscesses usually point. In caries of the os calcis the swelling is over the head of the os calcis, anterior to and below the internal malleolus, in which locality abscesses may also break ; or, as in a recent case of the writer's, the swelling and abscess may be localized below and posterior to the internal malleolus. In all cases of doubt an anæsthetic should be employed.

Prognosis. The course, duration, and termination are partly influenced by the age of the individual. In children recovery is the rule without surgical interference, but in adults, if the general health deteriorates, radical operations are indicated as life-saving measures.

Treatment. In children conservative methods accomplish the most good. The general constitution must be sustained and improved by every possible method suggested by modern medicine. The joint should be immobilized with a plaster-of-Paris cast or leather ankle-splint (Fig. 87), and with a high sole upon the sound foot and the use of crutches; the patient should secure the benefits of fresh air and sunshine. During the exacerbations the patient may be confined to bed. In addition to fixation of the joint, extension may also be secured by a modification of Foster's ankle-brace. (Fig.

88.) This consists of a racheted upright attached to a foot-piece, and secured to the leg above by a band, upon which are flat buckles for extension plasters applied to the leg. Counter-irritation may with advantage be employed, preferably with compound iodine ointment. Even should abscesses be present, recovery will often ensue, the average healing period by this, the so-called expectant plan, being,

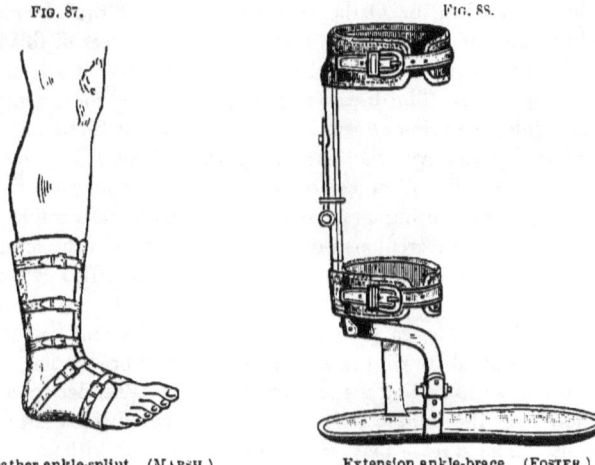

Fig. 87.   Fig. 88.

Leather ankle-splint. (MARSH.)   Extension ankle-brace. (FOSTER.)

according to Gibney, about 16.7 months. If the life of the patient be endangered by prolonged suppuration or amyloid disease, the object should be to remove thoroughly all tuberculous matter, since curetting of the ankle-joint is dangerous from general tuberculous or septic infection. Statistics would seem to indicate that while time is saved and walking feet are obtained after operative methods of treatment in children, the mortality is far greater than by conservative methods (Gibney, Shaffer, Willard, and others[1]). Thus, of 18 operative cases reported by Scudder 6 died—from tubercular meningitis (5) and septicæmia (1). In adults the same general plan of treatment is to be instituted during the early stage, with the hope that the disease may be arrested; but if the general health deteriorate, or abscesses appear, radical measures—either excision or amputation—must be at once performed. The choice of operation will be one of individual judgment, excision or erasion being always preferred, and strict antiseptic precautions being enforced.

[1] *Vide* Discussion, Orthop. Trans., vol. ii. p. 70.

CHAPTER VII.

OTHER MAJOR ARTICULATIONS: SHOULDER-, ELBOW-, AND WRIST-JOINT DISEASES.

Shoulder-joint.

OF all the major joints the shoulder is among the least frequently affected by tuberculous disease. Thus in 2292 cases of orthopedic affections treated in the Orthopedic Department of the Hospital of the University of Pennsylvania there were only 2 cases of this affection. Young persons are usually affected. Traumatism is usually the exciting cause. The disease may originate in a chronic synovitis, but, as a rule, the primary seat is in the head of the humerus. The glenoid cavity and even the coracoid process (Agnew) may be later involved. The dry form, *caries sicca*, is a common form in this locality—the head being gradually absorbed without the formation of abscess. The tuberculous process may extend from the joint into the medullary cavity of the shaft, producing the so-called *caries carnosa* (Koenig).

Symptoms. The early symptoms are stiffness and soreness in the joint, followed by pain in the joint, increased on motion, spasm of the muscles controlling the joint, and atrophy. Tender spots may be found over the joint capsule in front or behind the head of the humerus. The disease may run the entire course without abscess formation, but when suppuration occurs its advent is ushered in by rigors, fever, and sweating.

After a time the joint contents find their way along the long tendon of the biceps, to open on the surface of the arm, below the anterior fold of the axilla, or, perforating the joint capsule, escape beneath the deltoid muscle, to present themselves in the intermuscular septum between the deltoid and pectoralis major in front, or between the shoulder and scapula behind.

Diagnosis. This affection must be distinguished from primary bursitis and rheumatoid arthritis limited to the shoulder-joint. Inflammation of the bursæ beneath the tendons of the subscapularis and infraspinatus muscles is usually secondary to shoulder-joint disease, but when primary, pain will be elicited by drawing the arm away from the scapula and rotating it from side to side, and the essential symptoms of joint disease so often described will be absent. In primary bursitis of the sac beneath the deltoid muscle all the movements of the arm will be painful with the arm hanging by the side. Rheumatoid arthritis is a disease of advanced life, the head of the humerus is elevated and advanced forward, crepitation is present in the joint, without suppuration, rigors, fever, sweating, or other symptoms of tuberculous joint lesion.

Prognosis. Unless the disease be early arrested the destruction of the head of the humerus will be excessive and atrophy of the muscles great. The advent of suppuration increases the gravity of the prognosis. Under favorable circumstances caries sicca tends to spontaneous cure, by ankylosis, in from two to three years. The results of conservative treatment are excellent.

Treatment. Fixation and rest should be secured in the Velpeau position by an ordinary roller, starch bandage, or gypsum dressing.

Counter-irritation with flying blisters or tincture of iodine will be beneficial, and intra-articular and parenchymatous injections of iodoform have been highly recommended by Senn[1] and others.

The improvement of the general health by cod-liver oil, iron, quinine, malt and malt liquors, and other tonics, is of the highest importance.

Passive motion and electricity are valuable, but should not be resorted to until muscular spasm has ceased. False ankylosis may be overcome later by the use of forcible motion under an anæsthetic.

When suppuration has occurred free exit should be given to the pus, and antiseptic injections be faithfully persisted in. If the strength of the patient begins to fail as the result of prolonged suppuration, excision should be resorted to.

Elbow-joint Disease.

Tuberculosis of the elbow-joint is a more frequent affection than either shoulder- or wrist-joint disease, there being, in 2292 cases of orthopedic disease treated at the Orthopedic Department of the Hospital of the University of Pensylvania, 6 cases of this disease. It is essentially an affection of early adult life, the largest number of cases occurring between twenty and twenty-five years of age. The disease

---

[1] Tuberculosis of Bones and Joints, 1892, p. 898.

may begin in the synovial membrane, especially during childhood, but in the majority of cases the origin is primarily osseous. Thus in 137 cases Middeldorf[1] found the disease primarily synovial in 30, and primarily osseous in 107.

The humerus may be the first involved; but according to the same authority the relative distribution was as follows: Ulna, chiefly olecranon, 49; humerus, 33; external condyle, 4; humerus and ulna together, 18; radius, 3; all the bones, 2; and radius and ulna, 2. When the disease extends from the humerus to the bones of the forearm the ulna is usually implicated. The right side was affected in 57 cases, the left in 38, and both sides in 53. There were 53 males and 46 females.

Symptoms. The usual phenomena of chronic joint disease—rigidity, pain, swelling, atrophy, and abscess—are present. The arm is flexed and limited by muscular spasm and apprehension, and movement of the joint is attended with exquisite suffering. Muscular atrophy occurs early and usually is marked. The swelling appears first upon the posterior aspect of the arm on either side of the olecranon, destroying the natural configuration of the articulation, and extending in a spindle-shaped form up and down the arm.

When the pus breaks through the capsule, abscesses form and point upon the posterior aspect of the arm upon either side, usually the external side of the triceps muscle, or upon the anterior surface of the joint on the outer side of the biceps tendon.

When the disease is extensive sinuses may form at all points, communicating directly or by tortuous tracts with the necrosed osseous tissue within.

Prognosis. Unless arrested early the probabilities of false or even bony ankylosis are great, and the chances of the re-establishment of motion slight.

Treatment. Constitutional treatment is of the greatest importance and should be continued for a long period. Fixation and rest should be secured in a flexed position with the hand midway between pronation and supination. This position prevents the tendency of the head of the radius to dislocate backward and upward, and should ankylosis occur, will offer the most serviceable limb to the patient. A plaster-of-Paris cast or moulded leather splint from the metacarpophalangeal articulations to the axilla will fulfil the indications.

---

[1] Archiv f. klin. Chir., Bd. xxxiii., p. 226.

If the swelling be excessive the joint should be aspirated. Intra-articular injections of iodoform emulsion are strongly recommended, the joint being entered from the outer aspect, and strong iodoform ointment has been advised as a local remedy by Agnew.

When sinuses have formed, injections of bichloride of mercury, peroxide of hydrogen, and emulsion of iodoform should be employed.

When the disease has abated, ankylosis may be forcibly overcome, or gradual reduction may be made by the application of elastic force after the method of Weigel.[1]

When the disease is not arrested the caries or necrosis must be removed by an informal excision, the operation being limited to the affected bones. Where the destruction is more extensive and dis-organization of the entire limb is threatened, excision must be per-formed, the results being better than after similar operations upon any of the larger joints.[2]

If the health of the patient be gradually failing from the suppu-ration, and the disease upon examination be found to have extended far into the shafts of the bones, amputation will offer the best hope of recovery.

Wrist-joint Disease.

The radio-carpal articulation is affected with tuberculous osteitis about as frequently as the shoulder-joint, there being in 2292 cases of orthopedic disease treated in the Orthopedic Department of the Hospital of the University of Pennsylvania, 2 cases of this affection.

In 919 cases of tuberculous joint disease collected by Cheyne from the statistics of Jaffé, Schmalfuss, Billroth, Menzel, and his own records, the wrist and hand were affected in 6 cases.

Etiology. The majority of cases are primarily synovial, and, according to some writers, adults are most frequently affected, some of Senn's cases being from fifty to sixty years of age. Traumatism is usually the exciting cause, and the peculiar anatomical construc-tion of the joint—the numerous tendons passing over the anterior and posterior surface of the joint, with their sheaths and synovial membranes—renders the extension of the disease either to or from the joint peculiarly liable.

Symptoms. The usual signs of chronic joint disease are present; the joint is painful on motion; the hand is flexed; the back of the

---

[1] N. Y. Med. Journ., June 16, 1888.
[2] Senn: Tuberculosis of Bones and Joints, 1892, p. 416.

wrist swollen (Fig. 89), extending later to the finger-tips, and the arm muscles are atrophied. Abscesses travel up the arm and down

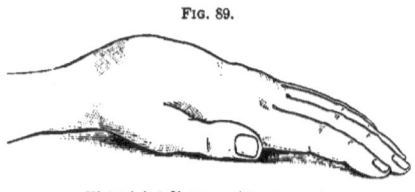

Fig. 89.

Wrist-joint disease. (Ashhurst.)

the hand, disorganizing the parts, and leading, in many instances, to backward dislocation.

Diagnosis. The affection is easily recognized when gonorrhœal arthritis, chronic rheumatoid arthritis, and teno-synovitis are excluded.

Non-articular disease of this articulation without acute inflammatory symptoms or history of gonorrhœa, is usually tubercular.

Prognosis. While the disorganization may be excessive, the tendency of this affection under favorable circumstances is toward spontaneous recovery, with ankylosis.

Treatment. Tonic constitutional treatment is to be supplemented by local measures.

The joint should be fixed in a semi-prone position by plaster-of-Paris, felt, or moulded leather splint.

Counter-irritation with iodine, strong iodoform ointment, or injections of iodoform emulsion should be employed. The needle can best be entered on the outer side below the styloid process of the ulna.

Abscesses must be early evacuated, either with the aspirator or lance, and sinuses should be irrigated with antiseptic solutions.

When disorganization is threatened a partial excision is necessary, but all writers agree that the functional result is more satisfactory after a cure by conservative methods than after excision. Amputation is seldom required.

CHAPTER VIII.

DIASTASIS.

DIASTASIS is a separation, pathological or spontaneous, of the epiphysis from the diaphysis. It is also known as separation at the epiphyseal juncture.

Etiology. Diastasis is an affection of youth and early adult life, and occurs usually as the result of accident or ulceration. One of the best examples of diastasis occurs between the first and second portions of the sternum, this region being also the seat of true dislocation, according to Brinton, Rivington, and Maissonneuve, who have observed a true joint in this situation.

True diastasis as the result of disease or traumatism may occur in any part of the skeleton, but in this connection only those which resemble or complicate joint disease in the extremities will be considered. The most common seats are the upper part of the femur, the upper part of the humerus, the lower part of the humerus,[1] and the upper part of the tibia. (Fig. 90.)

In some cases the epiphysis is pushed off by granulations, and in the case of the upper part of the femur the head of the bone may be found loose in the cavity of the joint. More frequently it results from direct or indirect traumatism, in the reduction of dislocations, of ankylosis, and in the correction of rhachitic deformities by manual or instrumental means.

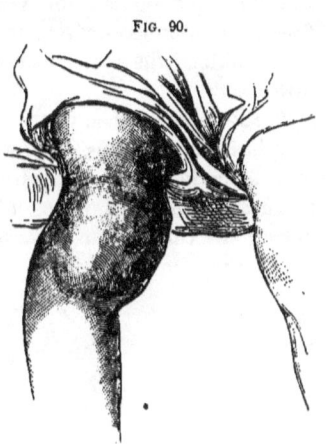

FIG. 90.

Caries necrotica of tibia (diastasis).
(BARWELL.)

Agnew, Barwell,[2] Marsh,[3] and others refer to this accident, and Hamilton[4] refers to three cases recorded by South, Parker, and Post.

---

[1] Watson : New York Med. Journ., Nov. 1858, p. 430.   [2] Diseases of Joints, 1881, p. 287
[3] Diseases of Joints, 1886, p. 127.   [4] Fractures and Dislocations, 1868, p. 339.

Separation of the head of the tibia has been recorded by Barwell, Marsh, and Senn.

Symptoms. Following accident or disease in a young individual there is immediate shortening, with great mobility, without crepitus, and flexion and rotation of the joint occasion no inconvenience. Severe inflammation may follow this accident, and swelling, redness, heat, and pain are followed by abscess formation and sinuses. Union is delayed, bony union may never occur, and shortening is an almost constant sequela.

Diagnosis. Diastasis has to be distinguished from shortening of the limb from joint disease. This may readily be done by observing that the shortening in diastasis occurs suddenly in the young, whereas in coxalgia, knee-joint disease, etc., it is a late event in the disease, and there are present the symptoms of articular disease. From fracture it is unnecessary to make a diagnosis, since, as pointed out by Hamilton,[1] the subsequent course and treatment will be identical with fracture.

Treatment. When this accident occurs during attempts to correct ankylosis, or rhachitic deformities, the use of a plaster cast in the most favorable position will be advisable. The surgeon should bear this accident in mind during all orthopedic manipulations, especially during the correction of rhachitic curvatures by osteoclasis or *brisement forcé.*

[1] Loc. cit., p. 28

CHAPTER IX.

ANKYLOSIS.

ANKYLOSIS consists of partial or entire immobility of an articulation in any position.

Synonyms. French, *Anchylose; Roideur Articulaire.* German, *Gelenkverwachsung; Gelenksteifigkeit.* Italian, *Anchilosi.*

The term ankylosis (ἀγκύλος, crooked) is derived from the Greek root ἀγκύλη (*ankule*), originally used to express articular rigidity in a flexed position, the term ὀρθόχωλος (*orthocholos*) being applied by Galen when the limb was fixed in a straight position. The word is now used interchangeably, the terms angular and straight being sometimes added, and the distinction being sometimes made of simple or complicated, associated with luxation. Two forms are recognized: true or complete, and false or incomplete. The ginglymoid or hinge joints are more frequently affected than the ball-and-socket joints.

Etiology. Congenital ankylosis is rare; but one case[1] has been recorded, of a child twenty-three months old, with complete ankylosis of the entire skeleton.

The affection is usually acquired, as a result of non-use, or as sequela of articular disease, fractures into the joints, traumatism, inflammatory processes, phlegmons, erysipelas, scalds and burns, in the neighborhood of joints.

As the result of non-use, ankylosis is not uncommon in the aged in every position in society, and interesting examples have been recorded by Paget, Manzel, and Reyher.

The tubercular joint inflammations furnish the largest number, but gonorrhœal, gouty, rheumatic, syphilitic, neurotic, and puerperal affections also lead to ankylosis.

The question of what produces and what prevents ankylosis has been a subject for discussion among surgeons for some time, some maintaining that if a normal joint be immovably fixed for a certain length of time, ankylosis will occur, and that motion is necessary

---

[1] Histoire de l'Académie des Sciences, 1716.

in order to preserve the normal integrity of the joint and to prevent ankylosis in injured and inflamed articulations; while others have asserted views diametrically opposite.

Both have used the fakirs of India, who for the sake of penance often assume one position for years, as an illustration; one claiming that their joints frequently become ankylosed in such an attitude (Schreiber), while another states that they, after holding their limbs in one position twenty years, quickly regain the normal use of their joints after their religious frenzy has passed (Thomas, quoted by Phelps).

The experiments of Phelps and Thompson[1] appear to have proved that neither of these statements about rest and motion is right, but that the question of ankylosis is determined by the severity, character, and duration of the inflammation, the presence of intra articular pressure, the subsequent cicatricial contraction of soft parts around the joints, the tissue involved, and the amount of destruction of bone and cartilage.

Pathology. In true ankylosis the articular ends of the bones are united by osseous material within the joint, without the joint as synostosis, or by both.

Preceding the union of the bones there has been destruction and absorption of synovial membrane, of cartilage, and the joint has been obliterated by cartilaginous or fibrous material.

In false ankylosis the intra-articular changes are largely synovial, consisting of false bands, cicatricial contractions, cohesion of the ligamentous capsule, or complete disorganization by inflammatory deposits. Cicatricial contractions of skin, fascia, tendons of muscles, are all of this false variety.

Diagnosis. A positive diagnosis between true and false ankylosis can only be determined under anæsthesia. When the ankylosis is osseous the articulation remains immovable under complete narcosis.

Prognosis. The prognosis at the present time is generally favorable under modern surgical methods.

Treatment. Treatment should not be undertaken until the original articular disease has disappeared or has been overcome by appropriate treatment, since too early manipulation may hasten what the surgeon is striving to overcome.

The reduction in mild cases of false ankylosis can be accomplished

---

[1] New York Med. Journ., May 17, 1890.

by massage, gradual manual and elastic traction. In more severe cases forcible reduction (*brisement forcé*) may be performed with or without anæsthesia. In doing this the rule employed by the bone-setter[1] should be followed accurately, to make firm pressure upon the tender spot at the moment the greatest force is exerted. The greatest caution and gentleness are at first necessary until the rigidity begins to yield, lest fracture result, as it frequently has done in the hands of the most skilful surgeons.[2]

The first movement should be in the direction of flexion, and forced flexion is often beneficial. When the false ankylosis is very firm, instrumental force must be resorted to, and osteoclasts of the preferred pattern may be used. Tenotomy is often required, and where necessary should be first performed.

In true ankylosis osteotomy or cuneiform resection is, as a rule, necessary; but amputation, as suggested by Bauer and others, is rarely warranted since the advent of antiseptic surgery.

The consideration of ankylosis of the individual joints will be found in their respective sections.

[1] Wharton P. Hood, M.D., M.R.C.S.: On Bone-setting. 1871.
[2] Agnew: Surgery, vol. ii. p. 143.

CHAPTER X.

SYNOVITIS.

SYNOVITIS includes inflammation of the synovial membrane from whatever cause.

Synonyms. *Sero-synovitis; Acute Serous Synovitis; Purulent Synovitis; Arthromeningitis; Hydrops Articulorum; Empyema Articulorum; Pyarthrosis; Tubercular Hydrops; Synovitis Hyperplastica Lœvis; S. Pannosa; S. Hyperplastica Granulosa; S. Hyperplastica Tuberosa, and Tubercular Empyema of Joints.*

Three principal forms may be distinguished, according to the pathological process: serous (hydrarthrosis), purulent (parenchymatous), fungoid.

The disease may be acute or chronic.

Etiology. The causes of synovitis are multiform, and may be included under traumatism, over-exertion, exposure to wet and cold, acute articular rheumatism, gout, syphilis, infectious diseases, rheumatoid arthritis, and tuberculosis.

Koenig states that in surgical clinics the surgeon will have one hundred cases of tuberculosis to deal with to one of the other varieties of inflammation (Senn); hence the importance which tuberculosis assumes in modern etiology and pathology of joint diseases.

The infectious diseases of the joints are largely pyæmic in nature and result as sequelæ of the following general affections: diphtheria, dysentery, erysipelas, gonorrhœa, malaria, measles, meningitis, pneumonia, pertussis, parotitis, pyæmia, puerperal fever, scarlet fever, septicæmia, smallpox, typhus and typhoid fevers, and varicella.

Acute serous synovitis begins as a hyperæmia of the synovial membrane, stasis of the blood-current, dilatation of the capillaries, and the ordinary phenomena of simple inflammation, the outwandering of leucocytes, and hypersecretion of synovia. The intra- and para-articular structures become thickened and softened, but should resolution occur all the tissues return to their normal condition without injury. Should the process continue unchecked, exfoliation of the cartilage occurs, fungous granulations invade the joint from the edge of the cartilage, and suppuration establishes a chronic synovitis.

Pathology. Primary synovial tuberculosis, both acute and chronic, is now a well-established affection, its classification being best made upon an anatomico-pathological basis in reference to the character of the inflammatory product.

Serous synovitis, hydrarthrosis, hydrops articuli, or tubercular hydrops resembles pathologically tubercular ascites, and is characterized by a copious effusion into the joint.

Fibrinous deposits may occur, and if abundant give rise to the title of hydrops fibrinosis, or *arthrite plastique ankylosante.* Later, thickening of the membrane may occur, with all the characteristics of chronic tubercular synovitis.

FIG. 91.

FIG. 92.

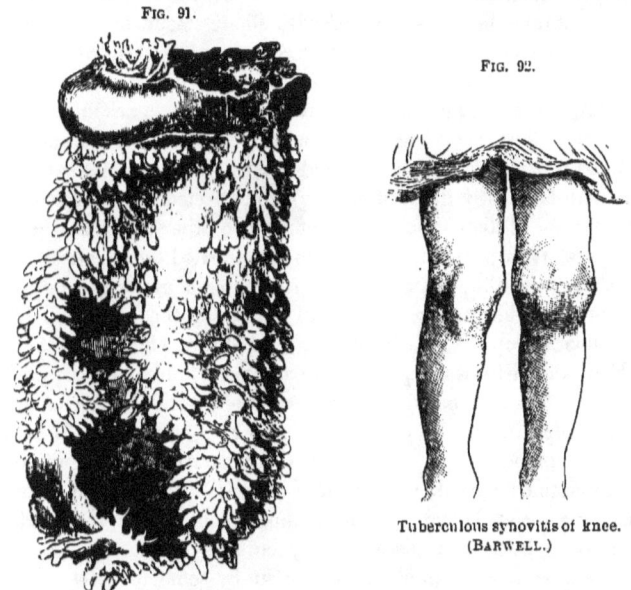

Tuberculous synovitis of knee.
(BARWELL.)

Fimbriated growths of the synovial membrane.

Purulent synovitis (parenchymatous) begins as synovitis hyperplastica, with thickening of the membrane, which is studded with minute granules, and a vascular layer of granulations invading the surface of the cartilage from the border toward its centre. Later the membrane becomes affected throughout, and constitutes the so-called synovitis hyperplastica granulosa. When the amount of suppuration is excessive a comparatively rare pathological condition results, resembling in its clinical history hydrops or acute suppurative synovitis, for which Koenig proposes the name *synovitis suppurativa tuberculosa.* It is also known as tubercular empyema of joints.

Fungoid synovitis is a more advanced stage of the preceding affection, being so called from the abundance of the granulations. This form may be primarily synovial, but is more frequently secondary to an osseous lesion.

Rarer forms are the synovitis hyperplastica tuberosa, characterized by the formation of circumscribed subsynovial masses, which either disappear spontaneously or the affection becomes more diffuse and the membrane becomes covered with small sessile or pedunculated growths, the so-called *rice bodies*, or corpora oryzoidea, this condition being also known as lipoma arborescens tuberculosum.

Symptoms. The symptoms are usually well marked and characteristic. The onset may be gradual, attended by soreness and stiffness in the articulation, or sudden without prodromal symptoms. The pain is usually diffused through the affected joint, accompanied by a feeling of distention, rarely referred to other joints, and in some instances, as in pyæmic synovitis, is complicated by hyperæsthesia of the skin. Swelling occurs early, producing a general enlargement of the part, obliterating the normal outline of the joint, and in the knee floating up the patella in a manner almost pathognomonic.

Heat and redness are present in superficial joints, and the position of the limb is the one which permits of the greatest degree of distention.[1] In acute attacks, especially during rheumatism, the local signs are accompanied by marked febrile disturbances. In chronic forms the symptoms resemble more closely osteitis, with which it is frequently associated, and for the symptoms of which the reader is referred to knee-joint disease. A peculiar symptom of the chronic form is a leathery crepitation, attributed by some to plastic deposits on the synovial membrane, but due in all probability to the formations of the subsynovial tuberous masses of tubercular origin. The advent of suppuration in the joint is marked by an increase of all the local symptoms, and by the occurrence of rigors, fever, and hectic.

---

[1] *Vide* Bonnet: Traité des Maladies des Articulations. 1845.

Treatment. The treatment of synovitis must be both constitutional and local. A saline purge, preceded in some cases by a mild dose of calomel, may be followed by the administration of nitrate of potash, salicylate of soda, iodide of potash, or colchicum, according to the origin of the affection.

Locally, rest either upon a pillow or in a plaster of-Paris cast, and usually flying-blisters applied about the border of the patella, or the use of iodine, will be attended by the best results. Massage may be early employed to reduce the swelling, and after the blisters have entirely healed, the remaining soreness may be removed by the use at night of the following original embrocation :

```
℞.—Spts.chloroformi  ⎫
              ⎬ aā ·              . ℥ss.
Tr. arnicæ    ⎭
    Ext. hamamelis fld. (witch-hazel) .        . ℥ij.—M.
    Apply with friction at bedtime.
```

Later, a roller flannel bandage must be worn for the relaxation of the joint, or, preferably, an elastic support.

In tubercular cases iodoform ointment locally is indicated, or injections into the cavity and peri-articular tissues of a sterilized emulsion of iodoform.

The treatment of chronic synovitis associated with osteitis will be found under osteitis of the individual joints.

CHAPTER XI.

LATERAL CURVATURE OF THE SPINE.

LATERAL curvature of the spine is "a permanent lateral deviation of the spinal column, or a part of it, from the natural physiological direction" (Drachmann), or "a distortion of the trunk in which an abduction and rotation of the spinal column is associated with a deformity of the thorax" (Vogt). This deviation is usually accompanied with rotation of the bodies of the vertebræ on their vertical axes, but as this is not essential it need not necessarily form a part of the definition of the deformity.

Synonyms. *Scoliosis.* German, *Seitliche Rückgratsverkrummung, oder Verbiegung; Bogenformige Deformität der Wirbelsäule; Seitliche Verbiegung.* French, *Scoliose; Déviation Latérale de la Taille.* All of these express clearly the deformity, though for scientific purposes the term *scoliosis*, which was given to this affection by Hippocrates about the year 400 B. C., is open to least objection, and is the one generally employed.

Lateral curvature is much more frequent than the ordinary observer would imagine, from the fact that by the proper arrangement of the clothing deformities of slight degree are readily concealed. It is also, without doubt, the most common of all orthopedic affections, although the broad estimate made by Werner, that there are 6500 scoliotic persons in Prussia, carries with it but little weight. Drachmann found in 28,125 children examined in the schools of Denmark 868 suffering with this affection. Combining, however, the 8000 cases of deformity examined by Schilling, Berend and Langaard, and Fischer, 2553 of whom suffered from scoliosis, we have a sufficiently large number to show the frequency of this affection. The relative frequency of this affection in surgical fracture is shown in the 2292 cases of orthopedic affections treated at the Dispensary of the Hospital of the University of Pennsylvania, 143 of which had scoliosis and 346 Pott's disease.

Taking the relative frequency among boys and girls, the larger percentage of the latter—about four or five girls to one boy—is

noticeable, possibly from the greater attention given to their development, and the greater likelihood of their early consulting the surgeon
upon the recognition of the deformity.

This relative proportion is found from the following table to
amount to 84.5 per cent.

	No. examined.	Boys.	Girls.
Kölliker .	. . 721	144	577
Roth . .	. . 200	17	183
Wildberger .	. . 120	19	101
Lonsdale .	. . 170	21	149
Ketch .	. . 220	40	180
Berend .	. . 896	123	773
Adams .	. . 173	22	151
	2509	386	2123

In very young children, under five years of age, the number of
males are said to equal or even exceed the females.

The aggravated forms are found in excess among males, though
severe cases occur among females; the most severe form the writer
has observed was in a female of thirty-five years. The relative
frequency at different periods of life is best shown by the analysis of
228 cases made by Ketch of the material of the New York Orthopedic Dispensary—the number of cases from the first to the twelfth
year being 52 per cent.; from the twelfth to the eighteenth year, 41
per cent.; and later than the eighteenth year, only 3 per cent.

Likewise in Eulenburg's 1000 collected cases, in 85.8 per cent. the
deformity began before the tenth year, in 96.5 per cent. before the
fourteenth year, and in only 3.5 per cent. after the fourteenth year.
His figures are as follows: 78 cases from birth to the sixth year;
216 from sixth to seventh; 564 from seventh to tenth; 107 from
tenth to fourteenth; 35 after the fourteenth. It therefore occurs
most frequently before puberty, and more particularly between the
eighth and twelfth years of age.

Scoliosis is of greater frequency among the enlightened; but that
it is a consequence of civilization, as has been stated by several
authorities, can hardly be granted, since some of the causes—congenital, traumatic, and pathological—occur among savages as among
the civilized. If, however, varieties only were considered—as
habitual, static, and professional—the statement would be accurate.

Six different forms of scoliosis may be considered, as follows:

1. Primary right dorsal scoliosis.
2. Primary left dorsal scoliosis.
3. Primary left lumbar scoliosis.

4. Primary right lumbar scoliosis.

5. Primary left convex total scoliosis.

6. Primary right convex total scoliosis.

From the anatomical construction of the vertebral column, scoliosis affects some portions of the spine more frequently than others, the right dorsal scoliosis being by far the most frequent as a primary affection, and left primary lumbar scoliosis next in order. The proportion of cases of primary right dorsal scoliosis has been variously estimated from 42.3 per cent. (Drachmann), 81 per cent. (Heine), 84 per cent. (Adams), to 92.7 per cent. (Eulenberg). It is probable that primary left lumbar scoliosis is relatively more frequent than has been supposed, Lorenz having found 38 per cent., Drachmann 47.7 per cent., and Klopsch 57 per cent. in recent investigations. Schmidt, Meyer, and Schreiber have referred to the same observation. As primary curves they rarely exist for any length of time, being sooner or later associated with a rotation of the bodies of the vertebræ, and a secondary or even tertiary compensatory curve to enable the column to maintain its equilibrium.

Etiology. Congenital scoliosis, though rare, does occur as a consequence of other deformity or as a primary malformation—a *vitium primæ formationis*—instances of which have been recorded by Willett, Busch,[1] Vogt, Schreiber,[2] Adams, and others. In Willett's case, an adult of thirty-five years, the deformity was believed to be due to an early embryological defect in the elements forming the lateral and vertebral plates. According to Redard, heredity is a frequent cause, especially in the scoliosis observed in young girls, 27 per cent. of the cases observed by him at the Furtado-Heine dispensary having histories of heredity. Eulenburg noted 25 per cent. in his 1000 cases, and Bouvier, Bouland, Dally, and others have referred to heredity as a cause. Hereditary examples of two or more members of a family, while the mother or father also present a similar deformity, are numerous and familiar to all.

By far the greater number, however, are acquired. As an acquired affection it arises from many and various causes, the most common among the predisposing being sex. Thus, the deformity is observed more frequently in girls than in boys—according to the statistics of Kölliker, of Leipzig, and Ketch, of New York, in about the proportion of four or five to one. Age is also a predisposing cause.

[1] Moderne Orthopädik, p. 135.   [2] Orthopädische Chirurgie, p 118.

General muscular debility in adult cases has but little weight, since the weak and delicate, unless as an acquired cachexia or diathesis the result of prolonged disease, do not suffer more frequently than the strong, muscular, and robust. This was particularly marked in two recent cases in unusually well-developed boys observed by the writer. In the majority of cases in the young, and especially in females, the muscular development is decidedly below par, the digestion is feeble, the circulation poor, and they are subject to cold hands and feet, intercostal neuralgias, and occipital headaches. This cannot be said, however, of the general debility resulting from rhachitis, which has been considered recently by one authority as one of the most common predisposing causes. When the number of bow-leg and knock-knee patients is considered, it is remarkable that lateral bending of the osseous spinal column does not with greater frequency yield to like causes. This, it may be suggested, may be due to the enlarged abdomen and ligamentous relaxation, tending toward a posterior curvature rather than a lateral deviation.

FIG. 93.

Production of scoliosis from carrying child. (REDARD.)

Later, as exciting causes, the direct results of rhachitis — bow-legs and knock-knees, anterior bowing of the femur and tibia, * flat-foot, etc.—contribute their quota to the number of curvature cases. The habitual position in which infants are carried is particularly liable to produce scoliosis, especially in rhachitic cases. (Fig. 93.)

Under exciting causes also may be reckoned all those influences which in any manner disturb the equilibrium of the spinal column, and which give to the muscles of one side an advantage over their antagonists. These may be conveniently classed in the order of importance and frequency, under habitual, static, professional, pathological, and traumatic.

The *habit* scoliosis which results from the partial or unequal use of the muscular system is equally, if not more common among the tailor-made society misses reared and instructed in the environment of a

Procrustesian atmosphere, and the factory slave unaffected by proprieties but compelled by necessity to assume cramped positions for long periods under the most unhygienic surroundings. Habitual faulty positions, therefore, constitute by far the larger number of cases. (Fig. 94.) Under this class belong also those cases resulting from

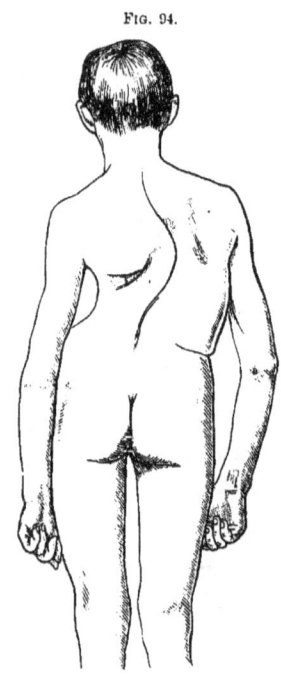

FIG. 94.

Severe right dorsal habit scoliosis.
(SCHREIBER.)

lateral posture assumed to relieve ovarian pain (usually left), and to relieve pain and distention from enlarged spleen, as in a recent case of Dr. John M. Keating's.

Static scoliosis, or inequality resulting from alterations of one extremity, is next in importance. Whatever produces a shortening of one lower extremity produces an obliquity of the pelvis in the opposite direction, and a primary deviation of the lumbar vertebræ. This may result from destructive changes in the joints, rhachitic curves and bowing, flat-foot, back-knee from ligamentous relaxation, or in consequence of excisions or amputations of either the lower or upper extremities. The exact production of scoliosis in cases of amputation is difficult to explain satisfactorily, since a person with an amputated arm may have scoliosis, while another with an amputated leg may not. The existence of a short leg from unilateral development is also very frequent, since Norton found in 513 healthy boys measured 272 with inequality in the length of their legs, and Garson, in London, in measuring the leg bones in 70 skeletons found only 10 per cent. with legs of equal length.

Professional scoliosis, so-called, results from oblique positions assumed during occupations. Particularly fruitful of deformity are the faulty positions assumed during writing, violin-playing, and the oblique attitudes required in bearing burdens. (Figs. 95 and 96.)

Pathological scoliosis is exceptional, but may result from certain

FIG. 95. FIG. 96.

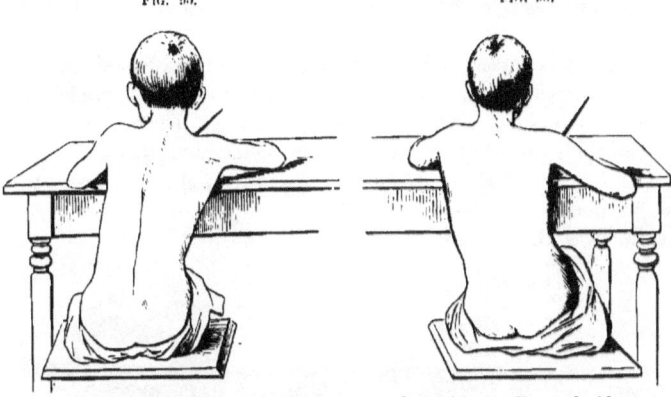

Faulty position tending to right-sided total scoliosis. (HOFFA.)

Faulty position tending to double curve. (HOFFA.)

inflammatory conditions, such as torti-collis;[1] cicatrices from burns; phleg-mons; caries or necrosis of ribs, etc.; pleuritic affections, especially empyema; certain muscular conditions, as spasms; unilateral muscular atrophy and hy-pertrophy; muscular rheumatism (myositis); sciatica and other nerve conditions; neurotic changes, either akinetic or hyperkinetic, especially the former, of which anterior polio-myelitis is the commonest example; morbid growths of the sides of the pelvis or trunk—as encephaloid, en-chondroma or sarcoma[2] — by their enormous weight; and sacro-iliac dis-ease from the habitual faulty position assumed to relieve suffering.

FIG. 97.

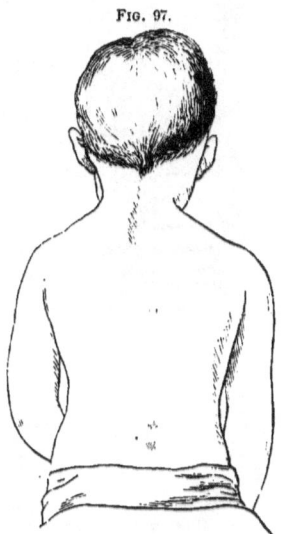

Scoliosis from infantile paralysis. (BRADFORD and LOVETT.)

Trauma is exceedingly rare as a cause, since fractures of the bodies or processes of the vertebræ are more liable to produce antero-posterior curvature. Non-reduced lateral dislocations would lead to

[1] *Vide* cases by writer: Medical News, February, 1890.
[2] Shattuck: Boston Med. and Surg. Journ., January 10, 1889.

permanent lateral deformity, as might also gunshot wounds of this region.

Pathology. The pathological anatomy of advanced scoliosis includes changes in the osseous, ligamentous, inter-cartilaginous, and muscular structures in the order of importance. In the early stages the change is probably first in the inter-cartilaginous disks.

The lesions which occur during the early stages are necessarily poorly understood, from the difficulty of obtaining post-mortem specimens at this period.

Fig. 98.   Fig. 99.

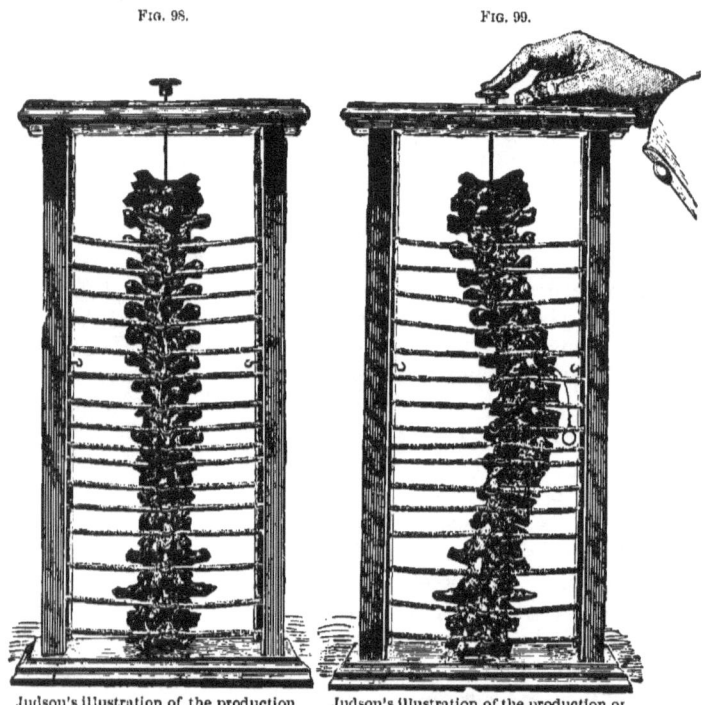

Judson's illustration of the production   Judson's illustration of the production of
of torsion: first part.   torsion: second part.

In a well-marked specimen the following changes are observed: The vertebræ are rotated horizontally, the excess of the displacement being in the anterior portion (the bodies). The manner in which the torsion occurs is well illustrated in the excellent experiment of Dr. Judson. (Figs. 98 and 99.) The spinous processes being comparatively fixed, any motion of the column is observed to result in a torsion or rotation of the bodies of the vertebræ. This is in all probability what occurs in man, since the human spine is endowed with but a slight degree of pure sidewise motion, so fully developed in amphibia, in reptiles, and in some mammals. The concave side of the body (Fig. 100) is atrophied, and in some cases ossified to the adjacent vertebra by a mild pressure ostitis, the result being a wedge-shaped body with the base toward the convexity. (Fig. 101.) The root of the arch is shortened on the side, the vertebral canal is ovoidal from pressure, and

FIG. 100.　　　　　　　　　FIG. 101.

Osseous changes in vertebral bodies.　　　Section of vertebra showing condensation
(HOFFA.)　　　　　　　　on concave side and rarefaction on con-
　　　　　　　　　　　vex side. (HOFFA.)

even the bony fibres of the body occupy a peculiar oblique instead of an upright position. The ribs, being attached to the vertebræ, are also much altered in general outline. On the concave side they are depressed, the angle is more acute, and may be united where they approximate by synostoses. On the convex side they are depressed, widely separated from each other, and the angle is obtuse. From these changes in the vertebræ and ribs, the axis of the cavity of the

12

thorax is oblique, the horizontal section representing an ellipsoid the greater axis of which is formed by the convex curvature, the lesser by the concave diameter.

The pelvis occupies an oblique position in the static cases, but in others it is more often horizontal than is generally believed. The changes occurring in the inter-vertebral disks take place early, and are marked. When we consider that the total thickness of all the inter-vertebral bodies forms about one-fourth of the spinal column exclusive of the first two vertebræ, and that the effect of the pressure of the body in the upright position for some time reduces them more than one-fourth of their proper size, the effect of lateral pressure long continued is readily appreciated—they become wedge-shaped and lose their elasticity and ability to return to a normal position.

The ligaments connecting the vertebræ (especially the inter-transverse and lateral) and connecting the ribs to the vertebræ are shortened on the concavity and lengthened on the convexity of the curve. The muscles on the convexity are relaxed, fatty degenerated, and atrophied, and on the concavity contracted, but altered to a less degree. The relations of the long dorsal spinal muscles are stretched and lengthened on the convex side, and contracted and shortened on the concave.

The flat muscles adapt themselves to the changes—the rhomboidei, trapezius, latissimus, and serratus major et minor, being thinned and atrophied over the bulging parts, and shortened and thickened over the depression.

The effect of lateral curvature is to cause displacements of the thoracic, abdominal, and pelvic viscera. The lungs are compressed on the side of the convexity, and the heart in severe cases is displaced toward the concave side. According to Adams, scoliosis leads to phthisis or a phthisical diathesis, but he admits the absence of statistics to prove a fact also contradicted by specialists upon lung disease, who may not see these cases, since the writer has himself lost cases of scoliosis from this cause. The stomach, intestines, and liver are displaced downward, and the spleen and the kidney upon the convex side are usually smaller than normal.

These pathological changes illustrate the results of the faulty position, but contribute little evidence as to the real etiology of the affection. The theories proposed are very numerous, and furnish, as Copeland facetiously remarks, material for a keen satire on the medical art; but the theory of superincumbent weight or pressure is applicable to the majority of cases, and is the one now generally accepted by writers.

They may, however, be included under five heads : the muscular theory, the ligamentous theory, the osseous theory, the theory of unilateral development, and the theory of pressure or superincumbent weight.

The muscular theory of unequal primary muscular action has been advanced and defended from time to time by a host of writers. The idea of an active muscular contraction identical with torticollis, supported by Guérin, Delpech, and others, was shown by Malgaigne to be based upon false premises. The unilateral relatively stronger action of the right serratus which formed the basis of Stromeyer's respiratory theory, supported by Barwell and Sayre, has been sufficiently refuted by the absence of anatomical and clinical facts, and the arguments of Werner. The modified form of the muscular theory advanced by Eulenberg has much to recommend it. It assumes that, inasmuch as continuous muscular action is necessary to maintain the erect position, if any muscle be weakened the spinal column will tend to bend, the convexity falling on the weaker side, over-stretching the muscles upon this side, and if continued for a time would cause permanent distortion.

Evidence of changed electrical reactions and primary weakness of the muscles either is wanting or has not been demonstrated in the early cases. While primary muscular relaxation may not be accepted as an actual cause, its importance as a predisposing factor in static cases cannot be overestimated.

Secondary muscular changes, such as atrophy due to degeneration of the cord referred to by Roth, or as the result of a disturbance of central trophic innervation as observed by Morvan and Lloyd, are accepted by all.

The ligamentous theory of a primary relaxation of the ligamentary apparatus was maintained by Adams and Malgaigne. The importance of the inter-vertebral ligaments in the etiology has never been appreciated. The highly elastic property of the yellow elastic tissue which composes some of these, especially the ligamenta subflava, serve to preserve the erect posture and to restore the spine after flexion. In static cases with feeble muscular system, when the spinal column is placed in the attitude of rest the ligaments sustain almost the entire weight and gradually yield and assume a curvature in the direction of the habitual position of the patient. This condition is identical with that which occurs in static cases of knock-knee, and while thoroughly tenable in all its parts has not received the support of authority.

The osseous theory, which attributes scoliosis to an insidious inflammatory softening, the patient instinctively assuming a lateral position to avoid pressure, has been maintained by V. Duval and Lorinser. Cases are rare in which the pathological findings correspond to this theory—but Schreiber mentions such an one—and most authorities doubt its existence or confound lateral curvature with lateral deviation in Pott's disease, and, as Reclard remarks, in scoliosis the cancellous tissue does not present a trace of inflammation or sclerosis. A modification of the osseous theory is met in Hueter's theory (Wachsthums-Theorie) of asymmetrical pressure from growth of the ribs. Numerous objections have been raised against this theory, and Lorenz and Dornblüh have proven its fallacy by the discovery of primary osseous deviations of other vertebræ, the absence of changes in the costal cartilages, the inability of the ribs to make pressure from before backward and from without inward, the frequency of other forms as cervical and lumbar, and from the researches of Lorenz, d'Hechenbach, and Albert, of the absence of the osseous deformations characteristic of the theory.

The theory of unilateral development, that scoliosis is only a pathological increase of what is physiological, has been supported by many eminent writers.

According to most anatomists the spine has a slight lateral curvature about the fourth or fifth dorsal vertebra, the convexity of which is directed toward the right side, being produced, as Bichat first explained, chiefly by muscular action, the right side being used by most persons by preference to the left, and Béclard having found in one or two individuals who were left-handed the lateral curvature directed to the left side. According to Sabatier and Bouvier, scoliosis results from an exaggeration of this, which the former writer and others (Cruveilhier, Sappey) considered to be due to the course of the aorta, while others (Vogt, Malgaigne, Busch, Volkmann) attributed it to the more rapid growth of the right half of the body, or to the increased weight of the organs on the right side (Struthers, Desruelles).

Albrecht has shown that the right upper extremity of the fœtus of all mammals has better blood-supply, which results in exaggeration of nutrition and physiological dorsal right curvature. If, as pointed out by Lorenz, Adams, and Woillez, it is not certain the right dorsal curvature is physiological in youth, if also we consider the frequency of primary cervical and dorsal curves and the occurrence of left scoliosis in persons who are not left-handed, but little importance can be accorded to the physiological theory of scoliosis.

The theory of pressure or superincumbent weight, which assumes the gradual transformation of the normal spine under mechanical influences falling obliquely upon it through faulty attitude, is accepted by most authors at the present time, and has been advocated especially by Volkmann, Roser, Vogt, Lorenz, Albert, Toldt, Bradford and Lovett, Schreiber, Redard, Hoffa, and others, and this the writer supports, giving especial importance to the ligaments in the secondary changes which occur.

Any external cause which destroys the equilibrium of the spinal column may be looked upon as a predisposing cause, and the superincumbent weight of the body as the direct etiological factor in the production of this affection.

Symptomatology. The lateral curvature or rotation of the vertebræ are rarely first observed. Usually the dressmaker or tailor discovers an inequality of one shoulder, or a growing out of one hip, and, these being further investigated by the surgeon, the true nature of the affection is at once revealed. They may be easily fatigued, complain of muscular weakness, and be disposed to recline or rest much of the time; or a boy complains of one suspender shoulder-strap constantly slipping, or a girl of her inability to keep the shoulder-straps in position. This should at once direct attention to the spine, and in slight cases the following changes will be noted:

In one of the most common varieties—the left primary lumbar curvature—the waist-line on the left is flattened or obliterated, and on the right deepened, and the waist-angle rendered more acute. The crest of the right hip becomes more prominent, and the crest of the left disappears. From a rotation of the vertebræ the left lumbar region becomes fuller, and the right flattened. Later, as the compensatory dorsal curve forms, the body placed in Adams' position (strongly flexed forward) reveals the serpentine course of the spinous processes, and the erector-spinæ mass of muscles is flattened on the left from the deviation forward of the transverse processes.

In this form, as in almost all cases of scoliosis, the first curve is known as the primary curve to distinguish it from the other curves which afterward form, and which are distinguished as secondary or compensatory curves.

Right convex dorsal scoliosis, very frequently a primary affection, gives symptoms still more pronounced. In the slight cases the ribs on the right side are more prominent, the scapula projects slightly backward, the left hip is a little more prominent, and spinous pro-

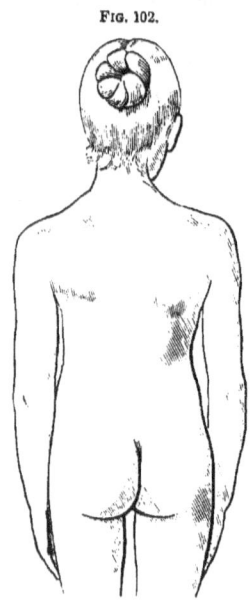

FIG. 102.

cesses in Adams' position are curved slightly (convexly) to the right. In severe cases these changes are increased and augmented by others.

The right costal angle is increased in extent and prominent, and the left flattened and decreased. The conformation of the anterior part of the chest is affected, but to a less degree than the posterior, and the breasts are unsymmetrically placed, the left being usually more prominent and higher. The abdomen is less prominent on the left, and the umbilicus is displaced to one side. The left scapula, more prominent, is elevated and projected backward, while the less prominent left scapula approaches the median line and is flattened. With the formation of a secondary compensating lumbar curve, the alterations before described under primary left lumbar scoliosis, are superadded. And later, with the formation of a compensatory left convex cervical, the outline of the shoulder and neck is altered, and the

Primary right dorsal scoliosis, showing so-called "high shoulder." (HOFFA.)

normal beautifully curved shoulder-line is changed; it is more curved on the right, the shoulder more prominent (Fig. 102), the neck longer; on the left it is flattened, the shoulder is less rounded, and the neck shorter. (Fig. 103.)

The exact process by which a cervical curvature from wry-neck induces a lateral curvature of the entire column is interesting. The deviation of the head to one side and the flexion of the chin upon the thorax by the unilateral contraction induce a bowing of the cervical vertebræ to that side and rotation of the bodies upon one another. There is at once produced a characteristic alteration in the outline of

the neck—the graceful double curve becomes flattened, the neck shortened and the shoulder less prominent on the one side, while the
curved line becomes more pronounced, the neck longer, and the
shoulder more prominent on the other side.

FIG. 103.

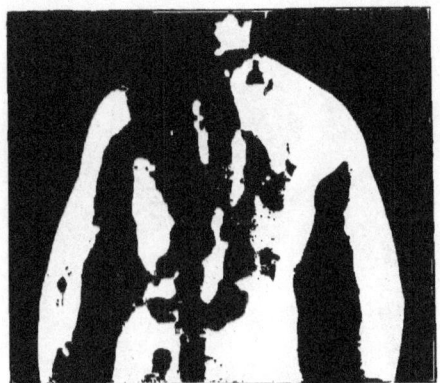

Right dorsal scoliosis in muscular subject, secondary to torticollis.

The natural balance of the column being disturbed, Nature
endeavors by a compensatory dorsal curve to restore the equilibrium.
This induces a rotation of the bodies of the dorsal vertebræ with all
its attendant deforming features: the flattening and twisting of the
scapula on one side, the deviation of the spinous processes in a
serpentine direction from the median line, and the elevation and
projection backward of the scapula on the other side. With the
development of a third or lumbar curve the crest of the hip becomes
obliterated on one side, the waist-line deepens, the hip becomes
prominent on the other side, and the lateral outlines become everywhere less symmetrical. (Fig. 104.)

It is popularly believed by the superstitious, that persons suffering
from lateral curvature possess great intelligence, bad tempers, and
intense venereal capabilities. Though these qualities, for obvious
reasons, are more or less frequently associated with caries of the vertebræ (Pott's disease of the spine), they do not apply to sufferers from
lateral curvature, who in other respects seldom differ from healthy
individuals.

In severe cases, in addition to the inconvenience and discomfort of the lateral deformity, actual pain in different parts of the spine and thorax from nervous pressure may be experienced ; and dyspnœa, emphysema, catarrhal bronchitis, pneumonia, and palpitation of the

FIG. 104.

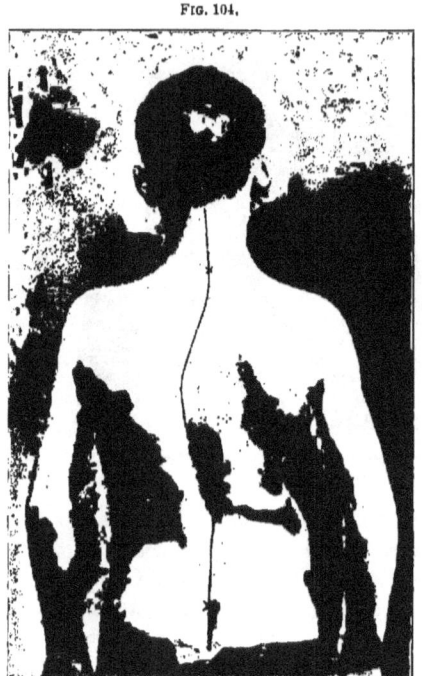

Photograph of case of left dorsal scoliosis.

heart may be complained of. The digestion may be impaired, the appetite capricious, and the patient nervous or neurasthenic ; while on the other hand, aside from the deformity, he may enjoy comparatively perfect health.

Three stages of the affection may be distinguished :

First stage, or initial stage.

Second stage, or stage of development.

Third stage, or stage of arrest.

The first stage in ordinary juvenile cases extends from the beginning of the affection—from the sixth to the twelfth year—to puberty ;

the second stage from puberty until the nineteenth to twenty-first year the period of the establishment of growth ; and the third stage includes the period after the completion of the developmental age.

Diagnosis. The importance of a correct early diagnosis cannot be overestimated, since it is the early stages which offer the most hope of perfect restoration. The examination should freely expose the trunk to a good light, and its contour and outlines be examined from different positions in front and behind. For this purpose the patient, if a child, should be freely exposed to below the trochanters, and the clothing conveniently secured by means of an ordinary roller bandage. The relation of the spinous processes should be carefully observed and marked while the patient, standing erect, bends far forward (Adams' position), or practises self-suspension from a bar or

Sayre apparatus. The inequality of the lower extremities may be tested by thin boards or books placed under the feet, while a rule or spirit-level estimates the amount of obliquity of the iliac spines. If doubt exists, or any inequality be discovered, the upper part of the body may be covered, and upon a hard couch the exact measurements of the lower extremities from the anterior superior spinous processes or umbilicus be taken. Many methods have been devised to measure and record the amount of deformity, by means of more or less complicated apparatus, of which Hoffa has recently collected a table of no less than thirty. One of the simplest of these is the scoliosometer of Mikulicz. (Fig. 105.) The ordinary cyrtometer used by clinicians to measure the chest, or an

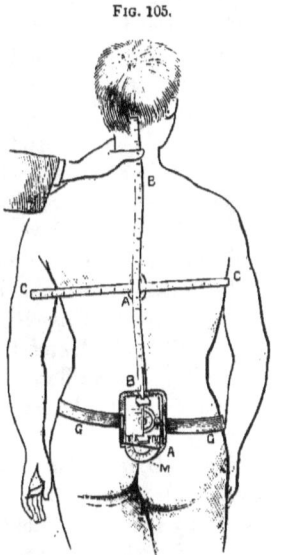

FIG. 105.

Mikulicz' scoliosometer applied.
(REEVES.)

ordinary lead measure with freehand sketching, has been found by the writer to answer all purposes, and this after using the most complicated of all—the Modellverband of Beely. The three stages or degrees of deformity should be carefully separated, as these bear directly upon the prognosis.

In the first degree, *the curvature can be readily restored* by suspension, lying, or slight manual pressure.

In the second degree, *the curvature cannot be entirely restored* by these measures.

In the third degree, *the curvature cannot be affected in the least* by these measures.

In this connection, particular attention should be directed to those cases of so-called lateral *bending* of the spine occurring in all classes and conditions of youth from general debility, in which the spine, though much deformed laterally, is without rotation of the bodies of the vertebræ, and can usually be restored by effort of the patient, who is quickly relieved by attention to the general health. From Pott's disease of the spine the differential diagnosis will be found under the appropriate section.

Progress and Prognosis. The progress, unless early arrested, is usually unfavorable; but the affection may be spontaneously arrested or become stationary at any stage. In the first degree, persistent and well-directed measures will permanently and surely restore the symmetry; in the second, the affection may with a degree of certainty be arrested, and, especially in primary lumbar scoliosis, be much improved, or even cured; but in the third stage nothing can be accomplished toward restoring the function of the spinal column, and measures must be directed to concealment of the deformity.

It has been asserted, but without positive proof, that scoliotic patients are short-lived—heart lesions and apoplexy being, according to Bouvier, the most frequent causes of death. The dangers of parturition in scoliotic women are unfounded, except when the degree is of the highest grade, or rhachitic deformity of the pelvis coexists.

Prophylaxis. The tendency to scoliosis should be guarded against by the introduction into schools of properly constructed benches and desks, by the care of the teacher to insist upon a correct sitting position, and perfect carriage while walking, and by the early correction of all deformities which tend to produce static curvature.

Treatment. This should be directed to the removal of the cause, the restoration of the health and strength, and the correction of the deformity.

The equalization of the extremities, change of occupation, assumption of a proper sitting position or carriage, and the administration of remedies for the correction of constitutional vice, such as preparations of cod-liver oil, iron, iodine, phosphates, etc., will all be useful in accomplishing a cure.

To correct or restore the deformity, properly directed gymnastic exercises are the most potent agents at our command. To assure a cure, the patient must devote herself (or himself) assiduously and persistently to the task for a long period. Self-suspension with a Sayre apparatus, double bar, horizontal bar, trapeze, or rings, should be frequently practised during the day, alternated with periods of recumbency. In all of these, where possible, the hand on the concave side should be uppermost. While in the prone position, a hard curled-hair pillow, a bag of sand, or an air-cushion should be placed beneath the convexity, or Wolf's suspensory cradle employed. Certain forward and lateral movements upon the chest-weights should be practised, or dumb-bell exercises employed. A very useful exercise is bending sideways supported by a strap (Figs. 106, 107), or

FIG. 106.

Self-correction apparatus for lateral curvature. (BRADFORD.)

over a padded bar ; or, better still, while sitting, the body may be inclined forward, resting the arms upon a table, and a padded belt to which a weight is attached be passed over the deformity.

By far the best exercises are the Swedish gymnastics. These must be administered by a person thoroughly familiar with the system as taught in Stockholm, and should consist principally of hanging and lying positions, and exercises intended to restore, and massage to strengthen, the weaker muscles. Care should be taken not to over-develop these, lest a curve result in the opposite direction, but all the muscles should be developed together. This method of treatment

FIG. 107.

Self-correction apparatus for lateral curvature. (BRADFORD.)

cannot be found in books, nor can it be taught from reading, and only certain general principles can be laid down and the general outline be described. To employ this method properly requires more ana-tomical knowledge than is the property of most physicians. For the general principles of the system the reader is referred to the author's former work.[1] In the special use of the Swedish move-ment for the cure of scoliosis two rules must always be observed : 1. *Retract only the retracted muscles, and only short of fatigue.* 2. *When exercising one side always isolate the other side.*

The special exercises for the three most common forms are as follows :

1. Dorsal Right Curve. Isolate the left arm by placing it in the neck-rest position. The right arm is extended up, with resist-

---

[1] " Physical Development," Keating's Cyclopædia, vol. iv.

ance. Bend forward with support about waist, four to six times, and rest.

2. Left Lumbar Curve. Contract rhomboidei muscles, and treat the special lumbar muscles.

3. Double Curve Left Side Neck. *a.* Exercise each arm separately. *b.* Place left hand to hip, exercise right side by flexion and extension of forearm with the arm extended, or bend forward on plane with massage of the dorsal and cervical muscles. *c.* The patient must take downward straight stride standing position four times *immediately*, so as not to allow the balance to be re-established. *d.* Follow by respiration movement.

Fig. 108.

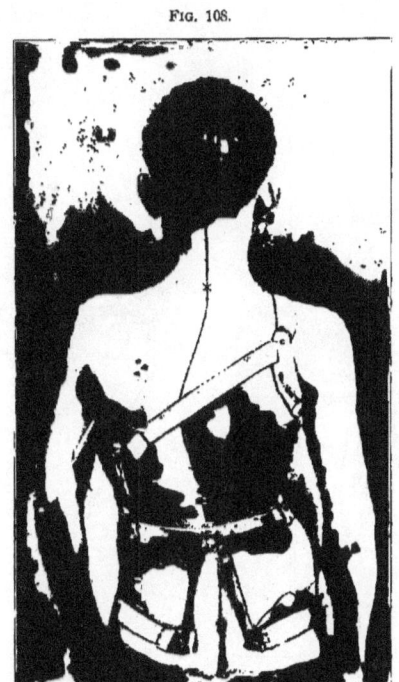

Shaffer spine brace applied to case of left dorsal scoliosis.

After a few lessons, a home prescription may be given, which can be followed if the parent or attendant coöperate; but where the patient can afford the expense, it is better to take these exercises three or four times a week for six or eight months, with additional exercises at home.   After the first few lessons, the exercises become a pleasant pastime, and the rapid and constant improvement leads the ambitious on to the acquisition of a perfect figure, with a buoyancy and vigor of health before unknown.

With these measures the best results have been accomplished, and the writer cannot speak too enthusiastically in their praise.

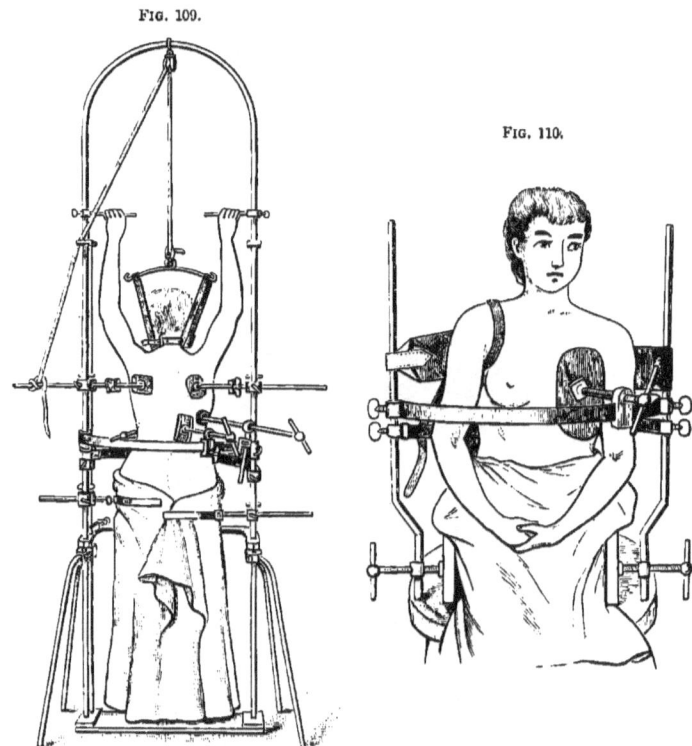

FIG. 109.

FIG. 110.

Patient in position in correcting apparatus.   Bradford's apparatus for forcible correction
(BRADFORD and BRACKETT.)   of scoliosis.

It will be observed that no mention has yet been made of the various apparatus still employed for correction and maintenance in this deformity.   The plaster-of-Paris jacket, myotomy, tenotomy, and forcible restoration under anæsthesia have, in the writer's opinion, no place in the treatment of lateral curvature. Of the mechanical supports, he has obtained the best results from, and has the greatest confidence in, a modified Shaffer brace. (Fig. 108.) This consists of a well-fitted steel waist-band, with diverging steel pieces in front and back attached together with webbing; from this waist-band is built up a crutch under the depressed shoulder, and a pad over the deformity. Strong webbing straps pass between the pad and the crutch in front and behind, and act laterally upon the prominence. The modifications made by the writer are a flattened axillary crutch after the pattern of Ernst, of London, and an increased number of straps of webbing below the band to insure greater stability. It is, however, to be clearly understood that this apparatus does not aim to correct the deformity by direct pressure, but is employed rather as a *reminder* to the patient to assume a correct position, and as a slight support to the body during the intervals of exercise and recumbency.

Mechanical Correction. In the second class of cases something more is needed to supplement the gymnastic methods, and in these

FIG. 111.

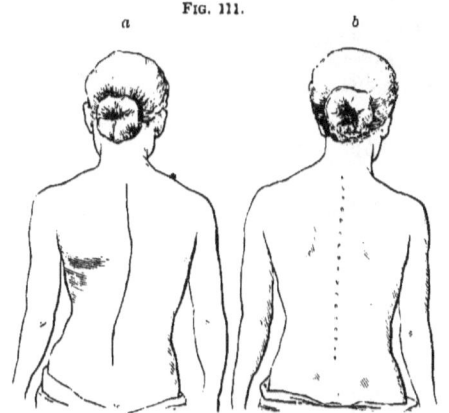

a. Case of right dorsal scoliosis before forcible correction. *b.* Same case after four months' treatment. (BRADFORD and BRACKETT.)

mechanical means for forcibly stretching contracted curves are of the greatest service. The principle of all is the same, and Beely, Lorenz, Redard, and Hoffa have shown how much force can be applied with advantage and without injury.

Two forms of appliances are employed : 1. Those which place the trunk in correcting attitudes. 2. Those exerting direct correcting pressure. The best form of the former is that of Beely—a square framework in which the patient is placed and in which weights and straps act as correcting agents. The appliance of Bradford and Brackett,[1] modified somewhat from that of Schede, of Hamburg, belongs to the second class, and is well adapted to the purpose for which it is designed. Its application, use, and results of treatment are well shown in the figures.

[1] Boston Med. and Surg. Journ., May 11, 1893.

CHAPTER XII.

INFANTILE SPINAL PARALYSIS.

INFANTILE paralysis is an acute inflammatory affection of the large multipolar ganglion cells in the anterior cornua of the cord, characterized by a sudden loss of power of the voluntary muscles unaccompanied by any sensory changes, and resulting in atrophy in one or more muscles or groups of muscles with deformity.

Synonyms. English, *Infantile Paralysis; Infantile Spinal Paralysis; Essential Paralysis; Atrophic Paralysis; Paralysis during Dentition* (Gull); *Regressive Paralysis; Myelitis of the Anterior Horns* (Seguin); *Tephromyelitis; Myogenic Paralysis* (Bouchert); *Poliomyelitis Anterior.* French, *Paralysie Infantile; Paralysie Spinale; Paralysie Atrophique Graisseuse de l'Enfance; Paralysie Essentielle; Tephromyélite Antérieure Aiguë* (Charcot). German, *Kinderlähmung; Spinale Kinderlähmung.*

Etiology. Infantile spinal paralysis is an affection almost peculiar to infancy and childhood, but may occur during adult life. The relative frequency of the affection may be inferred from the statement of Holmes Coote that out of 1000 children in the Royal Orthopedic Hospital 80, or 8 per cent., suffered from infantile paralysis. As to the possibility of the intra-uterine development of the paralysis the writer believes, with Gowers, that there is as yet no valid evidence, and certain it is that most of the cases of congenital club-foot (equino-varus) are not, as some writers believe, the result of an intra-uterine paralysis. It may come on soon after birth, as in the case of Duchenne, which was attacked on the twelfth day, and the case of Bramwell, which developed when the child was three weeks old; but it is infrequent during the first year. The great majority of cases occur between the ages of six months and three years or during the period of primary dentition. Thus, according to Gowers,[1] of all the cases under ten years, three-fifths occur during the first two years and four-fifths during the first three years. Of

---

[1] Diseases of the Nervous System, p. 254.

71 cases collected by Seeligmüller,[1] 90 per cent. occurred before three years; and of 350 cases collected by Sinkler,[2] in 335 of which the age of the patient was noted at the onset of the attack, 247 occurred under three years, and the average age of the attack in 244 cases was two years, one month, and two and one-fifth days. Occurring during the period of primary dentition, the disease has been attributed to dental irritation, especially by the early observers; but more recent and more thorough observations fail in most cases to reveal any direct connection. It is probable, as remarked by Meigs and Pepper,[3] that early age and dentition only act indirectly by inducing a remarkably susceptible condition of the entire spinal system. In this condition of exalted nervous irritability the spinal cord is especially susceptible to sudden changes of the surface temperature, as by a sudden cooling of the overheated surface. Sex appears to have no influence whatever upon its production, although it is stated by some writers that boys are more frequently attacked than girls. Of the 345 cases recorded by Sinkler 184 were boys and 161 were girls, and of the 63 cases collected by Barlow[4] 33 were males and 30 were females. The disease appears also to be almost as frequent among the children of the wealthy as among those of the poor, and Dr. Buzzard[5] thinks "it is more common than not for the disease to attack fine, grown, hearty children."

Heredity does not seem to have any influence, and cases in which more than one individual is affected in the same family appear usually to be the result of an epidemic; although Gowers[6] states that he has "been strongly impressed by two or three cases in which the other members of the family have suffered from other affections of the nervous system;" and Sinkler also refers to the history of nervous diseases in the family, especially chorea.

Long-continued ill-health seems to have no causative influence on the disease; but a number of cases are noted where the attack occurred during convalescence from some one of the acute exanthemata or other disease of childhood, as scarlet fever, measles, rheumatism, chorea, whooping-cough, cholera infantum, typhus and typhoid fever, and pneumonia.

[1] Gerhardt's Handbuch der Krankheiten, 188., v. p. 1.
[2] Keating's Cyclopædia of Diseases of Children, vol. iv. p. 685.
[3] Diseases of Children, p. 571.
[4] On Regressive Paralysis, Manchester, 1878, p. 4.
[5] Diseases of the Nervous System, p. 43.
[6] Brit. Med. Journ., May 20, 1882, p. 729.

It occurs with equal frequency in city and country, although the affection seems rare in some parts of the country, as in the locality in Ireland referred to by Bramwell,[1] where, during a period of thirty years not a single case has been met, nor are there any adults known in the district who are cripples in consequence of the disease.

The most conspicuous factor in the causation of this disease in children is the season of the year—a fact first emphasized by Sinkler some years ago.[2] In the 350 cases quoted by this authority,[3] in 270 the season in which the attack took place was recorded; of these there were 213, or 78.8 per cent., attacked in the hot months of the year—that is, from May to September, inclusive.

TABLE (FROM SINKLER).

		Cases.
Spring, 27 cases	March . . .	9
	April . . .	4
	May . . .	10
	Month not stated	4
Summer, 174 cases	June . . .	27
	July . . .	52
	August . .	65
	Month not stated	30
Autumn, 59 cases .	September . .	29
	October . .	24
	November . .	4
	Month not stated	1
Winter, 10 cases .	December . .	3
	January . .	4
	February . .	3
Total	. .	270

It will also be observed that the greatest number occur during August, when the temperature is but little lower than in July—a fact explained by Sinkler in two ways: "First, because the intense heat of July has prostrated the children to such an extent that they more readily succumb to the spells of heat which follow; and, secondly, in August the relative humidity is greater than in July, the figures being 72.1 and 68.6 respectively." Gowers states that two-thirds of his cases were attacked between June and September, and Barlow[4] has also confirmed the observation that a large proportion of the cases occur during the summer. The degree of humidity does not, according to Sinkler, show any influence on the number of cases unless associated with heat, nor does the range of temperature have any effect.

[1] Diseases of the Spinal Cord, p. 150.   [2] Amer. Journ. Med. Sci., April, 1875.
[3] Keating's Cyclopædia of Diseases of Children, vol. iv. p. 686.
[4] Regressive Paralysis, Manchester, 1878.

Exposure to great heat, long exposure to the sun, and great fatigue from over-exercise are mentioned as occasional causes of the attack. Exposure to cold and sudden chilling of the overheated body, as sitting on a stone step, sleeping in a newly-built house—as in a case of the writer's—or lying or sitting on the damp grass, undoubtedly have a marked influence upon the production of the disease.

Cases are frequently attributed to a fall, and traumatic hemorrhage into the substance of the cord and symptoms resembling this affection may and do occur; but in the majority of cases, inasmuch as falling is ordinarily a common symptom of the onset, and the interval between the particular fall and the onset of the disease is great, the association is of no importance.

Finally, it has been suggested[1] that this disease may be due to a microbe, and an analogy has been pointed out by Mathis,[2] between infantile paralysis and the infectious eruptive disease accompanied by paralysis which affects puppies, "maladie des jeunes chiens," in which a microbe has been demonstrated. The field for an ambitious investigator is still unoccupied, and further evidence is needed.

In conclusion, primary dentition may be considered a frequent predisposing cause, and sudden chilling of the surface of the overheated body during the season of the year when the *relative* humidity is greatest is the most common exciting cause, but the exact etiology at present remains uncertain.

Pathology. The morbid process of infantile paralysis consists of an acute destructive inflammatory lesion of the large multipolar ganglion cells in the anterior cornua of the gray matter in the cervical and lumbar enlargements of the spinal cord.

The exhaustive researches of Charcot and Joffroy, and the autopsies upon recent cases performed by D. Drummond,[3] Damaschino,[4] Charlewood Turner,[5] and Ashby,[6] have established beyond cavil the exact nature and location of the lesion.

The primary lesion is an inflammation which spreads over the greater portion of the cord, but is most intense and violent in the cervical and lumbar enlargements, and particularly in the anterior

[1] New York Med. Journ., Nov. 17, 1888, p. 530.
[2] Recueil de Méd. Vétérinaire, April 15, 1887.
[3] Quoted by Gowers, op. cit., p. 264.
[4] L'Union Médicale, 1883, quoted by Jacobi, op. cit.
[5] Ross: Diseases of the Nervous System, vol. ii. p. 121.
[6] Ashby and Wright: Diseases of the Nervous System, p. 446.

cornua, the most vascular part of the cord. This inflammation may in rare instances be preceded by a slight hemorrhage. The exudation of inflammatory material (exudation of leucocytes) leads to both temporary and permanent damage to the motor cells in this region. This exudation of leucocytes has been observed in the medulla and the white matter of the cord as well as the gray. When the inflammation subsides, absorption of the exudation occurs and a gradual improvement takes place in those areas where the destruction of gray matter has been incomplete, and perhaps new nerve fibres and cells

Fig. 112.

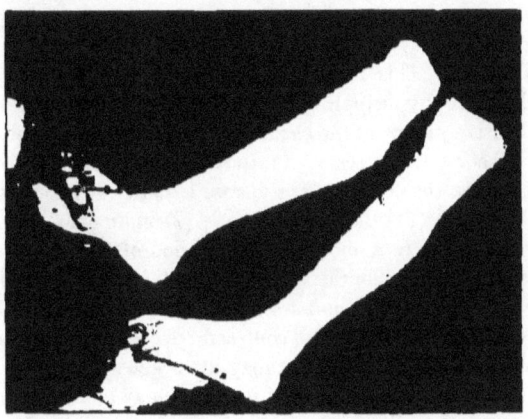

Photograph of a case of myelitis of the anterior cornua, showing atrophy of the leg type on the left side. (Gray.)

replace the damaged elements, and a certain amount of muscular power is regained. When the spinal cord is examined post-mortem some years after the onset of the paralysis, the destroyed areas in the anterior cornua are found replaced by cicatricial connective tissue. The bloodvessels in this region are enlarged and their walls thickened, and the débris of degeneration (corpora amylacea, pigment granules, etc.) are abundant. The large ganglion cells are found in different stages of degeneration and atrophy, or are entirely wanting. Degenerative changes take place likewise in the peripheral nerves connected with the damaged cord; the muscles waste; their fibrillar structure is replaced by fat-globules, and finally the muscular tissue loses its identity, and there is an absence of even fat-globules, it being replaced by hypertrophied connective tissue. According to Barlow, a muscle may, however, be much wasted in bulk without these microscopic changes being present. The muscles which oppose these paralyzed muscles are much wasted from disuse; the arteries are diminished in size; the tendons are elongated and atrophied, and the growth and development of the bones is greatly retarded, the medullary portion being increased. Changes also occur in the articulations, the articular extremities of the bones often being atrophied, the cartilages thin, and sometimes eroded. These changes are all atrophic.

Symptoms. This affection occurs in two forms: the acute and the subacute or chronic. The former, occurring most frequently in children and being the most fruitful of deformities, will be the only one considered. This disease is usually seen by the orthopedic surgeon when atrophy, crippling deformities, and contractures present a comprehensive picture of the chronic stage; but to thoroughly appreciate this affection, it is necessary to study the early stages also. According to Gowers, the course of the disease is most conveniently divided into four stages: (1) the initial stage ; (2) stationary period, which lasts from a day to a month; (3) a period of "regression," which lasts from one to six months, during which the motor function returns in certain of the affected muscles ; (4) a chronic stage, during which the degenerative changes and contractures occur, and during which slight and gradual improvement may intervene at any period.

1. The initial stage, or stage of invasion, may be characterized by a sudden paralysis, without any prodromal symptoms, but it is usually ushered in with high fever, general irritability, convulsions, vomiting, diarrhœa, muscular twitchings and hyperæsthesia, and great cerebral disturbance. These symptoms are usually overlooked by the friends, or attributed to dentition or general gastro-intestinal disorder, and it is not until the paralysis supervenes that the importance of these symptoms is recognized, and frequently the patient becomes convalescent and attempts to rise before even the paralysis is distinguished.

The pyrexia, though rarely high, may reach 104°. The general cerebral disturbance is so marked as to have been attributed to meningitis, both by the laity and the practitioner.

The paralysis usually reaches its height at once, or in a very few days—occurring within the first three days of the beginning of the general constitutional disturbance in 111 of the 254 cases recorded by Sinkler. A single group of muscles only may be paralyzed, or the paralysis may be monoplegic from the onset, but most systematic authors describe the paralysis as affecting all the extremities at the onset and rapid regression taking place. This has been the observation of the writer in a large number of cases. The severity of the initial attack seems to have no effect upon the extent of the paralysis which follows, since some of the most extensive and complete paralyses have supervened after the mildest initial stage. The paralysis is monoplegic in its distribution in more than half the cases, both at the onset and also at the time the patients applied for treatment, as will be seen in the following table of Sinkler:[1]

	Parts affected at onset.	Parts affected when examined.
Both legs	107	92
Right leg	63	90
Left leg	62	82
Left arm and left leg (hemiplegic form)	14	6
Right arm and right leg (hemiplegic form)	12	9
Left arm and right leg	...	1
Right arm and left leg	1	1
Both legs and right arm	6	6
Both legs and left arm	4	2
Both legs and one arm	...	1
Both arms	1	3
Right arm	5	4
Left arm	8	7
Arms and legs	35	16
Arms, legs, and trunk	22	9
Not stated	10	12
Total	350	350

The muscles most frequently paralyzed are those of one lower extremity, particularly those of the leg. Muscles of the trunk are only affected in the severer grades of paralysis (Fig. 113), and paralysis of the sphincters of the bladder and rectum are rare, but do occur. When single muscles are affected, the deltoid, according to Sinkler, suffers alone more frequently than any other muscle of the arm, and the tibialis anticus is paralyzed alone oftener than other muscles. The flexors of the foot and the extensors of the leg are next in the order of frequency. Paralysis of the facial muscles is rare, though cases are recorded by Sinkler, Gowers, and Barlow. Fortunately, certain muscles always escape paralysis, as the muscles of the eyeballs, ears, larynx, and those of respiration; the diaphragm and intercostals are only affected in the most exceptional instances. Universal paralysis, when it occurs, is quickly fatal from failure of respiration.

[1] Anterior Poliomyelitis, Keating's Cyclop. Dis. Children, vol. iv. p. 696.

2. The stationary stage is characterized by a period of from one to six weeks, or even four months, during which the muscle paresis remains stationary. As a rule, however, the regression occurs early, the first improvement usually taking place in the parts last affected, and extending until all the muscles have recovered, except those which are to remain permanently paralyzed. The reflexes, both superficial and deep, are lost, and in some cases sensation is diminished and even entirely lost. The circulation in the skin is interfered with by the cord lesion, and the temperature of the affected part is much reduced, there being frequently two or three degrees difference between the healthy and the affected limb, and in exceptional cases 10°, 20°, 30°, or even 40°. The circulation is sluggish, giving rise to a mottled, dusky purplish discoloration of the skin, easily excoriated under friction, but free from idiopathic ulcerations, sloughs, or severe atrophic changes. The muscle irritability is markedly changed during this early period, the reaction to the faradic current being usually entirely absent in those muscles that are permanently paralyzed.

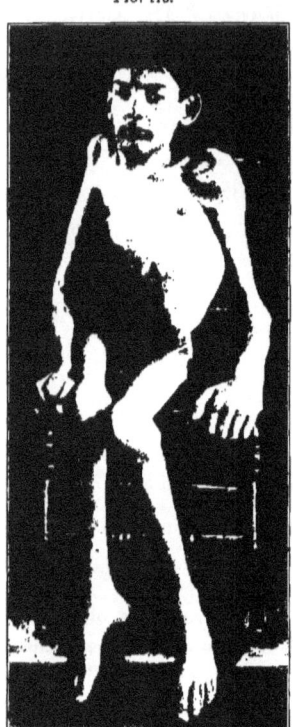

Fig. 113.

Photograph of severe case of infantile spinal paralysis.

The muscle irritability to the continuous current during this period is increased, but becomes gradually lessened later, and may entirely disappear. To the continuous galvanic current it still continues to respond, though the reaction differs essentially from that of healthy muscles, the change being the so-called "reaction of degeneration," due to the degeneration of the peripheral nerves. The electrical formula is reversed, a change to be referred to later under diagnosis.

3. The stage of regression, commencing after two or three days, or

after a delay of three or four months, continues till all the muscles have regained their power except those which are to remain permanently paralyzed. This includes a period which terminates only with the cessation of growth itself.

Recovery in the affected muscles is indicated by an increased response to the interrupted current, while a diminished response and progressive atrophy mark the advance of permanent paralysis in the muscles which are to remain permanently affected. This gradual regression of the original paralysis is a remarkable clinical observation, in a few instances entire recovery occurring in all the affected muscles, and in others recovery ensuing after one or more attacks and permanent paralysis finally resulting. During this period of regression the general health is usually excellent, all the bodily functions are perfectly performed, and nothing but the paralysis and atrophy remain to indicate the inflammatory stage passed.

FIG. 114.

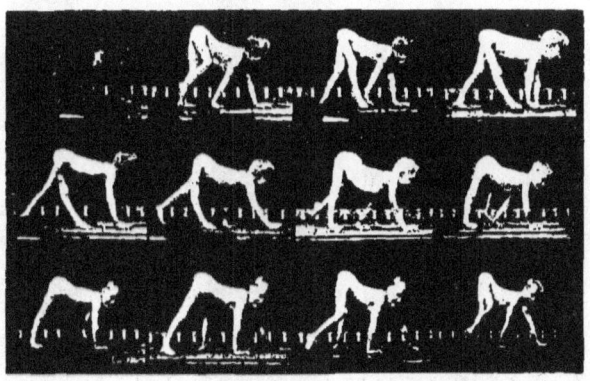

Quadrupedal position in walking. (WILLARD.)

4. The chronic stage, characterized by deformities and dislocations from atrophy and contractures, is of peculiar interest to the orthopedic surgeon. Six months after the onset of the disease marks the limit of rapid improvement, and the improvement, which may go on for years, is always slow and very slight after this period. (Fig. 114.) At this time may be estimated pretty accurately the amount and distribution of the permanent paralysis, which is always much less than

at first appeared. This permanent paralysis may affect only a single muscle or may include the muscles of all the extremities and some of the trunk muscles. (Figs. 115 and 116.)

Fig. 116.

Fig. 115.

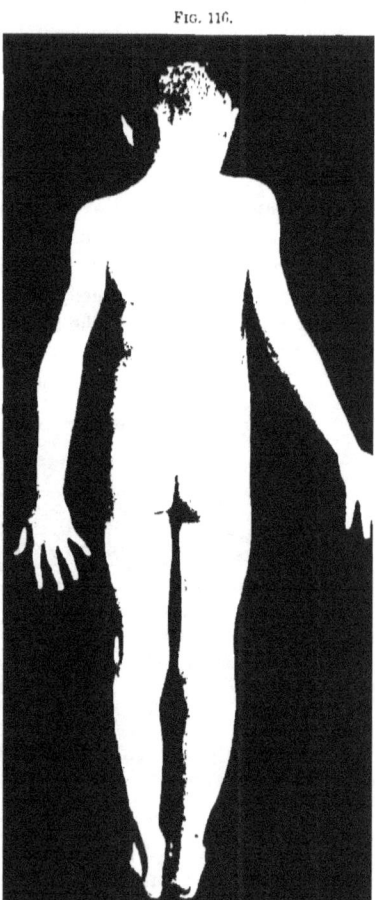

(FRONT VIEW.)   (REAR VIEW.)

Photographs of a hemiplegic boy of eight, showing difference between right (affected) and left (unaffected) sides five years after onset. (GRAY.)

The symptoms of the paralysis of individual muscles and groups of muscles are well illustrated in the following schedule from Jacobi:[1]

[1] A System of Practical Medicine, 1886, vol. v. p. 1124.

Upper Extremity. *Deltoid.* Absence of deformity, which is averted by weight of arm. Inability to raise arm. Sometimes subluxation. Frequent association with paralysis of biceps, brachialis anticus, and supinator longus.

Lower Extremity. *Ilio-psoas.* Rare except with total paralysis. Associated with paralysis sartorius. Loss of flexion of thigh. Limb extended (if glutei intact).

Glutei. Thigh adducted. Outward rotation lost. Lordosis on standing. Frequent association with paralysis of extensors of back.

Quadriceps extensor. Flexion and adduction of leg (if hamstrings intact). Loss of extension of leg. Frequent association with paralysis of tibialis anticus.

Tibialis anticus. Often concealed if extensor communis intact. If both paralyzed, then fall of point of foot in equinus. Dragging point of foot on ground in walking. Big toe in dorsal flexion (if extensor pollicis intact). The tendons prominent. Hollow sole of foot (if peroneus longus intact).

Extensor communis. Nearly always associated with that of tibialis anticus. Toes in forced flexion.

Peroneus longus. Sole of foot flattened. Point turned inward. Internal border elevated.

Sural muscles. Heel depressed. Foot in dorsal flexion (calcaneus). Sole hollowed if peroneus longus intact, flattened if paralyzed. Point turned outward (calcaneo-valgus).

Trunk. *Extensors of back.* Lordosis on standing. Projection backward of shoulders. Plumb-line falls behind sacrum (unilateral). Trunk curved to side. Trunk cannot be moved toward paralyzed side.

Abdominal muscles. Lordosis without projection backward of shoulders.

The deformities resulting from infantile paralysis have been ascribed to one of two causes: 1, the relaxation of the muscles and ligaments, and the undue action of their opponents; and 2, the result of growth upon the limb, the paralyzed parts remaining undeveloped. The first cause—the unequal action of the muscles—undoubtedly has its influence in certain instances, and is an important factor in the production of deformity, particularly where a single muscle or single group of muscles is involved; but the second cause—that they are due to atrophy and *arrest of growth,* and are not in any sense muscular—is by far the most important, and as such deserves emphasis. Nearly all these deformities may be accounted for on purely mechanical grounds. Hueter has demonstrated that the weight of the part in the position assumed at rest in paralysis, and the muscular insufficiency of the affected parts, which subjects the articular surfaces to excessive pressure when in use, cause the deformities.

FIG. 117.

Photograph of a case of myelitis of the anterior cornua, showing atrophy of the leg type on the left side. (GRAY.)

Thus in paralysis of the muscles of the anterior region of the leg (Fig. 117), the deformity is due entirely to the force of gravity, the foot falling into the position of equinus, and the anterior portion of the foot being adducted by its own weight. In these cases in the

early stages there is little or no contraction, and the deformity can
be readily reduced by manual pressure. Volkmann has also called
attention to the fact that the abnormal position assumed by the
affected limb eventually becomes permanent through the weight of
the body and abnormal growth, and not, as was formerly assumed,
from contraction. The apparent contractures which are met with in
this affection, as the muscles about the hip, knee, ankle, and foot,
are not, then, due to contracture of these structures, but are the result
of abnormal or normal growth of the bones and antagonizing muscles; these affected structures, from paralysis and atrophy, remaining
in their original condition. This corresponds to the so-called
"adapted atrophy" of Sir James Paget, the changes which ensue in
consequence of the mechanical relations of the foot to the leg. Thus,
in talipes varus resulting from infantile paralysis the plantar fascia
is not *per se* contracted, but its growth being retarded or abolished,
and the foot growing in length, the apparent contracture has resulted.

There is still one other factor at work which has been suggested as
the cause of the more severe paralysis of certain groups of muscles
over others, as in the anterior muscles of the thigh and leg, and this
is the stretching of these muscles by the weight of the foot and leg in
lying and sitting. This stretching of the muscles, if all the muscles
were equally affected at the onset, is sufficient to account for the
severe paralysis remaining after the other groups have recovered.

The deformities arising from infantile paralysis may be considered
under three groups: deformities of the upper extremity, deformities
of the lower extremity, and deformities of the trunk.

Deformities of the Upper Extremity.

The deltoid, though not a common seat of paralysis, suffers alone
more frequently than any other muscle of the upper extremity. In
addition to the inability to raise the arm there is a loss of rotund-
ity of the shoulder, and a prominence of the acromion process, the
shoulder presenting a flattened appearance, and sometimes a subluxation. Associated with paralysis of other muscles—the supra- and
infra-spinatus, biceps, and triceps—it constitutes the so-called "upper-
arm type" of Erb, a combination which differs from the "upper-arm
type" of Remak,[1] in which the supinator longus is affected along

---

[1] Archiv für Psychiatrie, 1879.

with the brachialis anticus, biceps, and deltoid. The trapezius, subscapularis, and serratus magnus are occasionally affected.

FIG. 118.

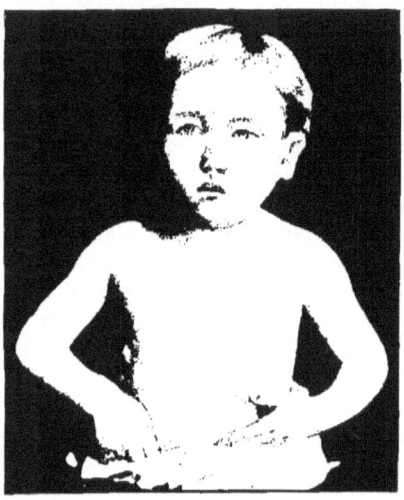

Photograph of a case of myelitis of the anterior cornua, showing the upper-arm type
on the left side, occurring twelve days after birth, in which there was absolute
paralysis for eleven months, and complete reaction of degeneration. (GRAY.)

Paralysis of supinators and extensors of the hand and the adductors of the thumb, when it occurs, may result in flexion of the hand
and fingers, with restricted mobility, and may prevent the apposition
of the thumb with the other digits. When the supinator longus
escapes, but the extensors of the wrist are affected, it constitutes the
so-called "forearm type" of Remak. A deformity amounting to
club-hand may likewise occur. Contraction of the shoulder, elbow,
and wrist resembles post-hemiplegic contractures.

Deformities of the Lower Extremity.

Paralysis and atrophy may affect the entire limb, and give rise to
the withered, useless, flail-like, doll-like limb known as "jambe de
polichinelle," or but a single extensor of the foot may be involved,
influencing locomotion but little or not at all. Fortunately, the
paralysis is more frequently partial than complete.

The extensor quadriceps and glutei are the muscles most frequently affected in the thigh. The limb flexed and adducted cannot be made sufficiently rigid to sustain the body-weight, and extension and outward rotation are lost.

Contraction of the sartorius, tensor vaginæ femoris, and, in rare cases, the ilio-psoas muscles, prevents the limbs being brought underneath the body, except by the forward rotation of the pelvis and the production of marked lordosis in the lumbar region of the back.

Dislocation of the hip may occur spontaneously from relaxation of the muscles and ligaments, or may result from the weight being improperly thrown upon it. Dislocation upon the dorsum of the ilium is the most frequent, but displacement in other directions may occur, and the relaxation of the joint may be very great. Shortening and great mobility, without pain or other symptoms, characterize the deformity, and if the dislocation is allowed to remain a new cavity for the head of the bone will in time be formed.

At the knee-joint, inability to extend the knee is the most common deformity, and great laxity with lateral mobility may occur, giving rise to the so-called "Schlottergelenk." Other deformities of the knee arise from use of the joint, the body falling improperly upon the relaxed ligaments. In this way occur the hyperextension of the knee, or recurvation, in which the knee is bent backward beyond the perpendicular, and the head of the tibia lies in a plane posterior to the line of the femur.

Permanent flexion from contraction of the hamstring tendons and knock-knee from elongation of the internal lateral ligament result in severe cases. Outward rotation of the tibia upon the condyles of the femur results from undue contraction of the biceps.

The most frequent varieties of club-foot are equino-varus and valgus. Equino-varus results from paralysis of the anterior tibial and peronei muscles. If, however, the peronei remain intact, equinus results. Talipes varus is rare, and talipes calcaneus is the rarest of all the deformities from poliomyelitis.

Talipes valgus acquisitus and talipes calcaneo-valgus result from the relaxation of the ligaments, and the improper transmission of the body-weight upon the ground, and the most severe forms of these deformities result from infantile paralysis.

In valgus the changes which ensue from the effects of growth upon the mechanical relations of the foot to the leg are observed, and in severe cases contraction of the peronei tendons is noted. Likewise in varus, contraction of the plantar fascia is of frequent occurrence.

True talipes calcaucus almost never occurs, and its existence has been doubted. What does occur, however, is known as pes cavus. The heel is lowered by relaxation ; the anterior part of the foot remaining in the normal plane gives an arched appearance to the sole of the foot, and an apparent elevation of the anterior portion of the foot.

According to Mr. Adams, these deformities occur in the following order of frequency : 1. Talipes equinus ; 2, equino-varus; 3, equino-valgus; 4, calcaneus, or calcaneo-valgus, and, 5, talipes varus. When both feet are affected, equino-varus of one foot is generally found with equino-valgus of the other.

Deformities of the Trunk.

Paralysis of the trunk muscles in severe cases gives rise to great distortion, and may render the patient perfectly helpless. (Fig. 119.)

Fig. 119.

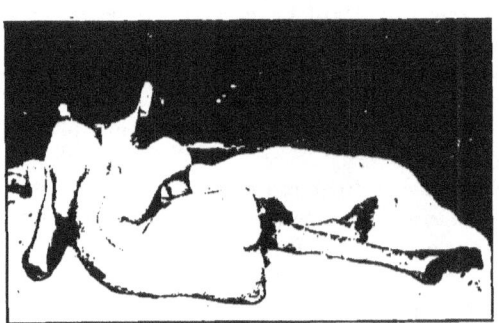

Excessive lateral curvature, with rotation and complete paraplegia from
infantile paralysis. (WILLARD.)

Paralysis of the extensors of the back, when bilateral, produces lordosis on standing, inability to sit erect, and projection backward of the shoulders. When unilateral, it results in lateral curvature, with inability to move the trunk toward the paralyzed side.

Paralysis of the abdominal muscles produces lordosis also, but without backward projection of the shoulders.

The lateral curvature of the spine, resulting from infantile paralysis, may be either (1) *static*, where the trunk muscles are unaffected but where the pelvis is tilted from inequality of one lower limb from paralysis and atrophy, or where faulty spinal attitudes are habitually assumed from paralysis and atrophy of the muscles of one upper extremity, or (2) *paralytic*, from unilateral paralysis of the intrinsic spinal muscles, that important group which controls the movements of the individual vertebræ, or the great erector spinæ mass of muscles, which controls the movements of the column as a whole.

Diagnosis. The symptoms in well-established infantile paralysis are so strikingly peculiar that a diagnosis is not difficult, and yet in the early stages, before the development of the paralysis, a correct diagnosis is most difficult to establish. The pyrexia, convulsions, and vomiting are frequently mistaken for cerebro-spinal meningitis, acute rheumatism, acute cold, indigestion, etc., and the paralysis which ensues is often mistaken for the prostration following some acute affection.

The characteristic symptoms upon which the practitioner must rely for a correct establishment of a diagnosis are :

1. Sudden onset.
2. Motor paralysis (sensation unaffected).
3. Regression of paralysis.
4. Change of electro-motor reaction.
5. Atrophy and deformities.

It should be recollected that the prostration following acute illness never amounts to complete loss of power, and that " forms of reflex irritation, such as ascarides, ådherent prepuce, and like peripheral conditions, may produce some of the symptoms, but their non-progressive character, and disappearance with the removal of the cause, should make the possibility of an error very remote." [1]

The only affection which cannot be differentiated from infantile paralysis by an electrical examination are the peripheral palsies, where localized paralysis results from traumatism or the presence of an enlarged gland or tumor on a nerve—in fact, wherever the muscles are cut off from the influence of the trophic centres.

The electrical reaction of the muscles furnishes by far the most important diagnostic test.

For the following practical description of the electrical reactions, the writer is indebted to his friend Dr. Charles S. Potts, instructor in electro-therapeutics in the University of Pennsylvania.

---

[1] A. L. Hamilton: Nervous Diseases, 1881, p. 285.

One of the most important symptoms of acute poliomyelitis is the changed behavior of the affected muscles and the nerves supplying them, when stimulated by the electrical current. This change is present to a greater or less degree whenever a muscle or nerve is cut off from its trophic centre, and has been termed by Erb the reaction of degeneration. In order to more fully comprehend what this alteration is, a brief account of how the muscles and nerves act when normal may not be out of place.

1. *Response to the faradic current.* When one of the electrodes is placed either over the muscle or the nerve supplying it, the other being placed at some indifferent point, there is a contraction of the muscle at each opening and closure of the circuit.

2. *Response to the galvanic current.* If the negative pole, termed the cathode, is placed over the nerve and the other at some indifferent point, the sternum preferably, and the current is gradually increased in strength and the circuit alternately opened and closed until a response be obtained, we will find that our first muscular contraction will take place when the circuit is closed; this is called the cathodal closure contraction. No contraction will take place when the circuit is opened; this can only be elicited by the most powerful current, and is for obvious reasons never obtained in a healthy human being.

If now for the cathode we substitute the positive pole, termed the anode, we will get no response until we increase the current strength, when contractions of about equal intensity will occur at both the opening and closing of the circuit; these are termed respectively the anodal opening and closure contractions.

These contractions have been conveniently formulated as follows:

Ca = cathode.
An = anode.
Cl = closure.
O = opening.
C = contraction.
C' = strong contraction.
C'' = very strong contraction.

Weakest current.
Ca Cl C.

Strong current.
Ca Cl C'.
An Cl C }
An O C } about equal.

Very strong current.

Ca Cl C''.

An Cl C'.

An O C'.

Ca O C (very weak contraction).

Practically the muscles when directly stimulated respond in a manner similar to that which ensues when the supplying nerve is stimulated.

When the muscle and nerve are separated from their trophic centres, the mode of reacting to the current undergoes a radical change. Further, the nerve and muscle each respond differently. If the case is seen immediately after the onset of the disease, which it rarely is, and the nerve or nerves influenced by the diseased cells are stimulated, we may find a stronger contraction of the respective muscles supplied by them taking place, both with the faradic and galvanic currents, than would take place if they were normal. This lasts for one or two days, when a progressive decrease commences which continues in marked cases until in the course of a week no response can be elicited by either form of current. The muscle when excited directly acts similarly with the faradic current, but with the galvanic current the following is noticed : During the first week a slight decrease in irritability will be noticed, to be succeeded by a marked increase lasting from three to six weeks; also, instead of the normal short, sharp contraction we will have one which is slow and long drawn out, in some cases almost tetanic in character, and in place of the Ca Cl C alone being excited by the weakest current that will cause a contraction we will find either the An Cl C equally prominent or in severe cases taking place first; this change in the mode of contraction is expressed by formula thus :

$$An\ Cl\ C = Ca\ Cl\ C,$$

or in severe cases,

$$An\ Cl\ C > Ca\ Cl\ C.$$

Ca O C can also be elicited and at times equals An O C, although it has never been observed greater; we express this change by formula as follows :

$$Ca\ O\ C = An\ O\ C.$$

At the end of this period the power of response gradually becomes weaker; stronger currents become necessary to produce contractions.

Ca O C disappears, then An O C, until in very severe cases we can only get a weak An Cl C, which, if the case goes from bad to worse, finally ceases.

In very mild cases we often observe what is termed a partial reaction of degeneration, the difference being that when the nerve is stimulated the muscle responds normally or with only a slight decrease in irritability. The muscles act when directly stimulated as described above.

To demonstrate these changes as detailed above takes skill and experience, and even then they cannot always be shown. For the practical purposes of diagnosis a knowledge of the following facts will suffice, and are really the essential features of the degenerative reaction.

1. The loss of the power of a muscle to contract when stimulated by a faradic current, and the response, if a recent case, of the muscle to a weaker galvanic current than that which normally causes a contraction.

2. The existence of the long wave-like contraction and An Cl C equalling or occurring before Ca Cl C.

Some valuable points in prognosis may be obtained by a study of these reactions. If after several weeks only the partial degenerative reaction is present we can predict a comparatively speedy recovery, and the prognosis is correspondingly worse the farther advanced the reaction is, and the longer it persists without any tendency to improvement. This improvement is manifested first by the gradual return of the nerves to the normal reaction, the changes evolved by direct stimulation of the muscles persisting for some time after the irritability of the nerves returns. If only An Cl C is present the chances of complete restoration are extremely slight and a prolonged course of treatment will be necessary to effect any. If no response can be elicited the case is practically hopeless.

Differential Diagnosis. Infantile paralysis must be distinguished from cerebral paralysis, myelitis, diphtheritic paralysis, rhachitic pseudo-paralysis, spastic paraplegia, cerebro-spinal meningitis, progressive muscular atrophy, pseudo-hypertrophic paralysis, birth or pressure palsies, hemorrhage into the cord, multiple neuritis, congenital dislocation of hip, and hip-, ankle-, and shoulder-joint disease. Of these, it is most frequently confounded with other forms of paralysis (spinal or cerebral) and progressive muscular atrophy and hip-joint disease.

Cerebral Paralysis.

It is with this affection, particularly in its earlier stages, that infantile paralysis is most frequently confounded, and from which, when the latter disease is hemiplegic with involvement of the facial nerve, it is difficult and almost impossible without an electrical examination to distinguish.

In cerebral paralysis the onset is usually sudden, convulsions frequently occur, and hemiplegia with facial paralysis results.

In the disease under consideration, when hemiplegia occurs the arm usually soon regains its power and the leg remains paralyzed, while the reverse occurs in cerebral paralysis. In the latter, also, the muscles of the affected part are frequently rigid; the tendency to atrophy and deformity (except post-hemiplegic contractures), the changed electrical reactions, and the lowering of the temperature of the affected part are all absent. Moreover, the causes of cerebral paralysis in children—meningitis, cerebral hemorrhage, and infectious diseases—add their own appropriate symptoms to the essential symptoms of this affection.

In the more difficult cases the electrical reactions alone will distinguish between the two affections.

Myelitis.

Acute transverse inflammation of the cord is characterized by complete loss of power, with marked loss of sensation, with diminished reflex excitability, and electro-muscular contractility and subsequent atrophy of the affected muscles. The loss of sensation, the tendency of the affection to grow progressively worse, and the grave character of the affection would serve to distinguish it.

In this connection, in speaking of the differential diagnosis from spinal paralysis, Mr. Alexander Shaw[1] says he has "never seen a case which it was possible to confound with spinal paralysis, and should suppose that a moderate amount of care would always suffice to prove the absence of any spinal affection."

Diphtheritic Paralysis.

Diphtheritic paralysis may be distinguished by the previous history, the association of paralysis of the palate and pharyngeal mus-

---

[1] Holmes' System of Surgery. Amer. ed., 1882, vol. ii. p. 876.

cles, the unchanged electrical reactions, and the absence of severe atrophy. Moreover, the paralysis of the extremities rarely arises suddenly, and, as a rule, according to Jacobi, they involve a series of muscles at the same time, improving in the same order as the individual muscles became affected.

Rhachitic Pseudo-paraplegia.

This is not so much a paralysis as an indisposition to use what muscular power is retained on account of the general tenderness. The loss of power may be almost as great as in infantile paralysis, but the gradual onset, the association of other evidences of rickets, the absence of atrophy, and the normal electrical reactions will serve to distinguish it.

Spastic Paraplegia.

The tetanoid rigid condition of the limbs, the gradual onset, the exaggerated reflexes, the absence of atrophy, and the unchanged condition of faradic irritability would readily distinguish it.

Cerebro-spinal Meningitis.

This may at the onset be confounded with infantile paralysis. The epidemic nature of the disease, the opisthotonus, the dorsal pain, hebetude, coma, and convulsions, the progressively grave character of the affection, are important in arriving at a correct diagnosis. Moreover, the paralysis, while frequently hemiplegic, is not permanent.

Progressive Muscular Atrophy.

" Wasting palsy " is of rare occurrence in infancy and childhood, and the gradual onset, corresponding to the increase of the atrophy, its progressive nature, and the presence of faradic contractility and the reflexes as long as any muscular fibre remains, will suffice to distinguish it in the earlier stages. Later, during the stage of contraction and deformity, the generalized atrophy of this disease will be sufficiently characteristic.

Pseudo-hypertrophic Paralysis.

Although motor weakness is the first symptom to attract attention in this affection, the increased size of the muscles, unattended by any marked electrical changes, is sufficiently characteristic; and subsequently, when atrophy sets in, its general distribution, the history, the deformities from muscular contractures, the disappearance of the knee-jerk as the disease advances, the entire absence of the reaction of degeneration, will readily distinguish this affection from the one under consideration.

Pressure or Traumatic Palsies.

Peripheral paralysis from nerve injury, as in birth palsies immediately after instrumental delivery, from tight bandaging, ligature, or from pressure of an enlarged gland or tumor on a nerve, is usually limited to the distribution of a single nerve and has associated loss of sensibility and trophic skin lesions. In all such the course of the great nerves of the part should be examined, for in such cases the electrical examination fails, since it is identical.

Hemorrhage into the Cord.

Traumatic hemorrhage into the anterior cornua is identical with anterior poliomyelitis, cases being recorded by Allbutt[1] and Turner.[2] Hemorrhage into the gray substance of the cord resembles the affection under consideration in its sudden onset, the subsequent atrophy, the absence of reflex action, and the loss of electrical irritability; but differs from it by the absence of the initial pyrexia, the almost instantaneous occurrence of paralysis, the association of sensory disturbances, the paralysis of the sphincters, bedsores, and other trophic changes.

Multiple Neuritis.

Of rare occurrence in childhood, it differs from poliomyelitis in the gradual onset, gradual loss of response to the faradic current, the absence of regression, the marked hyperæsthesia, and tenderness over the nerve trunks. Motion is painful.

Congenital Dislocation of Hip.

When slight this may resemble the disease under consideration, but may be distinguished by the slight atrophy, the normal electrical

[1] Allbutt : London Lancet, 1870, ii. p. 84.
[2] Turner: Trans. Path. Soc., 1879, xxx. p. 202.

reactions, the position of the trochanter above the Roser-Nélaton line, and the possible reduction of the deformity by traction. When the hip is dislocated in infantile paralysis the difference would be less, but the history of the case and the electrical changes would still be available for diagnostic purposes.

Hip-joint Disease.

When sudden and associated with joint pain and tenderness, severe muscular atrophy, and modified irritability to the faradic current,[1] the resemblance of this disease to anterior poliomyelitis is great, but the characteristic muscular fixation, the gradual onset, deformed position of the limb, night-cries, etc., would serve to distinguish it. The same symptoms, being those of osteitis in general, would serve in a measure to distinguish ankle- and shoulder-joint disease. The writer has recently had a case of ankle-joint disease which at first glance closely resembled infantile paralysis.

Prognosis. With modern therapeutic, mechanical, and operative means, the prospects of improvement in this affection are now exceedingly promising. The effect of treatment in this affection is most marked. Without treatment of any kind, after the stationary period the paralysis will usually improve for one or two months, more slowly for two to four months, after which it remains permanently or improves very slowly. After a time atrophy, contracture, relaxation, and malposition lead to crippling deformities, which locomotion rapidly and indefinitely increases. Treatment faithfully and persistently continued is frequently rewarded by the return of power and usefulness in an atrophied and helpless limb, while apparatus and surgical skill will not only correct deformities, but will hasten and increase the improvement.

During the initial stage the danger to life is exceedingly small, though in very rare cases the patient may succumb at the onset from respiratory failure, as in the case recorded by Sinkler.[2]

Cerebral complications add to the gravity of the prognosis. An attack of this affection may during convalescence render the patient less resistant to other and more fatal diseases. There is no evidence to prove that moderate deformity from this disease shortens the tenure of life. After the paralysis has become stationary, in about a week

---

[1] Shaffer: Archives of Medicine, N. Y.
[2] Keating's Cyclop. Dis. of Children, vol. iv. p. 710.

or ten days, it is not likely to increase, and the possible improvement or recovery of the affected muscles may be estimated by an electrical examination. Perfect recovery is rare, but improvement may be looked for even in the severest cases. The muscles which do not respond to the faradic current will probably remain permanently paralyzed. After a few days the muscles which at first gave no response may feebly respond, and these may be expected to recover partially or completely. When no loss of faradic contractility is observed after the paralysis has become permanent, recovery may be predicted in a few weeks or months. When the faradic contractility gradually fails, wasting and paralysis for an extended period may be predicted. When complete paralysis and marked atrophy are observed, within two or three months the permanency of the paralysis may be predicted. As long as the feeblest response remains to the faradic current improvement may be predicted, but if the " reaction of degeneration " is present the paralysis may be considered absolutely permanent. Even in the severest cases great improvement may be obtained by mechanical and operative treatment, the possibilities being only limited by the extent of the paralysis and amount of atrophy of the upper extremities. Only cases which are so extensively paralyzed as to be unable, with apparatus, to use a wheelcrutch, are not amenable to treatment; and frequently, under modern methods, children doomed to spend the rest of their life as helpless cripples upon the floor, are restored to health and usefulness.

Treatment. The treatment will depend upon the stage or period during which the disease is seen.

Prophylaxis. A word of warning may here be given, since certain simple prophylactic measures may prevent the advent of this disease with its terrible crippling deformities. Since its principal exciting cause is the sudden chilling of the overheated surface of the body during a season of the year when the greatest mean relative humidity of the atmosphere is accompanied with great heat, due precaution on the part of nurses and parents may prevent its occurrence. To this end children should not be allowed to become greatly overheated nor be exposed to the sun for long periods, and should not be allowed to lie on damp grass nor sit upon damp stones when in an overheated condition. The body, especially in delicate children, should be well, but lightly, covered, preferably with light natural wool or hygienic garments, and under no circumstances should a child be allowed to take a cold or tepid bath when greatly overheated.

Medical Treatment. *The stage of onset.* During the first stage, before the development of paralysis, the indications are to reduce the temperature, to remove the exciting cause, and to relieve the urgent symptoms present. Active measures at this time may limit the amount of destruction in the cord. A brisk purge, preferably of calomel, rhubarb, or magnesia, should be administered, and a febrifuge mixture be prescribed. If worms in the intestines are suspected as the cause, santonin should be added to the calomel. A general hot bath should be administered, and the child placed upon the side or face to limit stasis of blood in the cord; should have mustard plasters, tincture of iodine, or dry cups applied to the spine, especially over the cervical and lumbar enlargements.

For dental irritation the gums should be lanced at once. If genital irritation exist and phimosis be suspected as the cause, circumcision will be indicated as soon as the more acute symptoms have subsided, and may be followed by the speedy amelioration of the paralysis. But the writer now has a boy under his care in whom circumcision undertaken by another surgeon soon after the attack had no effect whatever upon the paralysis.

If the disease be ushered in with a convulsion, cold may be applied to the head. Ergot is strongly recommended, and ten drops of the fluid extract may be administered to infants of six months and one-half drachm to children between one and two years. The subcutaneous injection of ergotine is painful and is attended by some slight risk of abscesses. Ergotine may be given by suppositories when the patient cannot swallow. Belladonna, mercurial inunctions, hydrargyrum cum creta, and iodide of potassium have been recommended in the acute stage, but their value is doubtful. Rest and quiet, with a sterilized-milk diet, should be insisted upon, and a moderately cool temperature should be, if possible, secured. Nitrate of silver and bromide of potassium have been recommended, but are still on trial. Strychnine has been recommended both internally and by hypodermatic injection, but its value is doubtful and its use is positively contra-indicated while any signs of the initial irritation remain.

Stage of paralysis. During the acute and the early part of the stationary period the paralyzed parts are to be protected from strain and pressure, and measures are to be adopted which will diminish the amount of the deformity, hasten the regression, and secure, if possible, the maximum improvement.

The limbs should be enveloped in sheet lint or lintine and neatly bandaged; should be placed upon pillows and supported by sand-bags, pillows, etc., in such a manner as to avoid any undue tension and stretching upon the paralyzed muscles or relaxed ligaments. The weight of the bedclothes must be taken off the toes by a bed-hoop or other device. If the paralysis be more extensive a canvas-covered frame, a Phelps "Steh-bett," or a wire cuirass may be employed to secure perfect rest to the palsied parts, and enables the child to be carried about. The arm when paralyzed may be secured in a sling or preferably a well-applied roller bandage, pressure being carefully avoided. It is essential that the limb should be kept warm. The limb should be constantly incased in woollen, and in winter be additionally protected by chamois or buckskin. Dry heat may be applied several times a day by sitting before the fire with the par-alyzed limb placed through a hole in a sheet of heavy cardboard, or wood. Friction, rubbing, or massage is very useful, and is best applied by the bare hand, but sweet oil or vaseline may be used if the dry rubbing produce irritation. Rubbing may be employed by any intelligent person, and requires no especial skill. It is best applied at night after the limb has been thoroughly warmed.

Massage is best given by persons skilled in its employment, but where the expense is too great the rubbing may be substituted. Its use, at first, must be limited, and ten or fifteen minutes will be suffi-ciently long if only one extremity is affected, and the time may be gradually extended to twenty or thirty minutes. The massage is best applied in the following manner: First, firm stroking is applied over the entire limb from the foot to the hip; second, friction is then applied from below upward, and followed by stroking; third, the limb is then kneaded, the tissues being carefully separated; and, fourth, the manipulation is concluded by percussion over the mus-cular parts of the entire limb. "Stroking" is performed by holding the limb with one hand, while with the palm of the other the part is gently and quickly rubbed toward the centre of the body. In this manner the entire limb is gone over.

"Friction" consists of strong circular manipulations applied with the thumb or finger-tips, and is always followed by stroking as before given.

The thumbs apply pressure by a quick rotary motion, and sliding the thumb a little upward toward the centre of the body after each rotation.

"Kneading" may be applied with the palm of the hand, the tips of the thumbs, or the tips of the thumb and index finger. The palm or fingers grasp the muscle to be manipulated and knead them gently with firm pressure toward the centre of the body.

"Percussion" includes several movements, two only of which are serviceable in the treatment of infantile paralysis—clapping and hacking.

"Clapping" or slapping with the hands upon the skin will increase the circulation of the skin and stimulate its nutrition, and percussion with the ulnar border of the hand and fingers, or " hacking," as it is technically called, will also stimulate the nerves and improve the nutrition of the underlying muscles.

Finally, the foot, leg, and thigh should be rotated gently and firmly, and the joints should be carried through their normal movements several times.

A period of rest must follow these manipulations at all times, so that they are preferably performed at bedtime.

Exercise of the paralyzed limb, preferably on the principle of the Swedish movement cure, will form a valuable adjunct to the rubbing or massage.

Muscle and brain are developed by reciprocal action, and every movement of a composite nature develops equally the gray centres of the brain and cord. Active muscular exercise, however obtained, enables the weakened muscles to regain their power and daily places them at a greater advantage. For the general effect of exercise the reader may be referred to one of the writer's former papers upon " Physical Development in Children."[1]

In the treatment of this affection every movement of the paralyzed part which can be accomplished, provided the muscles are not overtaxed, may be employed. No exact series of exercises can be given, and the exercises must be varied to suit the individual case. To the same end apparatus should be early employed, to enable the child to walk as soon as possible, and so increase the amount of exercise.

Electricity is the most important remedy at our command to restore power to the paralyzed muscles. In the absence of fever and hyperæsthesia its use may be begun one week from the onset of the

[1] J. M. Keating and J. K. Young : Keating's Cyclop. Dis. of Children, vol. iv. p. 274.

paralysis, but for two or three weeks only the mildest currents may be employed and these only for a limited period. The rule in the administration of electricity is to use the current which will give the greatest amount of contraction with the least amount of current and the smallest amount of pain. The slowly interrupted faradic current is to be preferred, and according to Sinkler,[1] each muscle may be made to contract three or four times and the treatment should be applied daily. Galvanism is to be substituted when the muscles fail to respond to the faradic current. The good effect of the electricity is observed in the increasing size and power of the paralyzed part, and the improved circulation and appearance of the skin. Electricity is also useful, after the parts have reached their maximum development, to maintain the nutrition of the tissues. In the later stages massage and active muscular exercise will be of more importance than electricity.

Strychnine, if of any value in this affection, is probably of benefit in the chronic stage.

Mechanical Treatment. The principles involved in the mechanical treatment of infantile paralysis are : first, the support and protection of the limb in a manner which will render it most serviceable for progression, and at the same time exercise as much as possible the weakened muscles ; and, second, the prevention and correction of deformities. The lightest form of apparatus which will accomplish this should always be selected, and particular attention must be taken to have the bands thoroughly padded, and friction must be avoided, since the circulation of the part is poor. Rubber elastic bands are of service in some forms of apparatus to assist weakened and partially recovered muscles, but rigid appliances with locks and catches at the joints are to be preferred, the main use of braces being the support of the joints.

Mechanical appliances will be indicated whenever the limbs cannot support the weight of the body, and where in locomotion the weight of the body produces distortion and deformity of the limbs. No special form of apparatus can be given to meet all cases, and the changes and alterations required will tax to the utmost the ingenuity of the surgeon and the skill of the instrument-maker.

The conditions requiring mechanical appliances may be considered under four divisions :

1. Paralysis of the lower extremity.

---

[1] Cyclop. Dis. of Children, vol. iv. p. 713.

2. Paralysis of the trunk and lower extremity.

3. Paralysis of the trunk.

4. Paralysis of the upper extremity.

Paralysis of the Lower Extremity.

The application of mechanical appliances for the relief of paralysis of the lower extremity will include appliances for paralysis of the leg with associated varus, valgus, and calcaneus; appliances for paralysis of the thigh muscles; and appliances for complete paralysis of the lower extremity.

FIG. 120.

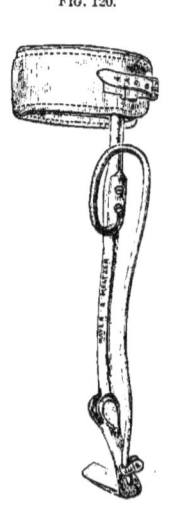

Apparatus for use after division of tendo Achillis for paralytic equinus.

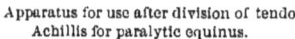

FIG. 121.

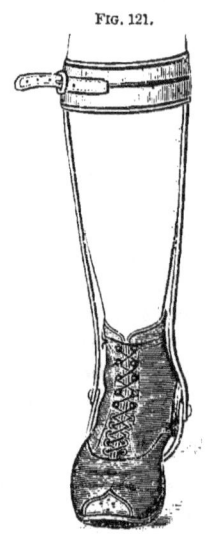

Ankle support for talipes valgus.

When simple calcaneus results from paralysis of the anterior muscles of the leg, the foot may be prevented from fully extending by attaching lateral steel uprights to the ordinary shoe, and the addition of a right-angle stop-joint opposite the ankle-joint. A single upright of this kind upon the outer side of the leg extending to a little below the knee may be used, but double uprights are preferable. (Fig. 120.)

The same result may be accomplished by means of a light, firm spring (Fig. 121) attached to the ankle-joint of the brace, which automatically elevates the toes, and which the weight of the body in stepping readily overcomes. The same result may be obtained from elastic straps attached to the outer and inner sides of the toes of the shoe, crossed in front of the ankle, and buckled or fastened to the leg-band above. In all these appliances in which an ordinary shoe is used, a steel plate must be inserted between the leather of the sole during the making of the shoe, or an inside steel plate must be attached to the shoe when the brace is attached. Without this the sole quickly loses its shape, and renders the surface uneven and insecure.

When varus coexists from the foot rolling inward from the weakness of the tendons about the ankle-joint, and the relaxation of the ligaments, a pad should be fastened to the outer ankle-joint of the apparatus, and a T-leather strap should be added to the outer side of the shoe, and buckled over the inner upright.

If, as is often the case, valgus is associated, this pad and T-strap should be attached to the inner side, and in addition a steel plantar spring, described under acquired club-foot, will afford additional support, and tend to invert the foot to its normal position. In many cases the T-strap may be dispensed with, and a strap of webbing may be attached to the outer upright just above the external malleolus.

These conditions of varus and valgus may also be corrected by means of the Taylor shoe, described under club-foot, the shoe being applied as before given for varus, but for valgus having the inner side elevated into a supporting plantar plate.

When the thigh muscles are paralyzed the limb drops forward, and is unable to sustain the weight of the body. If, however, the limb be thrown backward from relaxation of the posterior ligaments, the limb furnishes some support, but "back-knee" results.

For these conditions it is necessary to support the knee in a fully extended position. This is readily accomplished by carrying an outer and inner steel bar from the shank of the shoe to the upper part of the thigh, to be attached to a posterior steel band, and at the knee a broad leather band maintains the knee in position. (Figs. 122, 123.)

This is the principle upon which all paralysis leg braces are constructed, but for convenience in sitting they should be furnished with a lock or catch (Figs. 124, 125) at the knee. For this purpose numerous catches are in use. A useful and inexpensive one is the

FIG. 122.   FIG. 123.

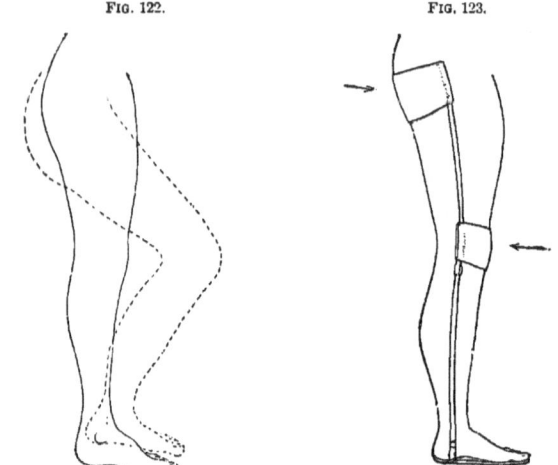

Supporting appliance in paralysis of the anterior thigh muscles.   (BRADFORD and LOVETT.)

FIG. 124.   FIG. 125.   FIG. 126.

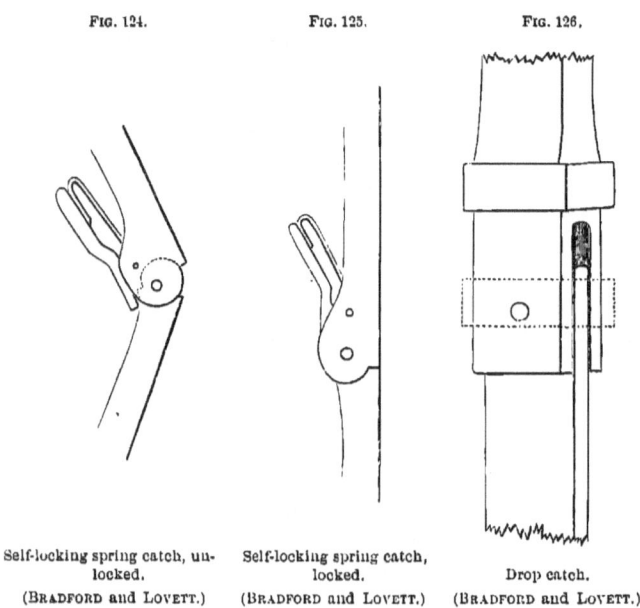

Self-locking spring catch, un-   Self-locking spring catch,   Drop catch.
locked.   locked.
(BRADFORD and LOVETT.)   (BRADFORD and LOVETT.)   (BRADFORD and LOVETT.)

drop catch (Fig. 126), a simple ring which falls when the limb is extended, and may be raised with ease to allow flexion. An excellent catch, and one the writer prefers, is known as the Congdon brace. This consists of semicircular pieces of steel attached to the knee-joints on either side, into which the ends of a steel bow play in such a manner as to automatically lock when the limb is extended, but which, upon being slightly elevated, permits the knee to flex.

To prevent hyperextension of the knee the knee-joints of the apparatus should have a stop-joint. This may consist of a simple pin or a piece of metal inserted just below the joint.

If the loss of power be not complete, the muscles may be assisted by elastic bands so attached to projecting bars as to supplement the palsied part, and the stop- or catch-joint is omitted.

Where contraction of the hamstring tendons produces deformity of the knee, this may be overcome by extension, bandaging to a splint, or in the manner described under deformity from knee-joint disease, and especially the Thomas knee splint and the Billroth splint.

A very useful brace for contraction of the knee is the splint employed by the writer after the correction of deformity from tumor albus. It consists of an ordinary paralysis brace extending from the sole of the foot to the upper part of the thigh, to which at the knee-joints are attached semicircular plates attached to the lower upright, and perforated near the border so as to be altered by screw to any angle. Pressure is applied by means of a leather knee-cap, provided with four straps closely perforated to fit the pins upon the lateral uprights. Instead of the semicircular pieces, a worm-screw and ratchet may be attached, and with a key the splint can be set at any angle.

Either of these braces should fit accurately, and this may be secured by broad leather bands above and below the knee, attached laterally to the brace and laced in front.

In many of these cases the affected limb will be found shorter than the sound one, and in applying the apparatus the length of the limbs must be equalized by applying a high cork patten on the short side. In some instances an insole will add the necessary height.

Paralysis of the Lower Extremity and Trunk.

In these cases, which without operation and apparatus are practically helpless, the entire body from the feet to the axillæ must be

15

made rigid, to support the weight of the body while the body is swung forward upon crutches. (Fig. 127.)

The body may be incased in a leather jacket, or a steel framework may be made to encircle the trunk, and to this the leg appliances can

FIG. 127.

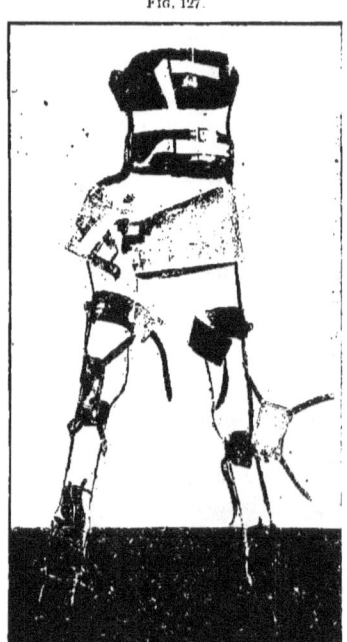

Rigid apparatus to give artificial legs. (WILLARD.)

be connected. To enable the patient to sit, locking and unlocking joints at the hips and knees are essential. If possible, crutches should be employed, but if these cannot be held securely, a steel-framed walking apparatus, a Darrach wheeled crutch, or the wheeled crutch used by Prof. Willard[1] will be found useful. (Fig. 128.)

These are simply light frames with wheels, the top of the framework being padded to fit under the axillæ, and straps or a padded

[1] *Vide* article in Trans. Amer. Orthop. Assoc., 1891; also Sinkler's article, *loc. cit.*, for description of Darrach wheeled crutch.

piece of metal for the hands to hold. By such apparatus the unaffected muscles are exercised, and the weakened ones are also brought

FIG. 128.

Wheeled crutch to assist locomotion. (WILLARD.)

into use. The construction and use of these apparatuses is well shown in the accompanying illustrations and need not here be described.

Paralysis of the Trunk.

Paralysis of the trunk muscles seldom occurs except in connection with extensive paralysis of one or both lower extremities.

Paralysis of the muscles of the back, when unilateral, gives rise to lateral curvature, and for its relief the mechanical appliance considered under the treatment of that affection can be employed. Bilateral paralysis renders the spine unable to maintain an erect position, and demands for its relief the use of corsets of leather, wood, wire, etc., as employed in the treatment of Pott's disease of the spine. In these cases crutches, or wheeled crutches, or a couch, will be required for locomotive purposes.

Paralysis of the abdominal muscles gives rise to a marked lordosis, with a protuberant abdomen. In mild cases abdominal corsets or supports will relieve the deformity, and in severer cases a well-fitting leather jacket will be serviceable.

Paralysis of the Upper Extremity.

The use of mechanical appliances in paralysis of the deltoid is the only condition calling for special consideration. Further relaxation may be prevented by the well-padded leather breast-ring encircling the base of the neck, with a leather capsule for the forearm attached to it by straps, as suggested by Schreiber.[1] More complicated apparatus may be employed, consisting of a well-fitting shoulder-piece and elevating springs attached to an arm-band encircling the part. In some cases a simple handkerchief sling will answer every purpose.

Operative Treatment. This is demanded when mechanical methods are inefficient or undesirable, and affords the most speedy and efficient result in the severest cases.

It includes (1) tenotomy, (2) aponeurotomy, (3) myotomy, (4) forcible straightening, (5) excision, (6) osteotomy, and (7) amputation. The first four procedures may be combined in one operation.

Tenotomy. This may be performed subcutaneously or as an open incision, the latter being the only safe method where the deep structures about the hip or knee have to be divided. In either case the technique is the same as already given, and strict antisepsis must be observed.

Immediate rectification should be accomplished, but over-correction is not necessary here as in congenital deformities. Contraction of the hip will require division of the sartorius, long head of the rectus femoris, and tensor vaginæ femoris, and by carefully avoiding the femoral artery and anterior crural nerve these may often be accomplished subcutaneously. If the psoas, the capsule of the joint, the external or internal rotators, or the adductors require division, a longitudinal open incision of considerable length will be necessary. Subcutaneous section of the psoas, although it has been successfully accomplished, is not a safe operation. As a conservative method, the superficial structures may be divided subcutaneously, the limb may be forcibly straightened, and bed extension may then be applied for a considerable length of time.

---

[1] Orthop. u. spec. Chir., p. 319.

Contraction of the knee in severe cases will require tenotomy of the hamstring tendons, and in some instances this had best be accomplished by the open method, since the parts are often greatly distorted and it is difficult to distinguish the structures. In this manner the dense bands in the region of the popliteal space can best be divided.

Resection of the quadriceps has been performed by Keetley[1] for paralysis and stretching with contraction of the knee.

After operation for contraction about the knee, horizontal bed extension with constant pressure over the knee are beneficial. Tenotomy about the foot for deformity is fully described under paralytic club-foot. Tenotomies are best performed subcutaneously. In talipes equino-varus the tibialis anticus, posticus, plantar fascia and ligaments will all require division. In talipes valgus the peroneals require division, an operation the writer[2] strongly advocates and one to which sufficient attention has not been given by surgeons. Forcible rectification will also be serviceable after tenotomy.

For severe calcaneus, resection or shortening of the tendo Achillis is sometimes desirable, and the methods for its performance are fully described under paralytic club-foot. To avoid removal of any piece of tendon the method of Walsham and Willett is to be preferred to that of Reeves. Where the peroneals are intact, Nicoladoni has adopted a surgical procedure which was followed by much benefit. It consisted in dividing the tendo Achillis and sewing the peronei tendons to it, so that the contraction of the peroneal tendons would elevate the heel.

An extreme amount of force is sometimes required, and may be employed with safety if only the surgeon's hands or sand-bags be used. Fracture is to be avoided ; but if it occurs it will heal readily and will not complicate the case. Forcible straightening under complete anæsthesia is advocated by Gibney.[3]

Excision of the knee- or ankle-joints may be required on account of extreme deformity or in order to secure a stiff joint. In the poorer classes resection in preference to the application of apparatus has been seriously considered by Ap Morgan Vance[4] and others. In in-knee and out-knee osteotomy and osteoclasis have been proposed, but resection is preferable, since it not only corrects the deformity, but leaves a stiff joint. Excision in order to secure a stiff joint, or

[1] Brit. Med. Journ., May 31, 1884, p. 1058.   [2] University Med. Magazine, 1891.
[3] N. Y. Med. Journ., April 3, 1886.
[4] Boston Med. and Surg. Journ., May 6, 1886, p. 416.

arthrodesis, as it is designated by Albert, was proposed by Zins-meister, and has been performed with satisfactory results by Albert,[1] Winiwarter,[2] J. Wolff,[3] Lesser,[4] and Rydiger.

In severe cases of deformity about the ankle and foot, if resection of the tarsal bones becomes necessary, the removal of the astragalus is to be preferred to tarsectomy, and both are to be considered as last resorts, for in most cases powerful force will accomplish all that can be gained by operative methods.

Amputation will, in exceptional instances, become necessary.

Treatment after Operation. The after-treatment will include bed extension and the thorough use of massage and electricity for a long period. After thorough correction by operative means mechanical appliances are often necessary. These are the same as have been fully described under the mechanical treatment of paralysis of the lower extremity and trunk. With these measures the most satis-factory results can be obtained, and a number of successful cases have been recorded by Dr. DeF. Willard,[5] in the mechanical treat-ment of some of which the writer has been associated. From a very extensive experience in the treatment of these severe cases his con-clusions are as follows:

1. Even the severe resultant deformities of infantile paralysis are capable of being benefited by the skilful employment of surgical measures and mechanical appliances. No case with fairly strong upper extremities should remain in helpless cripplehood, since even crutch locomotion is far preferable to a life upon the floor or upon the bed.

2. The deformities following infantile paralysis can be largely prevented by the early use of some form of apparatus.

3. Surgical measures in long-standing cases should usually precede mechanical appliances, since pain and time are thereby saved, and the resulting limbs are in nowise inferior to those obtained by the slower processes of mechanical rectification.

4. The surgical measures to be employed are tenotomy, myotomy, division of the fascia, application of force, and resection. Osteotomy and amputation are sometimes necessary.

5. Mechanical appliances should be used to retain the limb in proper position, but they should not interfere with the circulation of

1 Centralbl. f. Chir., Oct. 21, 1882.   2 Verhandl. des XIV. Chir. Cong., p. 141.
3 Berliner klin. Woch., xxiii., 1886.   4 Centralbl. f. Chir., 1887, No. 46.
5 Amer. Journ. Med. Sci., May, 1891.

the member. Crooked limbs can often be straightened so as to be made a part of the apparatus, and the muscles of these limbs should be compelled to do their full extent of work in supporting the body. The apparatus must frequently be made to support a large portion of the weight of the body, the helpless, flail-like limbs being accessories.

6. No case should be abandoned without the most careful and repeated attempts at rectification, as even feeble locomotion will in time become greatly improved by exercise in walking, and the health and happiness of the individual will thereby be greatly increased.

CHAPTER XIII.

INFANTILE CEREBRAL PALSIES.

THE cerebral palsies in children consist of certain spastic palsies due to cerebral defect, with atrophy, usually slight, of the affected muscles, and without marked electrical changes.

Although mentioned by Reil in 1812, Cazauvielh in 1827, Billiard in 1828, and described by Breschet, Lallemand, Rokitanski, and Cruveilhier from fifty to sixty years ago, the first accurate descriptions were given by Henoch in 1842, Little in 1853, Turner in 1858, and Von Heine in 1860. Since this time monographs have been numerous, and particularly since Strümpell's paper in 1884, which aroused special interest in this subject, many valuable papers have been contributed. For a complete bibliography reference should be made to the papers of Osler[1] and Sachs and Peterson.[2] The writer is also indebted to the valuable papers of McNutt, Gibney, Sinkler, Gray, Wood, Knapp, Hatfield, Bullard and Bradford, Seibert, Caillé, Smith, Lovett, and others.

Under the generic term cerebral palsies—the German *Cerebrale Kinderlähmung*—are included a large group of motor palsies, the result of a destructive lesion of the centres of the upper or *cortico-spinal* portion of the motor path sharply defined in their clinical features from the palsy—the common infantile paralysis—due to a lesion of the lower or *spino-muscular* portion of the motor tract. The relative frequency of these two groups is estimated to be about 1 to 4.16.

The cerebral palsies are classified, according as the distribution of the paralysis is unilateral, bilateral, or paraplegic, into three groups : 1, Hemiplegia ; 2, bilateral hemiplegia, or diplegia ; and, 3, para-plegia.

In 150 cases collected by Osler, 120 were hemiplegic, 19 were diplegic, and 11 were paraplegic. The sexes are about equally

---

[1] Medical News, July 14 to August 11, 1888.
[2] Journ. of Nervous and Mental Diseases, May, 1890.

affected. In Osler's 120 cases of hemiplegia, 57 were boys and 63 girls.

The age at which these different forms are most common is interesting. Of the hemiplegic cases the majority have their onset in the first three years of life, and a certain proportion (10 to 34+ per cent.) are congenital. Of the diplegic cases the large majority date from birth as the result of injury. Of the paraplegic cases the majority are congenital, a small proportion only occurring within the first years of life.

Etiology. The exact significance of many attendant or preceding diseases, and of environment and heredity, as etiological factors is difficult to estimate. At best the etiology is very obscure. Among the factors, however, may be mentioned fright or strong mental emotion during pregnancy; drunkenness at the time of conception; consanguineous marriages; difficult or abnormal labor; convulsions; asphyxia; injury with the forceps or by manipulations of the accoucheur; premature delivery; syphilis; cerebral traumata; thrombosis of the surface cerebral veins; defective nutrition; post-febrile processes, infectious or otherwise, from scarlet fever, measles, diphtheria, typhus, variola, whooping-cough, vaccinia, mumps, dysentery, meningitis, and violent vomiting.

Of the congenital cases fright and strong emotion during pregnancy and premature delivery, especially at the seventh month,[1] deserve especial prominence. The same may be said of traumata, particularly cerebral hemorrhages in birth palsies, and the influence of the infectious diseases in the palsies coming on during the first years of life.

Infantile Hemiplegia.

Synonyms. English, *Spastic Paralysis*. French, *Agénése Cérébrale; Sclérose Cérébrale; Atrophie Partielle Cérébrale.* German, *Hemiplegia Spastica Infantiles* (Bernhardt); *Hemiplegia Spastica Cerebralis* (Heine); *Acute Encephalitis der Kinder* (Strümpell); *Atrophische Cerebrallähmung* (Henoch).

Symptomatology. There is often present before the onset of paralysis a slight febrile attack, infectious or otherwise, an attack of some sort in cases resembling indigestion, a fall or blow on the head. The onset is usually characterized by convulsions and coma, though

---

[1] Young: Medical News, July 13, 1889, p. 38.

the disease may develop suddenly in apparently perfectly healthy children without spasms or loss of consciousness, or may develop insidiously without disturbance of any kind. Again, the onset may be accompanied by delirium or screaming-spells. A latent gradual onset is most common.

The convulsions are almost always accompanied by loss of consciousness, lasting from a few hours to many days. Fever, transient or persisting for weeks, is present in a large proportion of cases. Vomiting and general hyperæsthesia is sometimes observed. When consciousness is recovered the hemiplegia is usually complete; but occasionally loss of power, at first not complete, gradually becomes so. In about one-half the cases the face is paralyzed, and in these it is not complete, and rapid and complete disappearance is the rule.

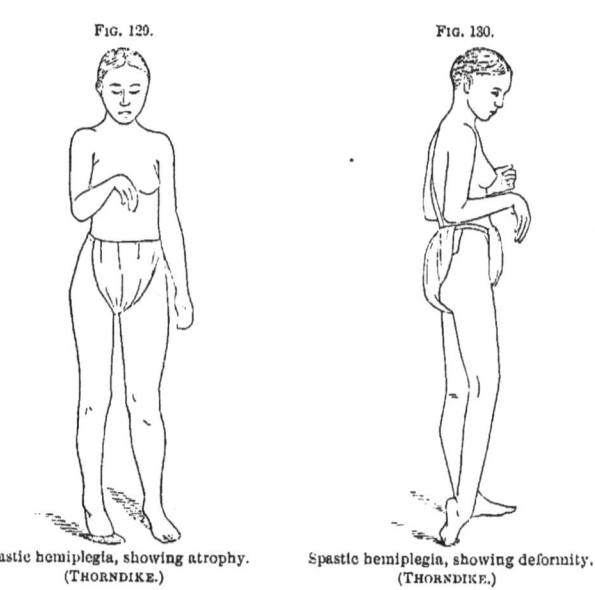

FIG. 129.   FIG. 130.

Spastic hemiplegia, showing atrophy.   Spastic hemiplegia, showing deformity.
(THORNDIKE.)   (THORNDIKE.)

The arm is more severely affected than the leg, recovers more slowly, and rarely recovers the more delicate movements. Arrest of development may occur, but a moderate degree of atrophy is usual. The leg, at first powerless, recovers more rapidly and more completely. Arrest of development is rare, and atrophy is never so marked as in the arms. (Fig. 129.) The paralyzed arm and leg may remain flaccid for ever, but rigidity is present in the large proportion of cases, and comes on at a variable period after the onset. The rigidity is increased by emotion or by attempts to use the limb or forcibly overcome the spasm, and disappears during sleep and under anæsthesia. (Fig. 130.)

The reflexes are usually increased, sensation is often unaffected, and vascular sluggishness is marked by coldness and blueness of the skin in severe cases. The electrical reactions remain practically unaffected.

Aphasia is observed in about one-third of all the cases; in some transitory, in others persistent. Post-hemiplegic movements in a certain proportion of the cases, about one-fifth, follow at a later stage. These have been referred to by Drs. Weir Mitchell, Greidenberg, Gowers, Knapp, and others as post-hemiplegic tremor, post-hemiplegic chorea, mobile spasm, athetosis, hemiataxia, and chorea spastica.

These consist of convulsive movements of the affected parts, incoordinate and choreiform in character. Mental defects, from complete idiocy to a feeble-minded condition, are frequently found associated with infantile hemiplegia, and three grades may be distinguished: idiocy, imbecility, and slight backwardness. The lesser degree of mental defect is met with in the majority of cases. Of 26 cases recorded by Bradford and Lovett only 6 had what was classed as average intelligence; in Gaudard's 80 cases 15 were feeble-minded and 19 idiotic children; in Wallenberg's 160 there were 50 with mental defects, and in 15 more imbecility followed the epilepsy; while in Osler's 120 cases only 12 of the infirmary series had idiocy or imbecility at the time of observation; and all of the 23 cases at the Pennsylvania Institution for Feeble-Minded Children had some degree of idiocy or imbecility.

Epilepsy, at first confined to the paralyzed side, but tending to become generalized, attacks sooner or later a considerable proportion, from a quarter to a half of all the cases. The most common form is the true Jacksonian epilepsy.

Post-hemiplegic contractions of a permanent nature are most marked in the upper extremity. In severe cases the arm is held to the side, the elbow flexed and semi-pronated, the hand flexed and the fingers contracted upon and embracing the thumb in the palm of the hand. The thigh is flexed and adducted at the hip, the knee flexed and rigid, and the foot held in a position of equinus or equinovarus. These contractures are, moreover, very firm and resistant.

Bilateral Spastic Hemiplegia.

Synonyms. English, *Spastic Rigidity; Spastic Rigidity of the Newborn* (Little); *Spastic Paralysis of Children* (Adams); *Tonic Contraction of Extremities; Spastic Diplegia* (Gee); *Essential Contraction.* French, *Spasme Musculaire Idiopathique* (Delpech). German, *Permanenter Kinder-Tetanus* (Stromeyer). In France the disease is sometimes spoken of as Little's disease (*maladie de Little*), and Gowers has applied the name birth palsies, since in the large proportion the trouble dates from birth.

Symptomatology. Bilateral hemiplegia is characterized by spastic condition of the extremities. This is observed immediately after or shortly succeeding birth, but occasionally it follows convulsions in one of the specific fevers.

Fig. 131.

Photograph of a case of congenital diplegia, with marked mental defect. (GRAY.)

The arms may be but slightly affected. The legs are rigidly extended, and the thighs firmly adducted—the so-called "clasp-knife rigidity," the feet are crossed and in a varus or equino-varus position.

On attempting to walk, the gait is awkward or impossible. The arms are flexed and held firmly to the sides, and the use of the hands is extremely awkward. The extremities are not atrophied. Sensation is unimpaired, the reflexes are increased, and the electrical reactions are unchanged.

Spasm of the affected parts with disordered movement, and variously described as *chorea spastica* and *double athetosis,* are not uncommon. Spasm of the muscles of the face and tongue is sometimes observed. Strabismus and nystagmus exist in some instances. The mental condition is seriously impaired, the patients being usually

Fig. 132.

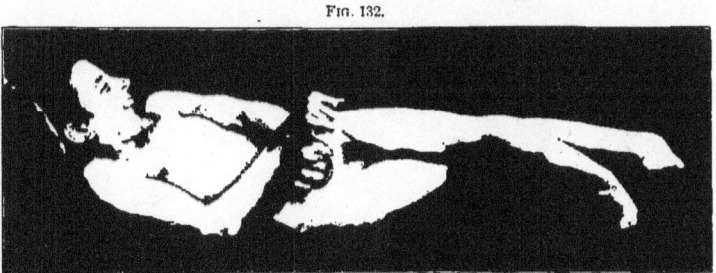

Photograph of case of cerebral palsy, showing spastic contractures of upper and lower extremities.

imbeciles or idiots. In some instances, however, intelligence is fair. Microcephalus, with or without asymmetry, is common. The facial expression is vacant, the teeth are defective,[1] and dribbling from the mouth is constant.

Epilepsy is not so frequent as in hemiplegia.

Spastic Paraplegia.

Synonyms. *Tetanoid Pseudo-paraplegia* (Seguin); *Deformity, with Rigid Muscles* (Adams); *Spastic Spinal Paralysis* (Erb); *Spastic Contractions* (Little). French, *Tabes Dorsalis Spasmodique*. (Charcot). German, *Paraplegia Cerebralis Spastica* (Heine).

Symptomatology. In children the condition usually appears at birth, or within the first year, although it may occur later. The lower limbs are drawn by the strongest muscles into a condition of spasmodic rigidity, the thighs flexed upon the pelvis and slightly

---

[1] Gollier: De l'Etat de la Dentition chez les Enfans, Idiots et Arriéres. Paris, 1887.

adducted, producing the " clasp-knife rigidity" (Fig. 133), the knees flexed and firmly adducted, so that they touch, or even overlap, the feet inverted and strongly extended in a position of extreme equino-varus. The contractions readily yield to firm, continuous manual pressure like a bit of lead, as Dr. Weir Mitchell has expressed it, but immediately return upon its removal. The child

FIG. 133.   FIG. 134.

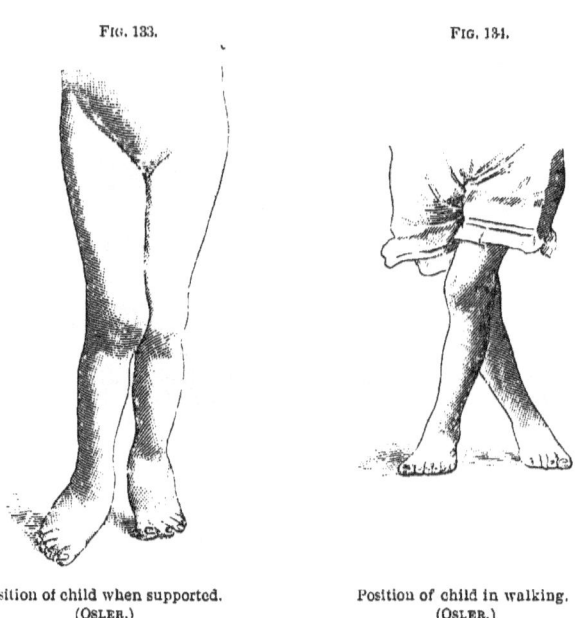

Position of child when supported.   Position of child in walking.
(OSLER.)   (OSLER.)

stands with assistance, resting upon the ball of the foot and toes, and walks, if at all, with a rapid, swinging, irregular gait, typical of the disease, and with the knees overlapping at every step, the so-called cross-legged progression. (Fig. 134.) The feet drag, wearing out the shoe caps, and the clothing at the inner side of the knees is worn by constant friction. (Fig. 185.) This stiffness is usually first discovered by the mother in washing and dressing the child, and the child is usually late and awkward in attempting to walk. (Fig. 136.) There is usually no anæsthesia, ataxia, atrophy, or any vesical or rectal incontinence. There is no loss of equilibrium in standing with the eyes closed, and the reflexes are exaggerated.

In the upper extremity the same condition exists, but to a less degree. The arms are approximated to the side, the forearm slightly flexed, and the hands flexed and pronated. The flexors of the trunk are sometimes contracted, and the head may be drawn forward and to one side (wry-neck). Strabismus may occur early, and occasion-

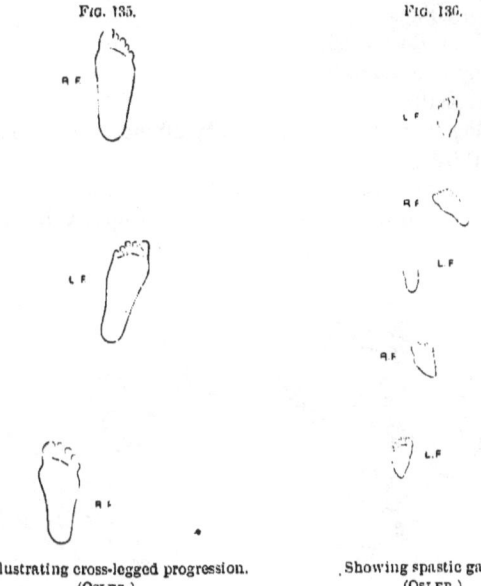

Fig. 135.

Fig. 136.

Illustrating cross-legged progression. (OSLER.)

Showing spastic gait. (OSLER.)

ally persist. The configuration of the head at times resembles that of the idiot, and with the half-silly, vacant stare, and an imperfect speech, presents the appearance of a lack of intelligence which does not always exist. In fact, in spastic paraplegia the cerebrum is less profoundly affected than in either of the other two forms, and some patients are intelligent and bright intellectually.

Pathology of Cerebral Palsies.

The pathological lesions are so variable, and the subject itself is so extensive, that it is only possible in this connection to give a *résumé* of the most recent matter. The pathology of infantile hemiplegia and diplegia have received much attention, but the morbid lesion in paraplegia remains at present unstudied. The morbid lesion in all three conditions is a destructive process in some part of the cortico-spinal segment.

In infantile hemiplegia. From a study of ninety autopsies, his own and abstracted from various sources, in cases occurring in infancy or childhood, Osler has grouped the lesions which occur under three headings:

1. Embolism, thrombosis, and hemorrhage.
2. Atrophy and sclerosis.
3. Porencephalus.

There being sixteen of the first, fifty of the second, and twenty-four of the third group.

Plugging of the sylvian artery (Fig. 137), usually embolic, occurred in seven, and hemorrhage in nine. Atrophy with induration,

FIG. 137.

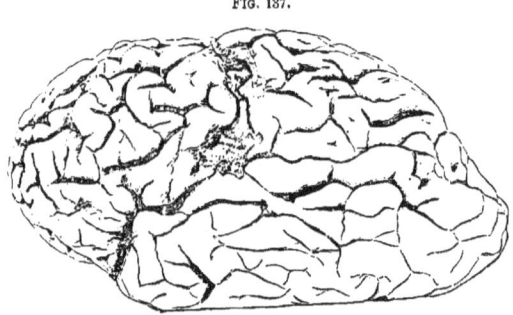

Left hemisphere, showing sclerosis in the Rolandic region. (OSLER.)

either of groups of convolutions, a lobe, or an entire hemisphere, were noted variously distributed over the cerebral cortex, and in all the recent cases a descending degeneration of the pyramidal tract was found. Porencephalus, meaning literally "a hole in the brain," represents a loss of substance of variable size, from a few convolutions to half a hemisphere, in the form of cysts or cavities full of liquid.

These lesions represent the final results of initial processes long past and originally a hemorrhage, an encephalitis, or other post-febrile change. (Fig. 138.)

It has been suggested by Strümpell,[1] that the lesion was the

---

[1] Jahrb. f. Kinderheilk., 1874, N. F., xxii. p. 173.

cerebral counterpart of infantile spinal palsy—a *poliencephalitis* of the motor areas of the cortex, analogous to poliomyelitis of the anterior horns—a very suggestive theory, but one which has not met with the favor it deserves, since anatomical proof is lacking.

FIG. 138.

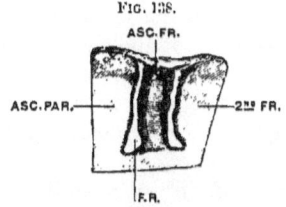

Section through the sclerosed area. (OSLER.)
ASC. FR. Ascending frontal gyrus. ASC. PAR. Ascending parietal.
2ND FR. Base of second frontal. F.R. Fissure of Rolando.

It is probable that infantile hemiplegia is the result of a variety of different processes, of which the following, according to Osler,[1] are the most important:

1. Hemorrhage, occurring during violent convulsions or during a paroxysm of whooping-cough.

2. Post-febrile processes: *a*, embolic; *b*, endo- and peri-arterial changes; and *c*, encephalitis.

3. Thrombosis of the cerebral veins.

In bilateral spastic hemiplegia the pathological lesion is in all probability bilateral atrophy, sclerosis, or porencephalous defect of the motor areas of the cortex cerebri.

In the reports of numerous autopsies this is the usual lesion observed, and, in many, descending degeneration in the pyramidal tracts has been found.

In infantile spastic paraplegia the pathological lesion remains to be worked out. It is not improbable that some cases may be due to spinal hemorrhage, and Ross has suggested that traction in feet presentation might injure the pyramidal tracts and be followed by sclerosis.

In the one recorded autopsy, that of Förster,[2] the cord changes were regarded as a descending degeneration, the consequence of cerebral lesion. The most plausible view is that the lesion, originally cerebral, has more or less completely disappeared and left the second-

[1] Medical News, August 11, 1888, p. 143.   [2] Jahrb. f. Kinderheilk., Bd. xv.

ary descending degenerative cord-changes as the most conspicuous feature. It has also been ascribed to primary sclerosis of the lateral columns (crossed pyramidal column), to various lesions of the spinal cord, chiefly myelitis,[1] and to genital or other peripheral irritation (Sayre).

Diagnosis. The characteristic signs of this group of cases are the spastic totanoid condition of the extremities, the distribution of the motor paralysis, the absence of atrophy in the affected parts, with increased reflexes, and unchanged electrical reactions, with or without mental defect. These form a comprehensive picture that can scarcely be mistaken for any other affection.

Differential Diagnosis. From infantile spinal paralysis, with which it is most frequently confounded, cerebral paralysis has been very carefully differentiated in the preceding chapter.

In infantile spinal paralysis the sudden paralysis is accompanied by early loss of reflexes, absence of rigidity, and marked electrical reaction changes, while in the infantile cerebral palsies the paralysis is accompanied with spasm and disordered movements, increased reflexes, gradual and slight wasting, and normal electrical reactions.

The hemiplegia must also be distinguished from obstetrical paralyses, cerebral tumor, and the di- and paraplegia from pseudo-paralytic rigidity met with in children.

Cerebral tumors sometimes produce symptoms identical with cerebral palsies and ordinarily cannot be distinguished. Such cases are reported by Osler[2] and Seeligmüller,[3] and a tumor producing bilateral rigidity has been recorded by Sharkey.[4]

Obstetrical paralysis of the face or upper extremity from forceps injury, described by Duchenne,[5] Nadaud,[6] and Budin,[7] are peripheral lesions and need not be confounded with palsies of cerebral origin.

Pseudo-paralytic rigidity, described also as idiopathic contraction with rigidity, and tonic rigidity of the extremities, is, according to Osler, apt to be confounded with true spastic paralysis. The fact that it is a painful carpo-pedal spasm, intermittent, transitory, often confined to the hands and arms, and associated with rickets or other constitutional disturbance, would serve to distinguish it.

---

[1] Seguin: Archives of Sci. and Prev. Med., N. Y., February, 1873, p. 100.
[2] Amer. Journ. Med. Sci., 1885.   [3] Jahrb. f. Kinderheilk., Bd. xiii.
[4] Spasm in Chronic Nerve Disease, 1886.
[5] Traité de l'Electrisation localisée, 3d ed.
[6] Des Paralysies Obstétricales des Nouveau-nés. Paris, 1872.
[7] Le Bulletin Méd., 1888, No. 20.

Spastic paraplegia in a mild form, sometimes resembles beginning pseudo-hypertrophic paralysis, and in such cases time alone can settle the diagnosis.

Prognosis. The prognosis will depend much upon the character and extent of the initial lesion, hemorrhage and vascular obstruction being more favorable than meningo-encephalitis, and the latter than intra-cranial growths.

In the traumatic cases it should not be forgotten that complete spontaneous recovery sometimes occurs, as in the two cases of hemiplegia referred to by Osler.[1]

In intra-cranial growths in only a very small number of cases, according to Knapp,[2] can we hope for a cure by surgical interference.

In regard to the bodily defect, while great improvement may be predicted, perfect recovery is rarely attained. Facial paralysis and aphasia usually disappear. Post-hemiplegic movements and epilepsy are bad prognostic signs. In hemi- and diplegic cases the mental enfeeblement is distressing and grave, and while training may do much to improve, the prognosis for many is bad, since they are liable late in life to become the subjects of psychoses. Many of these children live to be over twenty years of age. In spastic paraplegia, while perfect recovery is impossible, most children learn to walk and talk imperfectly, and operative interference is of great value, for as the researches of Rupprecht[3] demonstrate, tenotomy not only improves the position of the limbs, but also decidedly benefits the mental condition.

While the prospects of life are good, the vital resistance is not great, and there is a greater liability to death from intercurrent affections, and, according to some authorities, all cases of cerebral palsy are particularly liable to fatal meningitis.

Treatment. The convulsions and coma which characterize the onset require a hot bath with cold to the head, a calomel purge, and bromides and chloral. After the paralysis is established the indications are to favor the natural tendency to improve, to maintain the nutrition of the palsied parts, and correct the deformities by proper surgical and mechanical measures. Massage, warm clothing, and electricity will be of great benefit in maintaining the nutrition and temperature of the affected parts. The rubbing may be applied daily by the parent or nurse. The electricity must be applied faithfully

---

[1] Medical News, August 11, 1888, p. 144.
[2] Fiske Fund Prize Dessertation, 1891, No. 12, p. 110.
[3] Volkmann's Clinical Lectures.

and persistently for a long period. For the spastic rigidity and contractures, Dr. Weir Mitchell has recommended persistent massage with strong flexion and extension of the extremities.

Circumcision may be of service where genital irritation is evidenced by balanitis, painful micturition, and frequent erections. In a certain number of cases, however, it has no effect whatever, or is only temporarily beneficial, as pointed out by Eugene Dupuy,[1] referring to the cases reported by Sayre.[2]

Tenotomy offers the greatest hope in these cases. In diplegia little will be gained by operation, but in hemiplegia and paraplegia and in the post-hemiplegic contractions following, great benefit will ensue. The tendons that require division most frequently are the tendo Achillis, adductors of foot, hamstring, and adductors of the thigh. They should be performed as described under the treatment of club-foot; the parts should be immediately over-corrected and fixed. No attempt need be made to straighten the contracted parts by mechanical stretching, but tenotomy may be at once resorted to, since, as before remarked, the effect upon the mental condition is also beneficial. After operation an ordinary varus walking-shoe with stop joints at the ankle, or a Congden apparatus extending above the knee, with locking and unlocking joint to fix the knee, should be worn for some time, after which it may be permanently discarded.

The mental condition requires special training in all these cases, and the results attained at institutions specially devoted to the training of feeble-minded children demonstrate the advantages of such training in these particular cases.

Operations on the brain for the relief of epilepsy in cases of cerebral palsy have of late been rendered possible. Since the majority of cases are due either to (1) destructive vascular lesion, apoplectic, embolic, or thrombic, (2) sclerosis, or (3) porencephalus, all of which are beyond relief from surgical interference, there remains nothing but the intra-cranial growths, the melancholy prognosis of which has already been referred to.

In long-standing cases the possible existence of descending degeneration offers a serious objection, since the paralysis could not be benefited by operation. Surgical interference will be found useless in the majority of cases of cerebral paralysis, and when porencephalus is present, the paralysis due to a hole in the brain is not likely to be improved by making the hole larger.

---

[1] Journ. Nerv. and Ment. Dis., April 1887, p. 232.
[2] "On Reflex Paralysis, etc.," Trans. Amer. Med. Assoc., 1870.

CHAPTER XIV.

OTHER PARALYSES.

THERE are three other motor affections for which the orthopedic surgeon is frequently consulted, and which deserve brief notice here. These are:

1. Pseudo-hypertrophic muscular paralysis.
2. Progressive muscular atrophy.
3. Hereditary ataxia (Friedreich's disease).

Pseudo-hypertrophic Muscular Paralysis.

Pseudo-hypertrophic muscular paralysis is an affection of child-hood, characterized by an abnormal increase in the size of certain muscles with diminution in the size of others.

Synonyms. *Duchenne's Paralysis; Muscular Pseudo-hypertrophy; Lipomatous Muscular Atrophy; Diffuse Muscular Lipomatosis.* French, *Paralysie Myosclérosique; Paralysie Musculaire Pseudo-hypertrophique.* German, *Lipomatosis Luxurians Muscularis Progressiva* (Heller); *Myopachynsis Lipomatosa* (Uhde).

Etiology. The etiology is obscure. It usually appears between the ages of two and eight, and affects males more frequently than females in the proportion of about four or five males to one female. Congenital cases occur, and in rare cases this disease occurs as late as puberty. Heredity is an important factor, the morbid inheritance being usually maternal, and it is more apt to occur in families than in isolated cases.

Pathology. The pathological changes consist in degenerative changes in the muscular fibres, marked increase in the muscular connective tissue, an increased amount of fat between the muscular bundles and connective-tissue fibres. Various cord-changes have been described, but no constant lesion is found, and the disease is at present regarded as a primary muscular affection—a tendency to excessive growth of the muscular connective tissue.

Symptomatology. The first symptoms observed are motor weakness, difficulty in going up and down stairs, and a peculiar gait in walking. They fall frequently and experience difficulty in rising, from muscular weakness of the back and lower limbs.

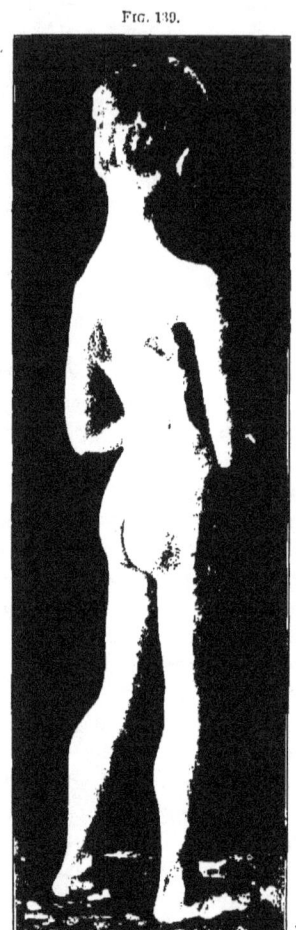

FIG. 139.

Muscular hypertrophy commences simultaneously with the motor weakness. The calf muscles, infra-spinati, and deltoids first become affected, but eventually, in from one to three years, all the voluntary muscles, with the exception of the pectorals, suffer. (Fig. 139.) This is succeeded by atrophy, the wasting appearing first and being most marked in the pectorals and latissimus dorsi. The posture and gait in walking are peculiar. Standing with the feet widely apart, the abdomen projects, the lumbar spine is arched forward, the buttocks are thrown back and the shoulders forward. (Fig. 140.)

In walking, the centre of gravity is brought well over each foot of the active limb to enable the passive leg to swing more easily forward, which gives a peculiar swaying waddle to the gait.

In advanced cases atrophy and contractions give rise to talipes equinus, flexion of knees and hip, and marked lordosis and scoliosis of the spine. The electrical contractility is lowered to both currents, knee-jerk disappears entirely in advanced cases, but sensation remains normal and rectal and vesical failure seldom occurs. Mental weakness is ofttimes associated.

Photograph of case of pseudo-hypertrophic muscular paralysis. Early stage.

Diagnosis. In well-marked cases the attitude, the peculiar walk, the enlarged calf muscles, and particularly the association of these symptoms with enlargement of the infra-spinati, with wasting of the latissimus dorsi and pectorals, are so characteristic as scarcely to be mistaken.

The gait resembles spastic para-plegia, the paraplegia of rickets, and of Pott's disease; and these will be referred to in their proper places. Progressive muscular atrophy is the affection with which it is most closely

FIG. 111.

FIG. 140.

Photograph of case of pseudo-hypertrophic muscular paralysis, showing the difference in size of the upper and lower extremities, and the helplessness. (GRAY.)

Photograph of case of pseudo-hypertrophic muscular paralysis, with absolute helplessness of hypertrophied lower extremities. (GRAY.)

allied, and from which it is most difficult, and in many instances impossible, to distinguish it. It is to be distinguished by the fact that it occurs later and the distribution of the atrophy is different. In pseudo-hypertrophy, the small muscles of the trunk escape; in progressive muscular atrophy, the affection usually begins in these muscles.

Prognosis. The prognosis in established cases is exceedingly grave. Under proper treatment the disease has been arrested,[1] and

[1] Donkin; Brit. Med. Journ., April 15, 1882.

two cases of recovery are recorded by Duchenne.[1] Death occurs from exhaustion or some intercurrent affection—especially pulmonary disease. The disease after attaining its maximum may remain stationary for a long time, and eventually prove rapidly fatal. According to Gowers, after the power of standing is lost, the patient will not probably live more than seven years.

Treatment. Cases of arrest are reported by Duchenne and Erb following the use of electricity. In conjunction with systematic exercise, gymnastics, and massage, electricity offers the best means of arresting the disease. Tenotomy of the tendo Achillis is of great service where marked equinus exists, and in some cases tenotomy of the hamstrings may be performed. Mechanical apparatus is of little or no value.

Progressive Muscular Atrophy.

Progressive muscular atrophy is a chronic disease characterized by progressive atrophy of individual muscles or groups of voluntary muscles, independent of any antecedent motor or sensory paralysis, or of any metallic poisoning.

Synonyms. *Wasting Palsy; Paralysis Atrophica; Cruveilhier's Atrophy; Cruveilhier's Paralysis.* French, *Progressive Muskelatrophie.* German, *Progressive Muskellähmung.*

Etiology. This is an affection of early and middle adult life, the earliest being two years, and the latest sixty-nine years of age. It affects more males than females, the proportion being about 6 to 1, possibly due to muscular occupation and greater exposure of males.

Two forms have been distinguished by some writers: a spinal muscular form, and an idiopathic muscular form, the former occurring after puberty from disease or violence affecting the spine, and the latter occurring congenitally, often hereditary, and sometimes influenced by consanguinity. The most frequent exciting causes are excessive muscular action, exposure to cold and wet, traumatism to head, neck, or spine; venereal excesses, constitutional syphilis, onanism, and antecedent febrile or zymotic affections.

Pathology. The pathological changes consist in a simple atrophy of the muscular fibres, with a hyperplastic growth of the perimysium, either primary, or secondary, to a destructive cord-lesion. Cord-changes may be entirely wanting, as atrophy and destruction of the

---

[1] Arch. gén. de Méd., 1868, i. pp. 5, 6.

ganglion cells of the anterior horns, with perhaps changes in the posterior columns and horns, may be present.

Symptomatology. The invasion is always gradual and insidious. Some weakness or inability to use the affected member first attracts attention to the part, which is usually discovered wasted and shrunken. In adults the ball of the thumb and hand, or the shoulder are first affected; in children the lumbar muscles, and later the leg muscles are first attacked.

FIG. 142.

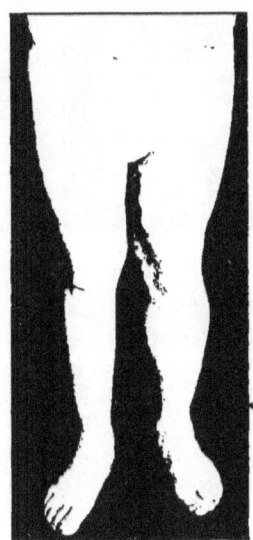

FIG. 143.

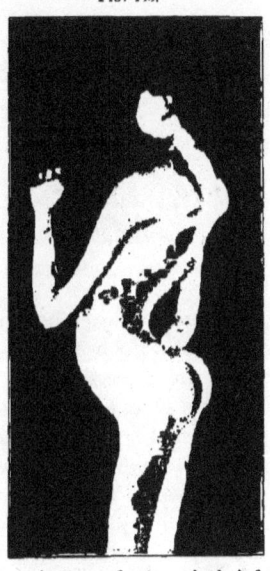

Case of progressive muscular atrophy, peroneal type. (GRAY.)

Photograph of case of extreme lordosis from advanced progressive muscular atrophy.

The disease may be arrested at any stage, or may invade other muscles, till finally every voluntary muscle may be involved, except those of the eyelids, eyeballs, and the muscles of mastication, a condition approaching in severe cases absolute helplessness. The atrophy of the muscles of the hands gives rise to the peculiar "claw-hand" characteristic of the affection. The affected muscles are soft and flabby. The expression is vacant and idiotic, from palsy and atrophy of the facial muscles, producing the so-called myopathic face. The affected muscles during the progress of the disease present constant fibrillary contractions visible under the skin, increased on exposure to cold. Electrical contractility gradually diminishes with the waste of tissue. Sensation is unimpaired, and pain of a neuralgic character is experienced in about one-half the cases. The mind remains unaffected to the end. Death is due to asphyxia from atrophy of the diaphragm and intercostals.

FIG. 141.

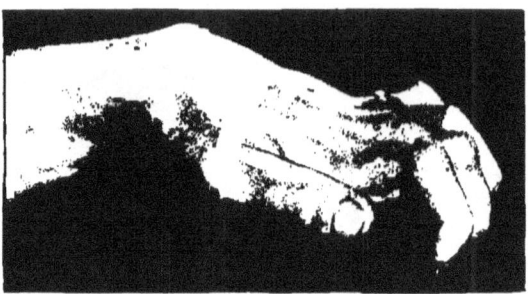

Photograph of the claw-hand. (GRAY.)

Progress is marked by repeated pauses extending over a few weeks to years, and the hereditary primary cases and those due to exposure to cold are more decidedly progressive to a fatal termination than those due to over-exercise of the muscles. The muscles of the bladder, rectum, and the heart are never implicated. Permanent contractions result in talipes equinus, flexions, lordosis, and scoliosis.

Diagnosis. The affection is most liable to be mistaken for peripheral palsy from disease or injury of a nerve, chronic lead-poisoning, and infantile paralysis.

In peripheral palsy, the history of injury, the sharply localized distribution, the sudden onset of paralysis prior to atrophy, and the impairment of sensation, would distinguish the nerve lesion.

In chronic lead-poisoning the history, symptoms, distribution of the palsy, and mode of progress will generally make the diagnosis evident.

In infantile paralysis the sudden onset of the paralysis preceding the atrophy, the loss of reflexes, and the "reaction of degeneration" are all of service in making a diagnosis.

Prognosis. The prognosis is influenced by early treatment, but when the affection becomes generalized it grows, as a rule, progressively worse. The congenital primary cases are less favorable than those due to fatigue from over-exercise, since the former are liable to become rapidly generalized.

Treatment. The general health requires attention, but local measures, especially electricity, offer the greatest hope. Strychnine ($\frac{1}{80}$ to $\frac{1}{40}$ of the nitrate) subcutaneously has been advised by Gowers. The continuous and interrupted currents are both of service; the more marked the atrophy, the more intense the current and the more rapid the interruptions. Frictions and massage begun early and continued persistently for a long time may delay the progress of the affection, and diminish the severity of the resulting deformity. When the atrophy is due to excessive use, rest is important.

Hereditary Ataxia.

Hereditary ataxia is a chronic degenerative disease of the posterior and lateral columns of the cord, characterized by serious motor disturbance, developed during childhood.

Synonyms. *Friedreich's Ataxia; Friedreich's Disease; Family Ataxia; Generic Ataxia.*

Etiology. The disease is inherited directly or indirectly, there being usually a family history of neuroses. Habitual intemperance, syphilis, consanguinity, and tuberculosis have been ascribed as etiological factors. Ataxia is rarely directly transmitted. The infectious fevers, especially scarlet fever and diphtheria, are usually the exciting causes. The sexes are about equally affected. It develops in infancy and childhood, at puberty, and in early adult life. The affection is comparatively rare. Dr. Griffith[1] collected 143 cases, and about 165 cases had been recorded up to 1890.[2]

Pathology. The lesion is similar to that of locomotor ataxia—a primary degeneration of the lateral columns with degenerative atrophy of the posterior columns. It commences in the lumbar region, and extends upward and downward, involving finally the medulla, and particularly the origin of the hypoglossal nerve. The difference is clinical and etiological rather than pathological.

Symptomatology. The earliest symptoms are weakness and uncertainty in walking, with slight numbness and pain in the lower extremities. The motions of the legs are improperly co-ordinated.

---

[1] Griffith: Monograph on Friedreich's Ataxia, 1889.
[2] Dana: Cyclop. Dis. of Child., vol. iv. p. 717.

Headache, vertigo, and impairment of speech, and nystagmus are often present.

The patellar reflex is early lost ; trophic disturbances are slight ; the sphincters are, as a rule, unaffected ; cutaneous anæsthesia is usually absent. Incoördinate or choreiform movements may develop in the extremities.

Deformities occur in the later stages—talipes varus or equino-varus, dorsal flexion of the toes, permanent flexion of the knees, and lateral curvature of the spine.

Diagnosis. The distinguishing characteristics are the development early in life of ataxia, paraplegic in character, with loss of reflexes, disturbances of speech, and deformities, without cutaneous anæsthesia, acute pain, or paralysis of sphincters, but with hereditary or family history of neuroses.

Prognosis. The tendency is progressively bad, death usually occurring from some intercurrent affection, the average duration of the disease being fifteen or twenty years.

Treatment. Improved hygiene with persistent regular massage is of the greatest service. Suspension, as in locomotor ataxia, should be tried, and cases are recorded in which it proved beneficial. The infectious diseases of childhood, especially scarlet fever and diphtheria, should be especially avoided, and injury by blows or falls should be guarded against.

Electricity may be employed to maintain the nutrition. Tonics, especially arsenic, are sometimes of temporary benefit. Flat-foot, club-foot, and lateral curvature require the appropriate treatment described in the other sections.

CHAPTER XV.

TORTICOLLIS.

TORTICOLLIS is a distortion of the neck, constant or intermittent in character, in which the head is drawn awry.

Synonyms. English, *Wry-neck; Caput Obstipum; Collum Distortum.* French, *Cou Tortu; Coutors.* German, *Schiefhals.* Italian, *Torticolli.*

Frequency. According to some writers, as Agnew, the sexes are about equally affected; but according to others, males are more frequently affected than females; and according to others, females are more frequently affected than males. Thus, in 264 cases collected by Whitman, 155 were females and 109 were males. The right side appears to be more liable than the left. Thus, in 37 cases collected by Dieffenbach, 5 only were on the left side, and in 29 cases recorded by Bouvier, 18 were on the right side. Of the 264 cases recorded by Whitman, in 187 of which the affected side was mentioned, 98 were on the right and 97 were upon the left side.

Etiology. Wry-neck is both congenital and acquired, the latter being much more common than the former.

Congenital. Cases are recorded of wry-neck in a stillborn infant,[1] and another of unilateral atrophy of the head,[2] but the majority of so-called congenital cases occur at birth from injury. In the true congenital variety the wry-neck depends upon a deficiency of the cervical vertebræ, from malposition of the fœtus *in utero*, from deficiency of liquor amnii, or possibly, as Petersen suggests, from an attachment between the skin of the face and the amnion in early embryological life. In congenital cases heredity appears to play a rôle, as in the case of mother and child reported by Dieffenbach, the seven children in one family recorded by Fisher, and the two sisters recorded by Petersen and Zehnder. The frequent association with other deformity in congenital cases points to malposition *in utero* as a cause. Little believes congenital wry-neck clearly orignates from pre-natal causes, acting through the nervous system).

---

[1] Archives de Tocologie, 1888.
[2] Boston Medical and Surgical Journal, June 1, 1882.

Acquired. The acquired form includes several varieties, which may be grouped into traumatic, tetanoid, paralytic, compensatory, cicatricial, and idiopathic.

Traumatic cases result from blows, twists of the neck, rupture of the sterno-mastoid muscle, and, as suggested by Stromeyer[1] and Dieffenbach, from violence received during delivery from forceps, pressure, or the manipulations of the accoucheur. Rupture of the muscular fibres, particularly of the sterno-mastoid, results, with subsequent induration and cicatricial contraction (Busch, Fisher, Volkmann, Vollert, Rust). This rupture may occur in a muscle shortened *in utero*,[2] as in the case reported by Bruns[3] at the last Congress of German Surgeons. Hæmatoma of the muscles is often present, but its relation to wry-neck is at present undetermined. Petersen believes hæmatoma is found in already existing torticollis, Fassbender has observed it upon the side opposite to the wry-neck, and the experiments of Witzel and Fabry fail to prove the associa-

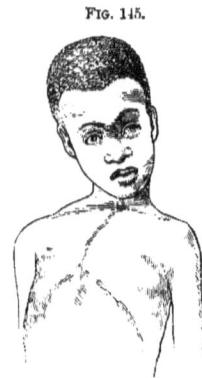

FIG. 145.

Attitude in cervical caries.

tion. Petersen[4] vigorously combats the theory of Stromeyer, Weiss[5] flatly denies it, and Whitman[6] believes that the majority of causes of congenital torticollis operate before rather that during delivery. In rare cases fracture and dislocation of the cervical vertebræ have been observed as causes.

Tetanoid or *spastic* contraction of the muscles of the neck, either tonic or clonic, results from a central lesion; a peripheral irritation, from inflammatory conditions of the bones, ligaments, or muscles (Fig. 146); or as a local manifestation of some central or spinal irritation. Choreiform spasm sometimes occurs. Peripheral irritation is a common cause of acquired torticollis, for of the cases collected by Whitman,[7] of 106 cases in which the cause seemed apparent, 50 per cent. were classed as from cervical gland irritation.

Paralytic wry-neck may arise from unilateral muscular action in paralysis of one side, either peripheral or central in origin.

[1] Handbuch d. Chir., Bd. ii. S. 426.   [2] Rennecke: Centralbl. f. Gyn., 1886, No. 22.
[3] Centralbl. f. Chir., 1891, No. 26.   [4] Archiv f. Chir., Bd. xxx. 1.
[5] Nouv. Dict. de Méd. et de Chir.   [6] Loc. cit., p. 477.
[7] Medical News, October 24, 1891, p. 475.

Compensatory torticollis is met with in lateral curvature, but in many instances of association of these two conditions, the wry-neck will be found to be primary.[1] Of a compensatory nature also are those cases of wry-neck developed from irregularity of the sight of

FIG. 146.

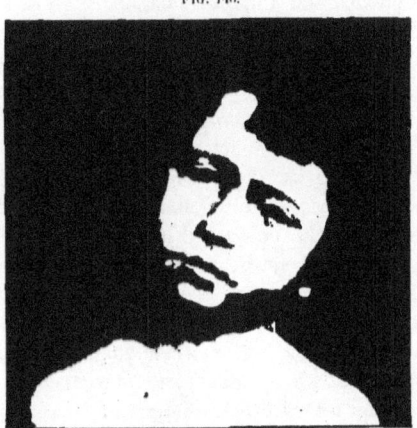

Photograph of a case of spasm of the trapezius. (GRAY.)

the two eyes—a condition termed by Quignet, *torticollis oculaire*—a not infrequent condition in the writer's experience.

Cicatricial wry-neck results from extensive burns and scalds, abscesses, cicatrices from suppurating cervical glands, or lupus of the neck.

Idiopathic cases occur in which no definite lesion is discoverable, but in which general nervous debility has had a local spasm engrafted upon it by some emotional cause, such as grief, fright, etc.; an acute torticollis, or as an accompaniment of hysteria. Among rare causes may be mentioned tumors of the sterno-cleido-mastoid muscle—sarcomatous, fibrous, or syphilitic, as in the cases reported by Graser,[2] Holmes,[3] Taylor,[4] and Hadra.[5]

Pathology. In the traumatic cases occurring at birth, the pathology consists of a laceration of the sterno-mastoid muscle from

[1] *Vide* cases recorded by the writer, Medical News, February 22, 1890, p. 204.
[2] Münch. med. Wochenschr., 1887, No. 13.   [3] Diseases of Childhood and Infancy.
[4] Trans. London Path. Soc., xxvi. p. 224.
[5] New York Medical Record, January 23, 1886.

torsion[1] of the neck, followed by hemorrhage,[2] succeeded later by an encapsulating inflammation,[3] and the characteristic hæmatoma. In the chronic forms alterations occur in the muscles, fasciæ, ligaments, and bones from malposition or disuse. The muscles most frequently involved are the sterno-cleido-mastoid, trapezius, splenius, scaleni, rarely the complexus, and the platysma, although in severe cases all the muscles of the neck participate. The affection involves the distribution of the spinal accessory nerve, and is usually due to central disease, lesion of this nerve, or the muscles themselves. The affected muscles undergo fibrous degeneration, becoming hard and unyielding. The contraction continuing, anatomical changes are produced which not only tend to fix the cervical vertebræ in their deformed relations but also to exaggerate the abnormal curves. These changes are chiefly atrophy of the bones and intervertebral cartilages on the concave side of the curve, with shortening of the ligaments and muscles, and hypertrophy of the bone and cartilage, with elongation of the ligaments and thickening of the muscles on the other side. Asymmetry of the face has been ascribed to gravitation of the blood to the more dependent side (Agnew); to muscular tension (Dieffenbach); pressure and retarded development of the vessels and nerves on the side of the concavity (Nélaton, Eulenberg); to tension of the soft parts, particularly the muscles of the sound side (Witzel); and to interference of function. Atrophy of the affected side and asymmetry of the skull have been variously ascribed by different authors to deficient activity of the muscles of inspiration on the affected side, as in ordinary scoliosis (Stromeyer); to the constant pull of the contracted sterno-mastoid muscles (Dieffenbach); to faulty assimilation on account of compression of the carotid artery and other great vessels of the neck on the affected side (Bouvier, Guérin, Mousson, Conillard-Labonnote, and Brock); to interference with the function of the part, causing atrophy from disuse, as in asymmetry of the face (Little and Busch); to hyperextension of the muscles on the sound side (Witzel); and to the misplacement of the epiphysis of the basilar portion of the occipital bone, and the effect of the weight of the skull upon the other cranial and face bones also (Nicoladoni, Falkenberg). Of these, the theory of Bouvier perhaps best explains the atrophy, and that of Nico-

---

[1] Küstner: Centralbl. f. Gyn., 1886, No. 9.
[2] Bouchut: Traité de Mal. de Nouv.-nés. Paris, 1873.
[3] Hüttenbrenner: Lehrb. f. Kinderheilk., 1886.

ladoni the asymmetry of the skull and also the osseous changes in the face.

Symptomatology. The affection may be acute or chronic, permanent or intermittent. In the acute form there is pain on motion, with tenderness of the affected muscles, and the head is held awry. The acute form may become chronic. The position of the head varies with the muscles affected. In the most common variety the head is drawn to one side, the face being rotated in an opposite and somewhat upward direction, and in severe cases the face may look directly to the sound shoulder and the neck on the affected side be buried out of sight. Asymmetry of the face, the oblique position of the nose, the irregularity of the eyes and commissures of the lips, the diminished size of the features on the affected side, are peculiar and characteristic. Asymmetry of the face is rarely lacking, whether the wry-neck be congenital or acquired. Marked atrophy of the affected side and asymmetry of the skull are present in severe grades of the affection. In chronic cases a compensatory lateral curvature of the spine results from the cervical curvature. Intermittent torticollis, either unilateral or bilateral, occurs in later life, and is often choreiform. Convulsive contractions vary from an occasional jerk to an almost constant spasm. It is more frequent in females, is usually clonic, but may become tonic. When unilateral it constitutes the so-called *tic giratoire* of Trousseau. Another variety, described by Dalby as occipito-atloidean, consists of a subluxation of the atlas on the occipital bone, producing a twisting of the head which resembles torticollis, and which is a bony ankylosis, a rheumatic sequela.

Diagnosis. While there is little difficulty in recognizing the deformity itself, it is most important to distinguish it from other conditions, such as cervical caries, simple cervical abscess and cervical adenitis, in which the head is held awry without muscular spasm or contraction. Ordinary stiff-neck from cold may at all times be easily distinguished by the transitory and shifting character of the affection, the pain and tenderness on pressure, its sudden onset, and acute character. Cervical caries can usually be distinguished by the greater rigidity, by more severe pain and its persistence even when the patient is recumbent, by dysphagia, the attitude of the patient, supporting the head with the hands, the facial expression of pain, and the co-existence of tubercular diathesis or lesions elsewhere. In anterior torticollis due to caries, the sterno-mastoid is not so permanently nor so prominently contracted. In posterior torticollis the symptoms are

17

almost identical with posterior idiopathic torticollis, but the fact that in spinal caries the head is rotated *toward* the contracted muscle, whereas in idiopathic cases of wry-neck the rotation takes place *away* from it, is of the greatest diagnostic value. The greatest importance attaches to the differential diagnosis between posterior idiopathic torticollis and torticollis due to cervical caries, since the writer has had related to him a case where an attempt to forcibly straighten a case of torticollis due to cervical caries resulted in instant death. In simple cervical abscess the symptoms are those of acute local inflammation, superficial in character and attended with fever, and in cervical adenitis there is circumscribed, deep-seated inflammation, without characteristic muscular spasm. Paralytic cases can be distinguished by the manner in which the unsupported head falls to one side.

Prognosis. Acute idiopathic muscular torticollis tends rapidly toward recovery, but may become chronic. Wry-neck due to peripheral irritation terminates usually upon removal of the cause. Paralytic cases require time and the recourse to mechanical appliances. Traumatic cases, the so-called congenital cases, and the ordinary chronic acquired affection are entirely curable by surgical means. The deaths reported following operation for torticollis have been due to hemorrhage, septicæmia, and injury to the pleura. The intermittent form may cease spontaneously, but usually remains unchanged for years, unless subjected to surgical interference, when encouraging success usually follows radical operation.

Treatment. The treatment will depend upon the variety, and may be divided into therapeutical, mechanical, and operative.

Therapeutical. In acute inflammatory forms much benefit will be obtained by the correction of the coincident constitutional disturbance. A gentle purge, followed by salicylate of soda, with the local application of warmth, the oleate of morphine, or atropine hypodermically, as employed by J. M. Da Costa, of this city, will be all that is required. In paralytic cases the endermic use of strychnine is recommended by Ashhurst. When dependent upon general debility, the health should be restored by means of tonics, massage, improved hygiene, and general bathing. When dependent upon ocular defect, the insufficiency should be corrected with prisms, tenotomy only being performed when the insufficiency amounts to twelve degrees or over. In a recent case of the writer's a one degree prism, base down, in the left eye corrected the insufficiency, and, with the proper correction for hypermetropia, relieved the torticollis. Electricity is of value and is sometimes followed by cure.[1] Gelsemium (fluid extract) in physiological or hyper-physiological doses, as suggested by Weir Mitchell, may be employed after failure to relieve the muscular spasm by galvanism or before resorting to nerve resection. In intermittent cases the actual cautery to the back of the neck has been recommended by Agnew and C. K. Mills, of this city. In chronic cases of posterior torticollis forcible straightening under an anæsthesia without tenotomy has been successfully employed by Delore,[2] Bradford, Lovett, and others. In these obstinate cases tenotomy of the deeper posterior muscles is impossible. The method of Delore consists in anæsthetizing the patient, and while an assistant firmly holds the shoulders, the surgeon grasps the head firmly with both hands and gradually but forcibly rotates it in all directions. When the deformity is over-corrected, the head, neck, shoulders, and trunk are fixed by a silicate of potash bandage. A positive diagnosis of posterior torticollis from caries of the cervical vertebræ must be made before resorting to this plan of treatment.

Mechanical. Mechanical appliances without operation are unsatisfactory, except in paralytic cases. In these, after the therapeutical means have failed to recall the lost muscular power, mechanical apparatus is useful. The mechanical principle involved is to obtain fixation on the trunk, from which counter-pressure is made upon the head. After operation apparatus should be worn from three to five months, when it may be permanently discarded. A simple and most efficient form of appliance after operation is that recommended by Ashhurst, and originally suggested by Little,[3] which consists of a broad adhesive strip around the forehead and occiput, and another around the waist, fastening the two together by means of a bandage carried from above the ear of the unaffected side across the chest to the opposite side of the trunk.

After operation the head may be secured in a slightly over-corrected position by incasing the shoulders erect and in a plaster-of-Paris cast, or, what the writer prefers in all operations about the head and neck, a starch bandage reinforced with strips of a wooden roller or aluminium.

More elaborate appliances are needed to twist the head, and these

[1] For method of application see case reported by Rockwell in Hare's System of Therapeutics, vol. i. p. 193.
[2] Gazette Hebd., March 22, 1878.   [3] Deformities of the Human Frame, 1853, p. 191.

are all constructed upon the principle of that of Jörg.[1] Of these a
great variety exist, and are well illustrated in the chin-rest used
(Fig. 168) in the treatment of cervical cases, in the accompanying
illustrations of that of Sayre,[2] recommended by the writer, and in
those figured in St. Germain[3] and Ernst.[4] The apparatus of Sayre

Fig. 148.

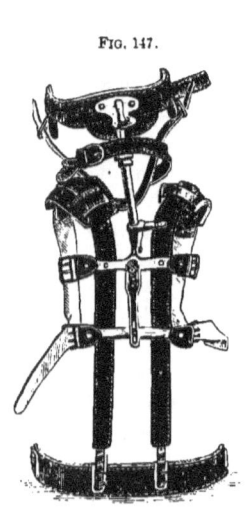

Fig. 147.

Sayre's brace for torticollis

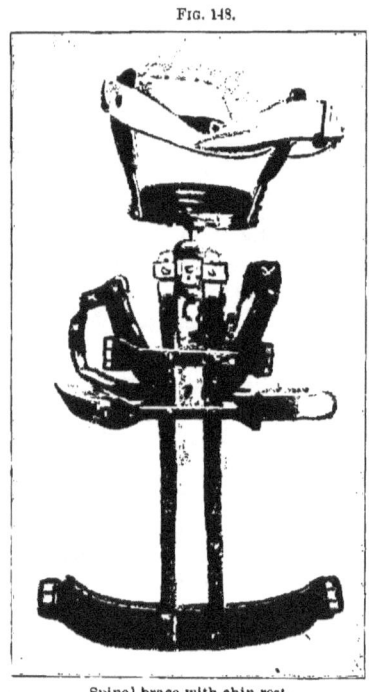

Spinal brace with chin-rest.

consists of two upright malleable steel bars which pass up the back
on either side of the spine and curve over the shoulders, on which
they rest. Below, they are attached to a band passing around the
pelvis. The head-piece is supported upon a rod by a universal joint
attached by means of a similar universal joint to the spinal uprights
at a point corresponding to the first dorsal vertebra, or, what is preferred, three Archimedean screws, since the former requires two

[1] J. C. G. Jörg : Ueber die Verkrummungen des menschlichen Körpers. Leipzig, 1810.
[2] Trans. Amer. Orthop. Assoc., vol. v. p. 253.　　　[3] Chir. Orthop., 1883, p. 239.
[4] A Guide to the Selection of Orthopedic Apparatus, p. 22.

persons for a proper adjustment. This rod has a ratchet and key movement for elongation. The head-piece, a padded piece of malleable steel, runs from the base of the skull forward and upward, curving over each ear. From the ends of the metal head-piece a leather strap, made adjustable by a buckle, passes across the forehead, and, by means of hooks, a strap is attached which passes under the chin. The apparatus of G. G. Davis,[1] while efficient, is most useful after operation.

Collars of pasteboard, felt, leather, and wire, sometimes spoken of as Minerva collars, *collier Minerve*, from the simplest to the most elaborate forms, are employed to fix the distorted head, but these are most efficient after operation.

Operative treatment. The operative procedures employed are tenotomy, nerve-stretching, nerve-division, and nerve-resection.

Tenotomy of the sterno-mastoid is most frequently required, and this may be performed subcutaneously, or by open incision. The operation of subcutaneous tenotomy is one of great delicacy, and not entirely free from risk, three fatal cases having been mentioned by Mr. Erichsen. Both the sternal and clavicular origin may require division. Anæsthesia is necessary, and the dorsal position is preferable, with the head well extended and rotated to the sound side. The sternal portion is best divided about half an inch above its origin, from within outward. Care being taken to avoid the anterior jugular vein, if present, a blunt-pointed tenotome should be introduced flatwise in front of the upper margin of the sternum anterior to the inner edge of the muscle, and be passed outward beneath the deep surface, the edge turned forward, and with a slight sawing motion, the tendon be divided. The clavicular portion may be divided through the same puncture, but is better and more safely accomplished by a second puncture between the two portions, the cautious insertion of the probe-pointed tenotome behind the tendon, and its division from behind forward. After the division the tenotome must be withdrawn flatwise, and the punctures be immediately sealed. The disadvantages of subcutaneous tenotomy are the difficulty of dividing the deep bands and the danger of wounding important vessels.

Open incision. In aggravated cases the deeper cicatricial bands may be preferably divided by the open incision of Volkmann. Under full aseptic precautions, a vertical incision along the inner border of the muscle, or preferably an incision parallel with and one inch above the clavicle, should be made. With forceps and the handle of

---

[1] University Med. Magazine, February, 1892, p. 377.

the scalpel, the platysma, the muscle sheath, and cicatricial bands may be separated and divided cautiously upon the grooved director. The deeper cicatricial bands should be firmly secured between two strong forceps before being divided. Great care must be exercised to avoid wounding the internal jugular vein, by relaxing the neck and not opening the deep cervical fascia.

The wound should be sutured with catgut (Heincke), and the head be secured in an over-corrected position, by plaster or silicate of soda dressing, or other mechanical means, for from ten to fourteen days, when the after-treatment should be begun. The orthopedic treatment consists in massage of all the muscles of the neck and the gradual but forcible manipulation of the head in such a manner as to stretch the formerly shortened muscles and to strengthen the weaker muscles. This may be entrusted to a skilled masseur, or may be performed by the patient swinging in an extension head-gear, the cord of which is not attached to the centre of the bar, but to the side of the contracted muscles. A weight should be held in the hand of the affected side.

Nerve operations. In intermittent torticollis, stretching of the spinal accessory nerve has been successfully employed by Hanson[1] (two cases), Southam, and Mosetig-Moorhof. Section has also proved temporarily beneficial in a case recorded by De Morgan,[2] in which resection of a piece of the nerve was afterward successful. Ewington,[3] Annandale,[4] Tillaux and Wood,[5] have also recorded successful cases of resection. The nerve may be conveniently exposed by an incision three inches in length along the posterior border of the sterno-mastoid muscle, the centre of which is half an inch above the centre of the muscle, or on a level with the upper border of the thyroid cartilage. After dividing the skin, superficial and deep fasciæ, the head may be flexed and the neck relaxed on the side being operated upon, the muscle may be separated with a grooved director and the finger, and turned forward, disclosing the nerve where it enters the muscle, or the bottom of the wound may be irritated with a director, as suggested by Richardson, the contraction of the muscle serving as a guide to the location of the nerve. In most cases where spasmodic torticollis of central origin requires operation, resection of a portion of the nerve appears to be the preferable procedure, and should be at once resorted to.

[1] Hospital Tidende, 1878, 2 R. v., No. 45.   [2] Brit. and For. Med.-Chir. Rev., July, 1866.
[3] Brit. Med. Journ., Feb. 8, 1879, p. 212.   [4] Lancet, 1879, i. p. 555.
[5] Ogle: Clin. Soc. Trans., vol. vi.

CHAPTER XVI.

NEUROMIMESIS.

NEUROMIMESIS (Paget[1]) includes certain functional or mimicked affections associated with the hysterical diathesis.

Synonyms. These affections have been variously recognized as simple hysteria (Brodie,[2] Skey[3]); as *Gelenkneurose* (Esmarch[4]); as the mimicry of disease (Weir Mitchell[5]); and as hysterical joint affections (Shaffer[6]).

Frequency. The frequency with which joint diseases are mimicked is very great. Brodie makes the assertion, fully indorsed by Esmarch, "that among the higher classes of society at least four-fifths of the female patients who are commonly supposed to labor under disease of the joints, labor under hysteria and nothing else." Skey and Shaffer assert its frequency also among the lower classes.

There is no type or form of organic paralysis, no contracture or spasm, and no articular deformity depending upon chronic inflammation or disturbed muscular action, which may not be simulated in hysteria.

Etiology. These affections are most common in women of a pronounced brunette type, and usually appear about the time of puberty or young womanhood. The manifestations may continue until the menopause or even later, and are often associated with ovarian or uterine disorders. Men and boys are not exempt. Debility, ill health, and neurasthenia are often associated. Heredity and education are the most important predisposing causes. Errors of refraction are sometimes the starting-point of a hysterical explosion. Traumatism is often an exciting cause. Real disease is often present, but the symptoms are all exaggerated, and hysteria is imposed upon an original organic pathological condition. The etiology, in fact, differs but little from the etiology of hysteria in general. The local

[1] Clinical Lecture and Essays. New York, 1875.
[2] Pathological and Surgical Observations on Disease of the Joints. 1818.
[3] Lecture on Hysteria. New York, 1867.   [4] Ueber Gelenkneurosen. Kiel, 1872.
[5] Lectures on Nervous Disorders. 1881.   [6] Archives of Medicine, vol. ii. p. 277.

manifestation is engrafted upon a state in which ideas control the body and produce morbid changes in its functions (Möbius).

Symptoms. These include three groups—the paralyses, spastic contractions, and joint disease.

Paralyses. These vary from slight transitory loss of power, which can hardly be termed palsy, to severe cases of long standing, which may be readily mistaken for lateral sclerosis, as in the case which Osler,[1] in 1879, frequently showed at the Montreal General Hospital as a typical example of lateral sclerosis, and which after persisting for eighteen months disappeared completely. In form they may be hemiplegic, paraplegic, or monoplegic, the paraplegic being most frequent. Left-sided hemiplegia is more common than right, Brique having observed 70 in the left to 20 in the right, and Weir Mitchell in the proportion of about 4 to 1. The reflexes may be increased, but the electrical reactions are normal and atrophy is absent. Hyperæsthesia, aphonia, and paralysis of the bladder are usually associated.

Spastic contractions. These may occur gradually, following paralysis, but usually they develop suddenly and disappear rapidly, at the termination of a variable period. Almost any group of voluntary muscles may be attacked, and the distribution may be hemiplegic, paraplegic, or monoplegic. The contracted part is rigid like an iron bar. Both flexors and extensors are in action, but the extensors usually predominate.

In hemiplegic cases there is often associated complete analgesia, deficiency of hearing, smell, taste, and retraction of the visual field on the affected side. Next to contraction of the arm, paraplegia is most common. In cases of long standing a certain degree of muscular atrophy may be present, but without the reaction of degeneration. True hysterical aura, beginning in the contracted part, terminating in globus hystericus, and followed by aphonia, are sometimes observed. Of this nature also is hysterical club-foot, described in another section.

Joint disease. Disease of any articulation may be mimicked, but it usually affects the knee or hip. The joint is usually hypersensitive, rigidly fixed or preternaturally mobile, and swollen. Atrophy occurs from disease only, but the electrical reactions are unchanged. Motion is painful, and the patient walks with a limp. The sleep is unaffected and "night pains" are absent. Superficial hyperæmia may increase the surface temperature.

---

[1] *Vide* case of Osler, Practice of Medicine, 1892, p. 970.

Organic disease may succeed the neuromimetic disturbance, as in the ankle-joint case of Esmarch, about which for many weeks he was undecided, which ultimately proved to be caries sicca, requiring amputation, and in the remarkable knee-joint case recorded in Weir Mitchell's lectures. In the latter the hysterical element was pronounced, but on account of its chronicity so eminent an authority as Billroth pronounced it organic, and Sands, upon operating, found the joint surfaces normal and non-tuberculous inflammatory thickening outside the joint.

Diagnosis. The presence of hysteria in any of its protean forms should put the practitioner on his guard in regard to the nature of any motor or joint affection.

The neuromimetic paralyses may be gradual or sudden in onset, but are usually associated with anæsthesia, paralysis of the bladder, aphonia, or other hysterical manifestation.

The neuromimetic contractions are very deceptive, but they disappear under full anæsthesia. The extension and rigidity of the part, the occurrence of areas of anæsthesia, deficiency of the special senses, particularly retraction of the visual field, and the hysterical aura, are of diagnostic value.

The neuromimetic joint affections have symptoms in common with both chronic synovitis and chronic osteitis of the joints. The absence of true reflex spasm, the normal electrical reaction of the muscles to the faradic current, the absence of a rise of temperature—either local or constitutional—and the association of globus hystericus, emotional attacks of weeping and crying, will serve to distinguish the hysterical affection from the true organic disease. The rigidity, moreover, yields to mild force if the attention is diverted, and wholly disappears under the usual doses of chloral or opium, and complete anæsthesia.

Treatment. The successful treatment of these affections requires a knowledge of human nature, and great tact, together with consummate skill. The general condition, and particularly the general *morale* of the patient, require attention. Tonics, nerve sedatives (with positive interdiction of morphia), electricity and massage are of great service. The method of Weir Mitchell offers the best advantages and is particularly applicable to these cases. It consists of free nourishment, isolation, rest, massage, and electricity. The assistance of an intelligent nurse is essential. In regard to the local affection—the neuromimesis—the treatment consists essentially in diverting the

attention from the affected part. In conducting this part of the
treatment a positive diagnosis is essential. This being assured, the
attention may be diverted abruptly or gradually. The first plan,
which is scarcely legitimate in regular medical practice, consists in
suddenly commanding the patient to use the affected part after the
method of charlatans and faith-healers. Hypnotism has of late been
extensively employed for the same purpose, but, as Osler remarks,
"Anyone who has seen the development of this method as practised
at present in France must feel that it is a two-edged sword, and that
the constant repetition in the same patient is fraught with danger."

The second and usual method consists in the gradual diversion of
the mind from the affected part, and the gradual use of the part.
The application of mechanical force is positively contra-indicated,
and the use of apparatus is entirely secondary. In spastic contractions and joint affections light traction in the line of deformity may
be applied for a time. The use of the limb is then encouraged, pain
being disregarded. Massage and electricity hasten the cure, the
massage being used daily as a substitute for exercise.

Operative treatment is usually contra-indicated, but if the contracture persist for a long period, despite thorough treatment, tenotomy
may be resorted to, as in the case in which Roberts, after consultation
with Weir Mitchell, divided the tendo Achillis for talipes equinus,
after which the deformity finally disappeared.

CHAPTER XVII.

SPINAL AND CEREBRAL ARTHROPATHIES.

THE neuro-arthropathies, or so-called Charcot's disease, are peculiar chronic deforming affections of the joints, due to some cerebral or spinal nervous lesion.

Synonyms. *Spinal Arthropathy; Arthropathy; Tabetic Arthropathy; Neural Arthropathy.*

While particular attention has been attracted to this subject by the writings of Charcot,[1] Allbutt, Raymond, Hitzig, Gull, and others, it is interesting to observe that to a distinguished American physician of this city, Dr. John K. Mitchell, belongs the long-forgotten credit of the first discovery that an "obvious spinal cause may produce a rheumatism characterized by heat, pain, redness, and tumefaction,[2] and the direct connection between Pott's disease and acute inflammations of the joints, and between traumatism and acute joint diseases."[3]

Etiology. The nervous lesions leading to arthropathies are injuries of the peripheral nerves,[4] acute myelitis, tumors of the gray substance of the cord, in Pott's disease, and in cerebral apoplexy. The most common spinal arthropathy is that dependent upon degeneration of the posterior columns, or locomotor ataxia—the one particularly referred to by Charcot. Hemiplegic arthropathies have been studied by Scott Alison,[5] Brown-Séquard, S. Weir Mitchell, and others, and according to Charcot similar lesions have been met in progressive muscular atrophy by Remak, Patruban, and Rosenthal. Two varieties have been recognized—an acute or subacute form and a chronic form. The disease is usually met with in adults, but cases are recorded as early in life as the sixth year. The right and left sides of the body are affected about equally, and all the major joints are affected and many of the smaller ones. Usually but one joint is affected, but occasionally two or more are diseased. Of all the large joints the knee-joint is most frequently affected.

---

[1] Leçons sur les Maladies du Système Nerveux. Paris, 1872-73, p. 100.
[2] Amer. Journ. of Med. Sci., 1831, p. 55. [3] Ibid., 1833, p. 360.
[4] Mitchell, Morehouse, and Keen: Nerve Injuries. [5] Lancet, 1846, vol. i. p. 227.

Thus, in the 169 cases analyzed by Weitzächer, the knee was affected in 78, the hip in 31, the shoulder in 21, etc. These occurred in 109 individuals, of whom 72 were men, and 37 women.

The exact mechanism of the production of these arthropathies has been a matter of much discussion. The discovery of arthropathies caused by injury of nerve trunks promised to simplify the research, but while it has been shown that they are caused neither by vascular palsy, vasal spasm, nor by inertia, as suggested by Mitchell, the local nerve irritation probably influences the centre, and through it and the entire nerve thread acts upon the joint to disturb its nutrition.

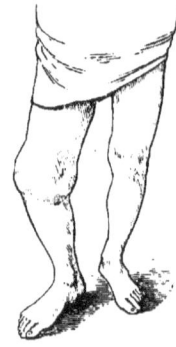

Fig. 149.

Ataxic knee-joint.

The theory that these arthropathies are only the result of constant traumatism upon joints whose sensitiveness was destroyed, as in conjunctival troubles following section of the facial nerve, is no longer tenable. Virchow's assertion that the distinction is due to faulty cellular or trophic change in the joint explains nothing. In some cases without peripheral irritation the disturbed centres are capable of producing joint lesions. The tabetic arthropathies occur independently or precede the active symptoms of spinal lession. In hemiplegic cases the arthropathy occurs from fifteen days to three months after the apoplectic attack, coincidently with the advent of "late rigidity," a symptom commonly ascribed to descending sclerosis. Occasionally the arthropathy may be observed as early as the day after an attack of hemiplegia, or on the third day.

Pathology. The pathological changes resemble those observed in arthritis deformans, but are more acute and destructive. Suppuration is never directly associated, unless incident upon direct injury.

The synovial membrane is the seat of chronic asthenic hyperæmia with hypersecretion of synovial fluid—hydrarthrosis. The cartilage is eroded in places, irregularly thickened in others. The epiphyses are hypertrophied and nodular, and osteophytes are abundant at the attachment of the capsule, and within the cavity of the articulation.

A. S. Roberts[1] systematically classifies the pathological changes as follows :

[1] Medical News, February 14, 1885.

1. A chronic asthenic hyperæmia of the synovial membrane—a hydrarthrosis.

2. An interstitial atrophy of the epiphyses.

3. A fungous or rarifying epiphyseal hypertrophy.

4. The formation of osteophytes and bony stalactites.

FIG. 150.

FIG. 151.

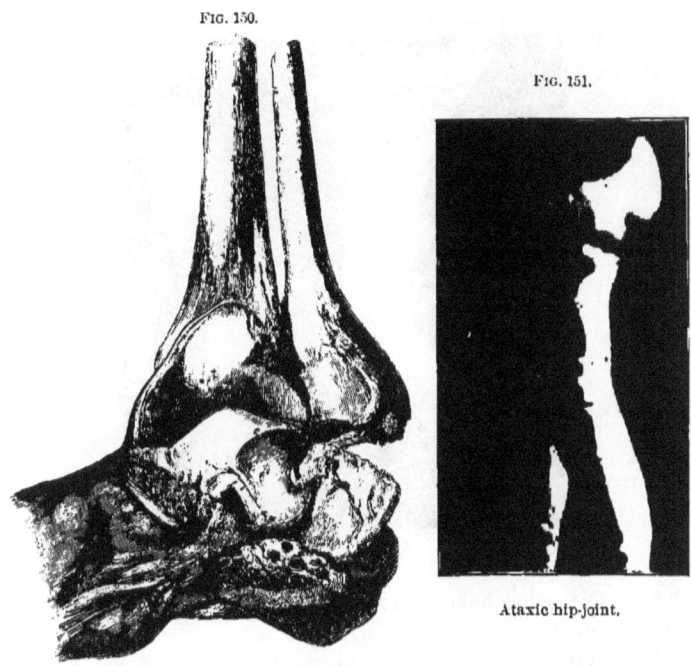

Ataxic hip-joint.

Pathological illustration of "tabetic foot."
(PAVLIDES.)

Symptoms. The local symptoms differ with the varieties, these being redness, swelling, and sometimes more or less violent pain in the acute or subacute form ; and a slow hydrarthrosis, with diffused, indurated swelling about the tissues of the joint, without the ordinary peculiarities of œdema, in the chronic and more usual form. The pain may be erratic, paroxysmal or constant, and exquisite tenderness to pressure. In spinal cases, pain and sensitiveness may be present over certain vertebræ with an aching distress over which ice develops a feeling of burning.

The course of the affection is varied and essentially chronic. With occasional exacerbations years may elapse before matured. Spontaneous recovery may be abrupt and rapid, or atrophy and absorption of the cartilages with proliferation of osteophytes, may result in total destruction of the articulation.

Fig. 152.

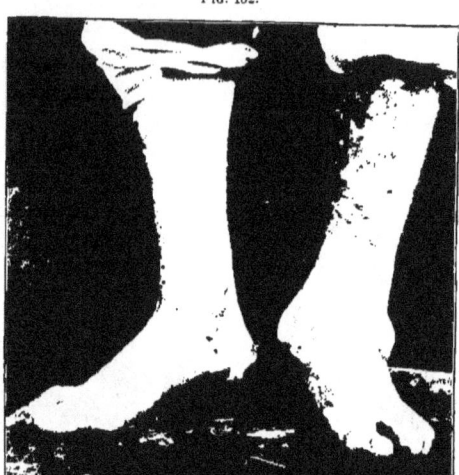

Ataxic ankle-joint.

Suppuration has been observed, but does not form an essential part of the disease.

The muscles become rapidly and extensively wasted, and respond only to galvanic currents of high power (thirty cells or more).

Diagnosis. In the earliest stages, before the appearance of the spinal or cerebral lesion, it is impossible for the most acute clinicians to distinguish these affections from some of the ordinary types of acute articular rheumatism. According to Charcot and others, they are to be recognized by the limitation of the affection to the joints of the palsied members, their relation in time to hemiplegia or other nervous lesion, and the coexistence of other trophic disturbances, as dystrophy of the nails, muscular atrophy, etc.

From chronic osteitis of the joints they may be distinguished by the absence of reflex pain, muscular spasm, and abscess formation;

and from malignant disease of the joints by the anamnesis and course of the disease, and of the presence of central or peripheral disturbance.

FIG. 153.

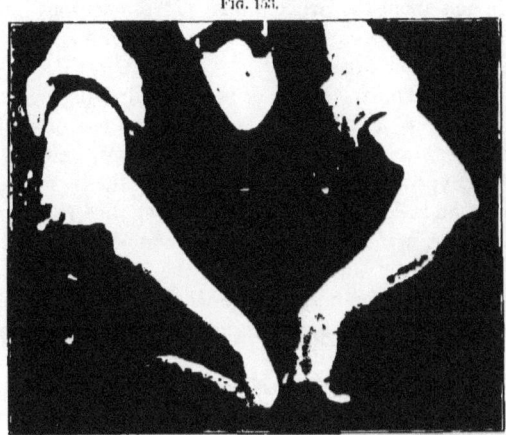

Ataxic elbow-joint.

FIG. 154.

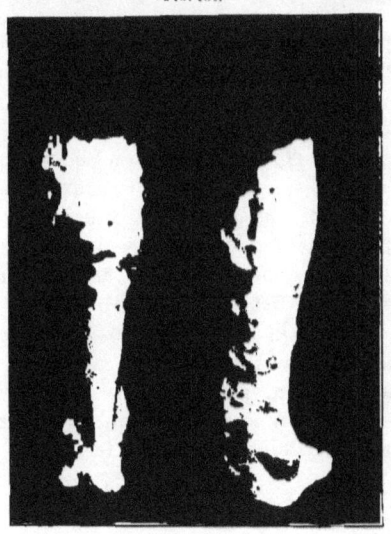

Ataxic ankle-joint.

Prognosis. Without treatment the affection is self-limited after a variable period, but repair to the lesion is impossible, and apparent diminution in the deformity is usually due to the disappearance of effusion in and about the articulation. If the nervous lesion can be controlled or cured, recovery may promptly follow, as in the cases recorded by S. Weir Mitchell.[1]

Treatment. The treatment consists in limiting or overcoming the central nervous lesion by absolute rest, the use of tonics, iodide of potassium or ammonium in moderate doses, and full doses of strychnia by hypodermatic injection. Locally the use of ice, massage, shampooing, and the application of a well-applied roller bandage or pressure sponges will afford the greatest comfort and relief.

Powerful galvanic currents through the joints and reversed galvanic currents have been recommended by Weir Mitchell. If the hydrarthrosis be excessive, aspiration or aseptic incision may be performed.

Excision of the knee-joint has been performed[2] with a favorable result, but the diminished reparative powers of the tissues seem hardly to warrant so serious an undertaking.

The Arthropathies of Syringomyelia.

The arthropathies of syringomyelia so often closely simulate those of locomotor ataxia that they deserve special mention. They usually occur after the thirtieth year, and are more frequent in males than in females, since syringomyelia attacks two males for one female. Thus, of the 35 cases of joint-affections in syringomyelia analyzed by Graf,[3] 26 were males. Arthropathies occur in not less than 10 per cent. of the cases, and whereas in tabes from 75 (Schrotter) to 80 (Graf) per cent. of the affected joints are those of the lower limbs, in syringomyelia the articulations of the upper limbs are generally affected. Thus, of 51 joints affected in the 35 collected cases, 39 were in the upper and 12 in the lower limbs, a difference readily explained by the fact that in syringomyelia the cervical and upper dorsal portions of the cord are those chiefly affected. The affection occurs most frequently in persons who use their arms to excess, as tailors, butchers, etc.

---

[1] Amer. Journ. of the Med. Sci., 1875, p. 339.
[2] Julius Wolff : Deutsch. med. Zeit., March 15, 1888.
[3] Graf: Beiträge zur klin. Chir. ; Bd. x., Heft 3. Brit. Med. Journ., Sept. 9, 1893.

Pathology. The changes found in the structure of the various articulations resemble those seen in the hypertrophic form of tabetic articular lesions, and Charcot[1] has distinguished as for the tabetic arthropathies two varieties—the atrophic form, which is rare, and the hypertrophic form, which is common. The latter are vegetating arthropathies characterized by asthenic hyperæmia, interstitial atrophy, leading to erosion of the articular surfaces, fungous epiphyseal hypertrophy and the formation of bony vegetations.

The central lesion of syringomyelia is characterized essentially by the formation of a cavity with gliomatous walls in the gray-substance of the spinal cord, or a simple dilatation of the central canal, which, as Sinkler[2] suggests, is probably a congenital condition and due to defective closure of the tube. The cord may present through its whole length an epitome of the gliomatous process in all its various stages, as in the case of Lloyd.[3]

Symptoms. The joint lesions of syringomyelia commence early, precisely in advance of the muscular atrophy and sensory disturbances, or, what is more ordinary, they are preceded or accompanied by thermo-analgesia, trophic cutaneous troubles, amyatrophy, etc. The development is generally spontaneous, abrupt, without history of traumatism, and without constitutional disturbance, pyrexia, pain, or muscular spasm. The articular and peri-articular tissues rapidly infiltrate, but extensive exudation is less frequent than in tabetic arthropathies. The ligaments soften and relax, and the articulations may be moved laterally or dislocated with crepitation but without pain.

One or more joints may be simultaneously affected. The course of the arthropathy is more chronic than it is in locomotor ataxy, and the initial disease may last for twenty or twenty-five years. The coexistence of arthropathy with syringomyelia will, it is thought by Graf, be accounted for by the discovery of post-mortem indications of neuritis and atrophy of the peripheral nerves.

Syringomyelia, says Bruhl,[4] has its own proper symptomatology from which a clinical diagnosis may be made.

The prognosis is absolutely unfavorable, and the pathological changes are unaffected by complete and prolonged rest. No other therapeutical suggestion has been made.

---

[1] Le Progrès Med., April 29, 1893.
[2] Sinkler: International Clinics, vol. iii., third series, p. 131.
[3] Lloyd, J. H.: Trans. Royal Coll. of Phys., 1893, p. 73.
[4] Contrib. à l'Etude de la Syringomyélie, par le Dr. I. Bruhl. Paris, 1890.

CHAPTER XVIII.

FROM the *Silpi Sastri*, or Treatise on the Fine Arts, the earliest known Sanscrit manuscript, to the most modern treatises upon anthropometry, the asymmetry of the human body has been recognized, and the attempts have been made to deduce the exact proportions of the perfect human form. Everything tends to establish that the human type of to-day is identical with that deduced from observation of the most symmetrical ancient statues. M. Quetelet recognized the existence of a central or typical form of man, *l'homme moyen*, as the *mean* result of large numbers of actual measurements of living men.[1] Any deviation from this may be considered as asymmetry. Atrophy is most common, although unilateral hypertrophy does sometimes occur (Strümpell).

The medico-legal importance of this subject has of late received much attention, particularly as it concerns inequality of the lower extremity. Measurements have been made by Hunt, Cox, Wight, Cobbendeck, and Morton, with a general result of from one-eighth to one inch and five-eighths difference in length in the lower extremities of healthy individuals in a large majority of those examined. Roberts, Dwight, and Garson, as the result of observations made upon skeletons, found asymmetry in regard to length in the majority of instances. In some cases there is also an increase in the volume of the part. Broca and Paget have described such cases. Asymmetry of the upper extremeties has also been observed by Hartwig and Poncet.

Etiology. Various causes have been assigned, among which may be mentioned a neuropathic disturbance (Strümpell), a former hemiplegia (Bradford and Lovett), a premature synostosis of the epiphyseal cartilages (Nicoladoni), a pathological lengthening following traumatism (Goodman). One side, or the right, may be *ab initio*

---

[1] For a complete review of this subject see article on "Physical Development," by J. M. Keating and the writer. Keating's Cyclop. Dis. of Children, vol. iv. p. 241.

the stronger (Hunt), or it may be due to the passage of a pure arterial current to one side (Albrecht).

Symptoms. The inequality of the lower extremities produces lumbar and pelvic pain, lateral curvature, and a limp which may be confused with coxalgia or infantile paralysis.

Treatment. The treatment consists in restoring symmetry and equilibrium by artificial means.

CHAPTER XIX.

RICKETS.

RICKETS is a disease of infants, acquired through malnutrition, characterized by impaired nutrition and alterations in the growing bones, and terminating spontaneously after a longer or shorter period.

Synonyms. *Rachitis; Rhachitis; Morbus Anglicus; Morbus Puerilis; Articuli Duplicati.* French, *Rachitisme; Novure; Maladie Anglaise.* German, *Englische Krankheit; Doppelte Glieder; Zwei-wuchs.* Italian, *Rachitide.*

The term rickets here used, the one popularly employed for nearly three centuries, is derived from the Saxon "rick," meaning a hump or elevation, from a Dorsetshire verb "rucket," to breathe laboriously, as suggested by Trousseau, or from the Norman word "riquets," applied to deformed persons. For scientific purposes, the terms rhachitis or rachitis are more correct, both terms being derived by the celebrated anatomist Glisson by slightly altering the orthography and adding a Greek root ($\dot{\rho}\acute{\alpha}\chi\iota\varsigma$). The spelling "rhachitis" has the preference.

Etiology. The disease is universal, but is most common among the poor in the larger cities. It is much less frequent in the large cities of this country than in Berlin, Vienna, or London, where from 50 to 80 per cent. of all children at the clinics are more or less affected. In this country it is very common among the negroes, Italians, and Portuguese. It is said to be comparatively rare in Canada, Greece, and Scandinavia; and, according to A. S. Ashmead,[1] it does not exist in Japan. It is sometimes observed in the country, and is not infrequent among the children of the wealthy, where it often assumes the form known as acute rickets, in reality a manifestation of scurvy. The severest cases occur in winter, from hygienic causes rather than from seasonal variation. Fischl refers to the greater severity of cases occurring in the winter in Munich. The disease affects male and female children about equally. From

---

[1] N. Y. Med. Record, October 11, 1890.

the obscurity of the etiology of this affection several theories have arisen. These may be included under four heads: the mineral theory, the acid theory, the vaso-nervous theory, and the primary inflammatory theory.

1. *The mineral theory*, which regards deficiency in the supply of the mineral constituents as the cause, was held by Virchow, who believed that the essential lesion was a deficiency of calcareous salts, an opinion which Jenner also held, but which has not been subsequently corroborated.

2. *The acid theory*, that the normally calcified bones are deprived of their lime by some substance, probably an acid, as carbonic or lactic acid, in the blood, was suggested by Fourcroy, and like Banquo's troublesome shade, seems altogether indisposed to "down," so often has it reappeared in different forms in literature. The essential lesion appears to be the deficiency in the deposition of lime salts, with increased absorption in the bone already formed.

3. *The vaso-nervous theory*, that it is due to some morbid condition of the nervous system, was proposed by Mayow,[1] who asserted that it was caused by some "obstruction" in the spinal cord, an opinion reiterated by Hoffman and Allen, and favored by Parry, who suggested a lesion of the vasomotor system.

4. *The inflammatory theory*, that the cause is a primary chronic inflammation beginning in the bone-forming tissues, while not new, has the indorsement of the most recent authority.

Glisson attributed rickets to disturbed nutrition of arterial blood, and the changes in the long bones to excessive vascularity; Rindfleisch and Jacobi observed and commented upon the augmented vascular supply, and Kassowitz has asserted the primary nature of the osseous changes, an opinion which is now generally accepted.

Rickets is usually an acquired affection due to certain known predisposing and exciting causes, under the influence of which almost any child may have rhachitis. The exceptional instances are those in which the tubercular or syphilitic diathesis is present and antagonistic. Rickets and tuberculosis are considered exclusive, but tubercular meningitis is, in rare instances, associated with rickets. The identity of the tubercular diathesis and rickets was supported by all the older writers. There is, however, no direct relationship between the two affections. A phthisical parent may bear rickety children,

---

[1] Tractatus duo, alter de Respiratione, alter de Rachitide. Leyden, 1671.

or rickety children may become tubercular, but children with marked tubercular diathesis rarely, if ever, become rickety, nor is it common to find rickets in a family where the other children are tubercular. The belief in the identity of rickets and syphilis is as ancient as the history of medicine itself. Rickets may affect the offspring of syphilitic persons, but severe rickets does not occur in syphilitic children. The belief that rickets is only a manifestation of congenital syphilis (Parrot) is certainly incorrect. Syphilis may act as a predisposing cause to rickets by impairing the constitution. The anatomical lesions in the two conditions are quite distinct. Syphilitic bones very rarely, if ever, present the spongy tissue peculiar to rickets, and rhachitic bones never exhibit the multiple osteophytes of syphilis. Rickets has been regarded by Oppenheimer as an effect of malaria. There is no evidence that the disease is ever transmitted. Anything which impairs the general health or seriously interferes with the assimilative power may be considered as a predisposing cause of rickets, and improper food is the exciting cause. Want of sunlight, impure air, and insufficient exercise are important factors, but given a healthy mother with an abundance of milk, rickets may be escaped, no matter what the character of the surroundings. Malnutrition through the use of improper food is the most common exciting cause. Whatever unfits the mother's milk for the nourishment of her child requires consideration. Gold cannot buy wisdom, and wealth, through ignorance and indolence, vies with poverty, through need and necessity, in producing milk which is poor in quality and deficient in quantity. Jenner called attention to weakness, and especially anæmia, of the mother, and Tilbury Fox gave especial prominence to the influence of menstruation during lactation. Other discharges, especially leucorrhœa, are special agents in its production. Too early weaning is not so common a cause as too late weaning. Through some false ideas of preventing conception, or through absolute ignorance, children nursed from eighteen to twenty-eight months are particularly liable. The occurrence of pregnancy during lactation, and too frequent pregnancies, are likewise frequent causes. The possibility of the occurrence of rickets after any acute disease, particularly malarial fevers, should not be overlooked.

Four varieties of rickets may be distinguished :

1. Intra-uterine rhachitis.
2. Infantile rhachitis.

3. Adolescent rhachitis.

4. Senile rhachitis.

Since Ritter described and figured the first authentic specimen of congenital rickets, Winkler, Hink, Jacobi, Ballantyne, Gueniot, Shattuck, Henoch, Lewis Smith, and others, have recorded similar examples; Beduar believes it to be common, and Virchow refers to cases of the kind.

Two varieties of intra-uterine rickets are recognized by Spiegelberg and other writers—*fœtal* rickets and *congenital* rickets, to which

FIG. 155.

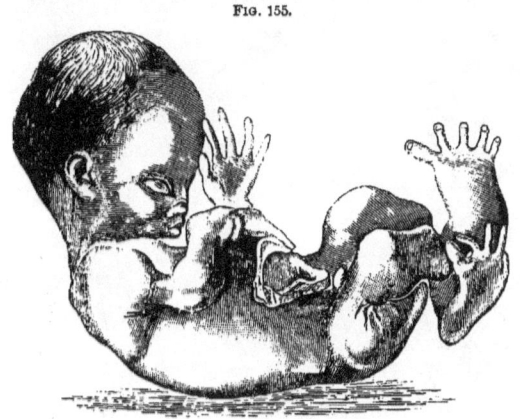

Intra-uterine Rickets. (BALLANTYNE.)

Winkler has respectively given the names of *micromelic rickets* and *annular rickets*. Indeed, slight rhachitic changes are far from uncommon in newborn infants whose parents have been living in bad hygienic surroundings during pregnancy, and Schwarz, out of 500 infants born at the Second Vienna Obstetric Clinic, found 80.6 per cent. with typical rhachitic changes. Doubt has been expressed as to the identity of the lesions found with the post-natal disease. Fœtal or micromelic rickets is in some obscure manner allied to the conditions known as cretinism and achrondroplasia (Parrot), and the opinion has been expressed by Barlow[1] and others that the cases of rickets which have been described are more properly to be regarded as examples of fœtal cretinism. Fœtal rickets appears in the later months of pregnancy and develops subsequent to birth, while achron-

---

[1] Cyclop. Diseases of Children, vol. ii. p. 258.

droplasia is an osteomyelitis of early intra-uterine development.
Porak[1] gives an excellent description of the latter condition.   That
they are cases of true rickets would seem to be established by the
specimen recently recorded by Ballantyne,[2] which likewise tends to
confirm the opinion of Spiegelberg, Lauro, and others, that the histological characters of intra- and extra-uterine rickets are identical,
by exhibiting in the same fœtus characters peculiar to both fœtal and
congenital rickets.

Infantile rickets (Fig. 156) seldom appears before the seventh or
ninth month, most frequently between the seventh and eighteenth

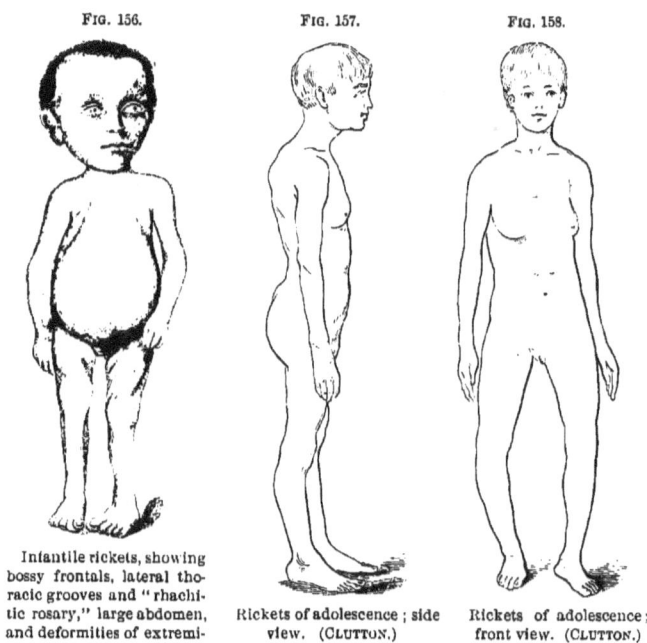

FIG. 156.   FIG. 157.   FIG. 158.

Infantile rickets, showing
bossy frontals, lateral thoracic grooves and "rhachi-
tic rosary," large abdomen,   Rickets of adolescence ; side   Rickets of adolescence ;
and deformities of extremi-   view. (CLUTTON.)   front view. (CLUTTON.)
ties.

month, and rarely after the second year.   In the 1876 cases collected
by Bradford and Lovett from various sources, 1541 occurred within
this period.

Adolescent rickets affects persons about puberty and is usually
associated with albuminuria.   In the cases reported by Clutton and

[1] La Pratique Médicale, Aug. 12, 1890.   [2] Edinburgh Med. Journ., June, 1890.

Drewett, the affection was identical with the infantile variety. (Fig. 157.) This form of rickets is fairly common, and the writer has observed it commencing as late as the eighteenth year. (Fig. 157.)

Senile rickets, first described by Reeves, occurs during adult and advanced life. Durham has collected 145 cases, most of which were associated with childbirth. Czerny, Scoutteten, Solly, Mosetig-Moorhof, Weinlechner, and others have reported cases. It is rarely met outside of the Rhine provinces, and is frequently confounded with *osteitis deformans.*

Pathology. The most important pathological lesions are in the bones, and especially in the epiphyseal junctures of the long bones and the ribs. The cartilage at these localities between the shaft and epiphysis, normally represented by two narrow bands one or two millimetres in thickness, is greatly thickened, from five to fifteen millimetres in width, reddish-gray, translucent, irregular in outline and softer, and the entire epiphysis is enlarged and softened. The periosteum is thickened and infiltrated, as is also the underlying bone structure, with spongoid, jelly-like fluid. The result of these changes is imperfect or delayed ossification, and the deposit of lime salts is arrested. The relation of organic to inorganic matter in the bones is very much increased, the diminution of calcareous salts being as low as 25 to 35 per cent. In the cranial bones large areas of imperfect ossification are met, giving rise later to areas of atrophy and premature hyperostosis. Later, the deposit of lime recommences and advances with undue rapidity, producing a condition of eburnation or petrifaction in the bones. Kassowitz, a leading authority to-day on the pathological anatomy of rickets, regards hyperæmia of all the bone-forming structures as the primary lesion. The spleen, liver, and sometimes the mesenteric glands, are enlarged. Catarrh of the alimentary and respiratory passages is usually present. The brain is usually hypertrophied, and, with the membranes, is unusually vascular. The voluntary muscles are pale, soft, and flabby, and the ligaments are soft and relaxed.

Symptoms. The symptoms of rickets are well marked and characteristic, and may for convenience be included under three divisions: 1. Incubation; 2. Deformation; and 3. Recovery.

Incubation. Preceding the period of bone change, there are certain characteristic symptoms which are liable to be overlooked. Loss of appetite, occasional vomiting, impaired digestion, flatulence, constipation alternating with diarrhœa, the stools being mucous, green, often frothy and extremely offensive, indicate disturbances of digestion which frequently precede but are not invariable precursors of this affection. The skin is pale and at first generally moist, afterward hot and dry; the flesh is plump, though flabby, the abdomen distended, and the anterior fontanelle depressed. Associated with these digestive disorders are two symptoms which are characteristic of the first stage—local sweatings and nocturnal fever. The local sweatings are chiefly confined to the head, neck, and upper part of the chest and back. These portions are cold and damp, in marked contrast to the rest of the skin, which is hot and dry. They occur usually during sleep, but may be induced by motion or exercise. Upon the forehead, face, and neck the perspiration appears in large, clear drops, and is often associated with a copious eruption of miliary vesicles. Associated with this local sweating the skin is hot, dry, and uncomfortable, leading to great restlessness and a disposition to kick the clothes off at night, even in the coldest weather. The pillow is thus often wet with perspiration, while the limbs are hot, dry, and uncovered.

Deformation. The commencement of morbid changes in the bones is marked by general hyperæsthesia or tenderness, first exhibited by signs of uneasiness when danced up and down ; the tenderness increases until the slightest involuntary movement causes pain, even the approach of persons produces fear and aversion, and the child prefers to remain quiet, motionless, and alone. The veins upon the head and scalp are enlarged and prominent, general hyperæmia of the cranium and scalp exists, and the carotid arteries and jugular veins are said to be disproportionately large. The digestive derangement continues; the stools consist of undigested food, whitish, curdy-looking, offensive mucus ; constipation is not uncommon, and diarrhœa and constipation frequently alternate; the urine is abundant and loaded with phosphates, the appetite is voracious, the abdomen is distended, the flesh is soft, flabby, and in some instances emaciated, and the skin pale.

These prodromal symptoms are prolonged into, and may even increase in severity during the stage of bone lesion. The changes in the bones which lead to deformity may be divided into three stages :

    1. The stage of congestion or invasion.

    2. The stage of softening or deformity.

    3. The stage of hardening or sclerosis.

In children in whom the disease has commenced at the beginning of teething, about the sixth or eighth month, the final stage is usually reached about the third year.

All the bones are simultaneously affected, but the order in which deformity usually appears is, first, the trunk; second, the head; third, the extremities.

Trunk. Among the first bony lesions are the changes in the epiphyseal junction of the ribs, forming the so-called rickety rosary, a series of bead-like enlargements easily felt beneath the skin. Important changes occur in the thorax. The transverse diameter being sometimes less than the antero-posterior, the bodies of the ribs are bent, forming a groove just outside the junction of the cartilages; the lower border of the thorax is pushed out by the large liver and spleen, forming a transverse groove, the so-called Harrison's groove, passing outward just below the fold of the pectoral muscle. The sternum projects forward like the prow of a ship, forming the so-called pigeon- or chicken-breast. Hypertrophy of the tonsils is frequently associated, and these changes in the thorax have been ascribed to them and other causes which interfere with the free inspiration of air, but according to some authors the two are the results of one cause. Changes also occur in the spinal column; an antero-posterior curve, a simple exaggeration of the normal curves through weakness and long-continued sedentary position, being most common. (Fig. 159.) Lateral curvature is also a common deformity, and lordosis, as a compensatory affection, sometimes occurs and may persist. The appearance of lordosis is increased by the large size of the abdomen. Changes in the pelvis occur from the sedentary position and from the pressure of the heads of the thigh bones. Illustrations of this are common in all obstetrical treatises. The abdomen is greatly enlarged from several causes—contraction and depression of the diaphragm, from the diminished capacity of the thorax, increased shallowness of the pelvis, tympany, and albuminoid degeneration of the liver and spleen.

Head. The head of a rickety child is large and misshapen, and the fontanelles, particularly the anterior, which normally should close about the eighteenth month, remain open until the end of the third or fourth year and ossification may not be complete until the termination of the sixth or even the ninth year. There are two varieties of skull typical of rickets—the oblong head, such as is met with in negro children, and the square head, the more common variety, which seen from above is rectangular, the *caput quadratum.* The forehead is high, the frontal and parietal protuberances prominent, and the comparative smallness of the face is characteristic. Another peculiarity is imperfect ossification, particularly in the parieto-occipital regions, the so-called soft occiput, or *cranio tabes.* This deficiency of ossification is preceded by thickening and softening of the bone. They are soft, round or oval spots, from one to twenty-five in number, situated just within the sutural margins, and give the sensation of an orifice closed by cartridge paper. The association of these spots in some cases with syphilis—in 47 of the 100 cases recently studied by George Carpenter—renders their exact relation to rickets somewhat

Fig. 159.

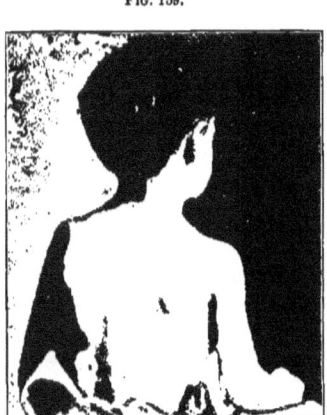

Fig. 160.

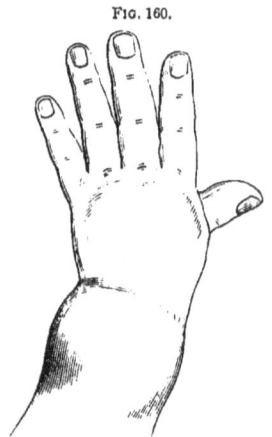

Photograph of case of rhachitic spine.

Rhachitic hand, showing the epiphyseal enlargement. (BRADFORD and LOVETT.)

doubtful. The occurrence of a systolic murmur over the anterior fontanelle was believed by Fisher, who first described it, to be peculiar to rickets, but the fallacy of this has been proven by Osler. The normal process of dentition is ordinarily retarded, and the eruption is irregular in time of appearance and position, and attended with great pain. Teeth which have appeared decay early, and at eighteen months or two years very few teeth may remain. In rare instances the teeth may present the characteristics of the so-called Hutchinson's teeth. Cerebral development in rickets is usually retarded; the intellect is dull.

Extremities. The alterations in the long bones occur early, commencing first and affecting most seriously those parts which have a thin covering of soft parts, such as the wrist (Fig. 160), elbow, knee, and ankle. The increase in size of the epiphyses is real, and the length and circumference of the shaft are also less than normal. From softening and imperfect ossification of the bones, any or all of the long bones may be bent or twisted. These deformities have been ascribed to muscular action, gravity, the weight of the body, and atmospheric resistance. Most of these deformities may be accounted for by the weight of the body, gravity, and the peculiar position occupied by the child. Flat-foot is present in almost all cases of rickets. It occurs usually before the seventh year, either alone or associated with knock-knee. Associated in certain cases with the deformity is a pseudo-paraplegia, known as Parrot's disease, which prevents the child from walking, and sometimes from standing. The muscles are atrophied and weak, but there is no permanent nervous lesion. The electrical reactions are normal, the reflexes are not exaggerated, and the pseudo-paralysis is due to a periosteal tenderness of the muscular insertions and muscular weakness.

Complications. Children affected with marked rickets are especially predisposed to certain diseases, to one of which death is usually directly attributed. These include laryngismus stridulus, convulsions, chronic hydrocephalus, bronchitis, and diarrhœa.

Laryngismus stridulus is more frequent in rickety children, though not invariably associated with this disease. Associated with it, and perhaps due, as Gee suggests, to the same condition of general malnutrition, general convulsive paroxysms are frequent and are often combined. Of 50 cases of laryngismus stridulus observed by Gee, 48 were rickety and 19 had general convulsions; and Jenner remarked the close relationship between infantile rickets and convulsions, particularly in those which occur after the sixth month.

Chronic hydrocephalus as a complication is most common, according to Merei, between the eighth and eighteenth months.

Bronchitis is extremely fatal in rickets, owing largely to the deformity and softening of the ribs.

Diarrhœa is frequent and dangerous on account of the extreme sensitiveness to atmospheric changes and the unhealthy condition of the alimentary canal.

Recovery. Though death often occurs from some intercurrent affection, the disease frequently terminates favorably, the recovery being perfect provided there is no thoracic complication. All the symptoms subside and gradually disappear, the bones become exceedingly thick and very hard, the muscles short, thick, and very firm; the epiphyseal enlargements diminish, and serious deformities of the long bones and throat remarkably diminish, and may entirely disappear within a few months. The growth, in length, usually remains stunted, the mean height of the body being, according to v. Rittershain, at a given age, never attained. Such patients may subsequently become marvels of endurance and strength[1] and live to a great age.[2]

The arrest of growth and deformity in the pelvis is important from a surgical aspect. In Cæsarean section the mortality of the operation is said[3] to be influenced by the rhachitic diathesis, and in lithotomy, in males, the contraction of the outlet of the pelvis may interfere with operations upon children, as in the case of Sir Henry Thompson.[4] In some cases the changes in the bones consolidate rapidly, leaving the bones thick and very hard, with great permanent deformity. This is particularly true where spontaneous cures result without treatment.

Prognosis. The tendency of uncomplicated rickets is to spontaneous recovery after a variable period. The danger lies in the complications—degeneration of the viscera, catarrhal affections of the respiratory passages, diarrhœa, hydrocephalus, and, in very rare cases, laryngismus stridulus. Under efficient treatment the disease is very amenable, and the prognosis as to life is good, unless some complications arise. If the disease be early arrested the bony deformities and the kyphosis and lordosis all tend to diminish, but the full stature is seldom attained by adults who have been affected in early life. The dangers attending operative procedures for rhachitic deformities have been reduced to a minimum.

Treatment. The treatment of rickets should be hygienic, dietetic, and medicinal.

The hygienic treatment should include the selection of a proper diet and the improvement of the entire surroundings of the child. A daily bath in tepid salt-water, with vigorous rubbing, should be given, and the child, warmly clad in warm woollen clothing and carefully wrapped, should spend the greater part of the day in the fresh air and sunshine on all except windy days. A flannel bandage about the abdomen should always be worn, summer and winter.

[1] Hunter, Lawrence: Lancet, June 5, 1829. Sir Charles Bell: Lancet, March 15, 1834.
[2] Stanley: Diseases of the Bones, London, 1849, p. 230.
[3] Harris: Amer. Journ. of Obstet., Nov. 1871.
[4] Med. Times and Gazette, Dec. 5, 1863.

Frequent pregnancies and nursing a child during pregnancy are important etiological factors to be overcome. If a nursing woman become pregnant, or the mother is unhealthy, or cannot, from other cause, nurse the child, a wet-nurse should be provided, or the child be artificially fed. If the child be bottle-fed the most rigid system of cleanliness must be enforced to maintain asepsis of the food. A sterilizer of the most improved pattern is important, and a bottle, smooth inside and easily cleaned, with frequent changes of plain rubber nipples, will best serve the purpose. If the patient reside in a cold, damp locality, a change to the dry, bracing air of mountain, country, and especially seashore, is often of great benefit. Seashore hospitals, established both in this country and abroad, are excellent resorts for rhachitic children, and the "day nurseries" in our large cities accomplish much good by the proper feeding and care of the children. The bowels should be regulated by castor oil, olive oil, "rhubarb and soda mixture," or compound licorice powder. The diet of a rhachitic child is most important. For very young children, artificially fed, cow's milk will, properly diluted to suit the age, constitute the chief food. The proportion, to begin with, should be two-thirds water, increased gradually to half-and-half from one to two months of age; at five or six months two-thirds milk, gradually giving the milk pure after twelve or fourteen months. The stools should be carefully examined for undigested casein. The quantity should be a pint during twenty-four hours for an infant, gradually increased. To this, in older children, barley-water and thin, strained oatmeal gruel may be added. For full details of proper diet, the reader should consult works upon diseases of children.[1]

Good fresh butter spread upon home-made bread may be early employed, and fish should be largely used by nursing mothers. Indeed, Ashmead ascribes the absence of rickets in Japan, among other reasons, to the large use of fish, crustaceal and iodized seaweed, oils of fish, blubber of whale, and especially loach, by nursing women, the children being all breast-fed and suckled for a very long period, the Japanese women having also an enormously large supply of milk. The milk of the average American woman is not rich after twelve months; notwithstanding which fact, the writer has frequently had

---

[1] Meigs and Pepper: Dis. of Children, article "Thrush," Keating's Cyclop. Dis. of Children, vol. ii, p. 257.

to order the discontinuance of nursing at the twentieth to the twenty-seventh month, in some cases where pregnancy also existed.

Medicinal treatment is secondary in importance to the foregoing. The general condition should be improved by cod-liver oil, which is considered by many a specific, and which has long been a common remedy on the shores of the Baltic.[1] It may be given pure, in tea-spoonful doses, or, what the writer prefers, in an emulsion with the lacto-phosphate of lime, as recommended by Parry and Lewis Smith. The latter formula is

R.—Ol. morrhuæ . . . . ʒvj.
 Syr. calcis lacto-phosphatis } āā . . ʒiij.—M.
 Aq. calcis
Sig.—ʒj or ʒij three times a day.

Another excellent combination of the writer's is with maltine, or maltine and syr. calcis lacto-phosphatis. It may also be rubbed into the legs, arms, and abdomen at bedtime. Syrup of hypophosphites is another excellent remedy, and may be used in summer to alternate with the oil. Iron is highly recommended by all authorities, and may be given in the form of hypophosphite, the wine of iron, the citrate of iron and quinine, or preferably, the syrup of the iodide of iron. Of late phosphorus has been recommended by Kassowitz and also advised by Jacobi, and reports are favorable in this, as in all wasting diseases of the bones. It is usually given to children in $\frac{1}{120}$ grain doses, dissolved in oil, three times a day. A more palatable form, recommended by Berg,[2] is Thompson's solution (Hazard & Hazard, New York), the formula for which is as follows:

R.—Phosphori . . . . gr. j.
 Alcohol. absolut. . . . . . . mcccl.
 Spt. menth. pip. . . . . mx.
 Glycerinæ . . . . fʒij.
Sig.—For a child two to four years, 6 minims, t. i. d., increased to 10 minims. Strength gr. 1/20 to ʒj.

Hartwig has advised intermissions in the treatment for a month at a time to avoid degeneration of the liver, a danger suggested by Wagner. The results of phosphorus treatment are contradictory, and this form of treatment may still be considered *sub judice*. For diarrhœa, which is met with in most cases, if the cod-liver oil does not diminish the number of stools, drop doses of tincture of opium in some aromatic water is a favorite prescription.

1 London Lancet, 1851, vol. ii. p. 546. 2 N. Y. Med. Rec., Nov. 16, 1889.

CHAPTER XX.

KNOCK-KNEE.

KNOCK-KNEE, or genu valgum, is that deformity in which the knee is thrown inside a perpendicular line drawn from the head of the femur to a point midway between the malleoli; in other words, where the bones of the leg form an abnormal angle opening outward with the bones of the thigh—an abduction contracture.

Synonyms. English, *In-knee.* French, *Genou en dans; Genou Cagneux.* German, *X-Bein; Knickbein; Ziegenbein; Bäckerbein; Knieng; Kniebohrer; Schemmelbein.* Italian, *Ginocchio Torto all' Indentro.* Latin, *Genu Introrsum.* Greek, *Entogonyancon.*

Occurrence. The deformity is frequently met with in surgical practice, but is not so common as bow-legs. Thus, of 6400 cases of general surgical disease in children treated at the New York Orthopedic Hospital and Dispensary, there were 270 cases of knock-knee and 400 cases of bow-legs. In 2292 cases of general orthopedic affections treated at the Orthopedic Dispensary of the Hospital of the University of Pennsylvania, 42 were cases of knock-knee, and 121 were cases of bow-legs. Boys appear to be more frequently affected than girls.

Congenital cases, though rare, do occur, and both knock-knee and bow-legs have been observed in the newborn. The extremely rare variety of genu valgum which follows on congenital luxation of the patella (Middledorpf, Maas), belongs under this head.

The deformity usually appears when the child begins to walk, between the ages of two and four—*genu valgum rhachiticum infantum;* or about adolescence, between the ages of twelve and eighteen—*genu valgum adolescentium.*

The congenital form is usually the expression of general rickets.

The acquired forms present several varieties, which may be grouped into: 1. Rhachitic. 2. Atonic. 3. Paralytic. 4. Arthritic; and 5. Traumatic.

1. The rhachitic variety includes almost all the cases occurring

19

during the first period, and Mikulicz and others would include some of those also which occur at puberty, and consider the process a local rhachitic one—a form of "latent rickets."

Rickets softens the bones, weakens the muscles and relaxes the ligaments (Kassowitz), and the superincumbent weight of the body accomplishes the rest. Genu valgum in these cases is simply the local manifestation of a constitutional diathesis.

2. The atonic or statical variety affects individuals of feeble physique about the time of puberty—*genu valgum staticum sive adolescentium*—whose occupations compel them to stand most of the time. Carpenters, waiters, cooks, young bricklayers, and especially bakers, who work in a warm, moist atmosphere, and carry heavy loads of bread, are most liable to be affected. They exhibit no evidence of rickets, but suffer from relaxation of the ligaments and muscles.

The production of this form of genu valgum is usually the result of faulty positions assumed from weakness or fatigue.

The normal human individual stands erect with a certain amount of knock-knee, the femurs form an angle with each other of 15 degrees or more, and an imaginary line drawn from the head of the femur falls outside the centre of the knee-joint. To compensate for this the normal internal condyle of the femur is from one-quarter (Clark) to one-half inch (Holden) longer than the external.

Children and adults of feeble physique instinctively assume a valgoid position with the knees extended and the feet everted and separated, the so-called "attitude of rest," a position in which ligamentous is substituted for muscular support, and the limbs placed in a position most favorable for the production of this deformity. Flat-foot is often associated in this form as well as in the rhachitic, sometimes as causative, as a secondary condition, or both may result from the same faulty attitude.

Reeves has suggested that asymmetry may be an important factor in the genesis of this form of knock-knee.

3. The paralytic varieties are met with in connection with poliomyelitis anterior (infantile spinal paralysis) and in spastic paraplegia.

Trophic disturbances affect the nutrition of the bones, ligaments, and muscles. The bones are more curved, thinner, and softer than normal; the ligaments are relaxed, and the muscles have their equilibrium destroyed, as Volkmann has shown, not from loss of muscular antagonism, but from growth while the part remains in an abnormal position.

4. The arthritic variety includes the few cases which result from destructive disease of the joint, the so-called *genu valgum inflammatorium*, from osteitis, rheumatoid arthritis, osteomalacia, osteoarthritis, etc.

5. The traumatic form occurs from fractures of condyles of the femur or of the articular facets of the tibia, and in rare cases from over-correction after osteotomy for genu varum.[1]

The chief theories which have been advocated to explain this deformity may be included under three heads : 1. The ligamentous theory. 2. The muscular theory. 3. The osseous theory.

The ligamentous theory considered either that the internal lateral ligament was primarily relaxed, permitting lateral and downward hypertrophy of the internal condyle of the femur (Stromeyer, Malgaigne, Guérin, Blasius) ; or that the external lateral ligament was primarily shortened, producing pressure atrophy or deficiency of growth of the outer condyle. According to Roberts, the *internal* ligaments of the joint, principally the crucial, are largely concerned as causative agents.

The muscular theory assumes either a primary shortening of the biceps, popliteus, and tensor vaginæ femoris muscles, or a primary relaxation of these same structures (Bonnet, Verneuil, Duchenne, Guérin, Little, Brodhurst, Adams).

The osseous theory assumes a primary rhachitic, inflammatory, or other osseous changes in the epiphysis or lower portion of the femur or the upper part of the tibia, producing hypertrophy of the inner portion of the joint, with or without atrophy of the outer portion (Ogston, Hueter, Annandale, Macewen, Chiene, Gosselin, Tillaux).

The defective growth of the external condyle has been ascribed by Gosselin and Ollier to premature inflammatory synostosis of the outer part of the epiphyseal cartilage from excess of pressure. Malnutrition of the epiphyses, the result of central changes having their expression in the epiphyseal cartilages of the knee-joint, has been suggested by Roberts as a cause.

These theories have been formulated to explain the pathological findings in genu valgum, but in the majority of instances no one theory will account for the production of this affection. In many cases rhachitis is the sole cause ; in a large number the statical conditions, with or without rickets, are causative ; in others paralysis,

---

[1] *Vide* Bradford and Lovett, loc. cit., p. 652.

local disease, or traumatism, are etiological factors. In all cases, after the condition is once established, the superincumbent weight of the body is an important factor in increasing the deformity.

Pathology. The morbid anatomy will depend upon the stage of the affection and the degree of deformity.

In the rhachitic form the most important changes are in the bones forming the knee-joint.

The elongation of the internal condyle is both apparent and real (Guéniot, Santi, Lannelongue, Chiari); the external condyle is atrophied and sclerosed, and the entire lower epiphysis of the femur is broadened, shortened, and obliquely placed upon the shaft, from the lengthening of the inner side of the lower part of the diaphysis.

In some cases the lower epiphysis of the femur is twisted or rotated out (Volkmann), while in others the upper epiphysis of the tibia is oblique and rotated generally outward, in rare cases inward, and the femur is apparently normal (Noble Smith). The shafts of the femur and tibia are bent above and below the joint, and in some cases at the upper part of the shaft of the femur and the lower part of the shaft of the tibia.

The density of the osseous structure depends upon the stage of the morbid process, being a little softer than normal bones during the first stage (stage of vascularity), as soft as cheese during the second stage (stage of softening), and as hard as ivory in the third stage (stage of consolidation).

The articular cartilages are hypertrophied on the outer side and atrophied upon the inner side.

The internal lateral ligament is elongated and relaxed and hypertrophied. The external lateral ligament is contracted, and in some cases the crucial ligaments are atrophied or entirely absent (Lannelongue).

The muscles upon the outer side are shortened, and those upon the inner side are relaxed and elongated.

In the paralytic cases the primary changes are muscular, the action of which has already been described.

In the pathological and traumatic cases the changes are local and primarily osseous, the other structures becoming subsequently involved.

Symptoms. The symptoms of genu valgum are the deformity about the knee, the peculiar gait in walking, and the secondary deformities which complicate this affection. In standing, the knees are more or less unduly prominent upon the inner aspect, the leg projects outward, and the feet are separated to a varying degree. The in-knee should be examined in the fully extended erect position, or better, in the extended recumbent position. Preternatural lateral mobility of the articulation is characteristic. Pain on the inner side is not constant, but when present suggests a local inflammatory process. Flexion of the knee causes almost entire disappearance of the angular deformity, a peculiarity explained by two factors: the obliquity of the articular surface of the condyles and consequent oblique axis of rotation, as in a Charnier joint obliquely placed (Bosch, Reeves); and also the outward rotation of the femur upon its own axis (Albert, Mikulicz). A simpler and more satisfactory explanation is found in the fact that the increase in the internal condyle is only in length and not antero-posteriorly, and that in flexion the facets of the tibia come in contact with the posterior normal condylar surface (Guéniot).

In unilateral cases obliquity of the pelvis downward upon the affected side, and flexion of the thigh of the sound side, diminishes the inequality of the limbs and modifies what would otherwise be a limp.

In bilateral cases the feet are widely separated, the knees slightly flexed to prevent them from striking, and with each step the knee deformity is increased, producing a mild half-jerking, half-rolling gait that is characteristic. Flatfoot, lateral curvature, hyperextension of the knee (back-knee) are associated as secondary or coincident deformities dependent upon the same cause.

FIG. 161.

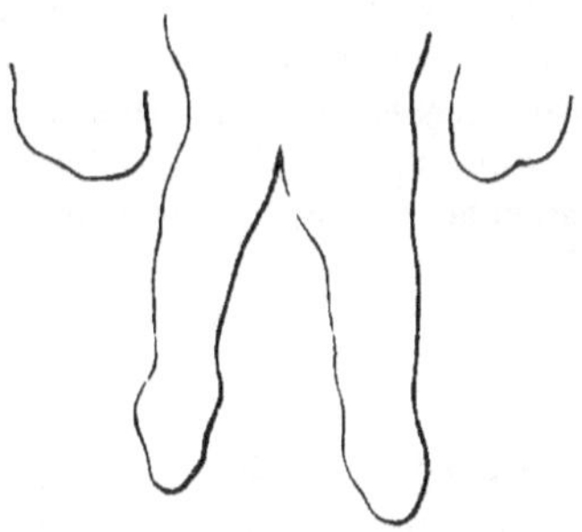

Tracings of a case of knock-knee, with outlines of condyles. (BRADFORD and LOVETT.)

In rare and severe cases, in persons whose muscles are not weak, a compensatory supination or adduction of the foot may develop, and, notwithstanding the oblique position of the leg, the sole may be set flat upon the ground. (Schreiber.)

In single knock-knee a compensatory or accommodative bow-leg of the opposite leg may occur. Methods for estimating the amount of deformity have been proposed by Mikulicz, Roberts, Schreiber, and Reeves. The simplest and best method of recording the deformity consists in making an outline tracing of both limbs with the patient seated with the limbs extended upon a sheet of paper (Fig. 161), or a lead tracing of the inner side of each limb may be made.

Diagnosis. The diagnostic points have already been given under the symptoms. It is necessary also to distinguish the varieties. In children the majority will be found to be rhachitic, and during adolescence, while the majority are statical, the rhachitic diathesis may often be discovered. In paralytic knock-knee the associated nervous lesions distinguish the cause. In severe inflammatory cases the local symptoms of tumor albus will distinguish, and in the traumatic cases the history of fracture or operation and the osseous changes will render the diagnosis clear.

Prognosis. Parents are frequently assured that children will "grow out of" this deformity, but such is only the truth in the mild degrees. In moderate and severe cases the probabilities are that the affection will remain stationary or grow progressively worse unless treated. Robust health, strong muscles, and recovery from the rhachitic diathesis are all favorable to such a termination, but the large number of adults seen in public places who are knock-kneed would seem to prove conclusively that many never recover. Such individuals do not apply at clinics because the deformity has practically caused them no serious inconvenience.

Since the introduction of antiseptic osteotomy cases requiring operation are uniformly successful. In arthritis deformans the prognosis is unfavorable, and the limb may become useless.

Treatment. The treatment of knock-knee may be considered under three heads: 1. Hygienic; 2. Mechanical; 3. Operative.

Hygienic treatment. The hygienic method of treatment is intended to assist Nature's efforts to correct the deformity, and is most valuable in infantile rhachitic cases and in statical cases occurring during adolescence. The constitutional treatment for rickets, or general tonic treatment, should be begun, and every effort should be made to improve the hygienic surroundings of the patient. General gymnastic treatment is of the highest importance, and a change to country or mountain air will often work marvels. The entire lower extremities should be rubbed with bathing whiskey, and manipulated every night before retiring. To accomplish the most good the limb should be grasped above and below the knee and the thumbs be applied upon the inner side of the knee. (Fig. 162.) In this position strong traction should be made with the hands while firm pressure is exerted upon the prominent in-knee. This last may be applied after the child has fallen to sleep. The pressure should be firm, forcible,

FIG. 162.

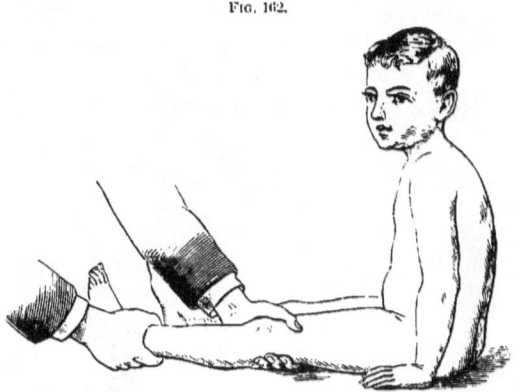

Manual correction of knock-knee. (HOFFA.)

repeated several times for a few seconds each time, and should not be severe enough to awaken the child or make it cry. This manipulation is of the greatest value, with or without the use of apparatus.

Mechanical treatment. The treatment by apparatus is employed to limit the progress of the deformity by taking the weight off the limbs, or to gradually correct the deformity by making counterpressure against the internal condyle. Two methods are employed: that which confines the patient to bed, and that which encourages locomotion. The former has been almost entirely abandoned in this country. Apparatus is usually employed before the end of the third year, when, in rhachitic cases, the stage of hardening usually occurs, though cures have also been affected in adolescents. The condition of the bones can usually be determined by the age, and also by the elastic or springy feel when gradual manual pressure is applied. In cases in which doubt exists, mechanical treatment can be tried for several months before resorting to operation, or if time be important an exploratory incision may be made and the bone be bored into with the cog-wheel drill of Colin, of Paris, as suggested by A. S. Roberts. In the sclerotic stage bones bore like ivory, and mechanical treatment is useless; in the soft stage the bones are entered with as little resistance as muscular tissue, and operative treatment is unnecessary. Plaster-of-Paris furnishes the simplest and cheapest mechanical appliance.

Hueter advocates retaining the knee in a flexed position for a long period by means of a plaster cast, but König, Waitz, and others failed to obtain good results by its use. Heine employs a plaster bandage from which an elliptical piece has been removed from the inner side of the knee, and with a splint upon the outer side makes traction. (Fig. 163.) German writers speak favorably of it. Mikulicz

FIG. 163.   FIG. 164.

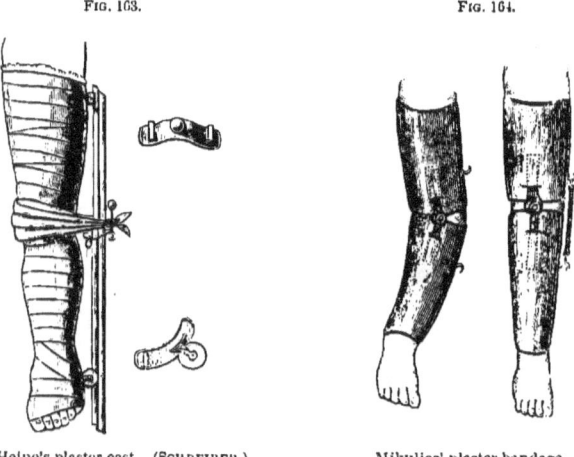

Heine's plaster cast. (SCHREIBER.)   Mikulicz' plaster bandage. (SCHREIBER.)

(Fig. 164), Vogt (Fig. 165), Billroth, and others incorporate metal hinges in the plaster dressing, and after removing a section from the outer and inner sides of the cast, apply elastic traction by means of rubber bands. These cheap methods are all useful where better apparatus cannot be secured. It is unnecessary to describe the various forms of appliances devised for the relief of this deformity. The principle of all are identical, the application of counter-pressure upon the hypertrophied internal condyle, but the method of application is different. (See Fig. 166.)

Constant pressure is impossible to obtain, and attempts to produce it result in the application of intermittent pressure, an undesirable factor, since pathologists agree that while "constant pressure on a part produces absorption, occasional pressure, especially if combined

with friction, produces thickening or hypertrophy."[1] Elastic traction
or pressure produces absorption and destruction of bone—a fact well
illustrated in aneurism of the aorta. (See Figs. 167 and 168.)

FIG. 165.

Vogt's plaster bandage with elastic
traction. (SCHREIBER.)

FIG. 166.

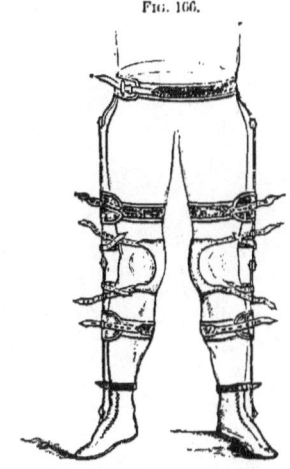

Apparatus for knock-knee. (ERICHSEN.)

FIG. 167.

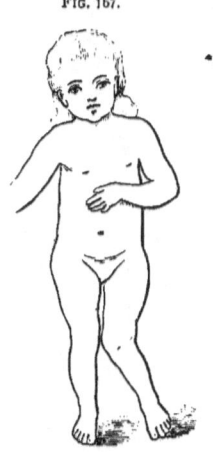

Before application of elastic
traction. (SCHREIBER.)

FIG. 168.

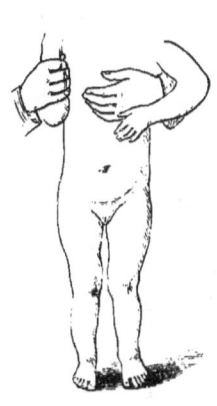

Result of application of elastic
traction. (SCHREIBER.)

[1] Sir James Paget: Lectures on Pathology.

Upon this principle of elastic traction, A. S. Roberts[1] has constructed an efficient brace (Fig. 169) to be worn at night as a substitute for the retention braces to be worn during the day. It consists of a tempered steel upright applied to the external aspect of

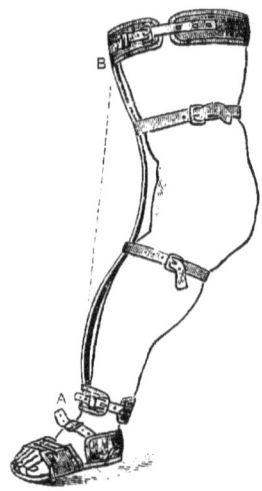

FIG. 169.

Roberts' elastic traction knock-knee brace.

the limb, extending from a foot-piece well up the thigh, and secured just above and just below the knee-joint so that in its efforts to straighten itself it will make continuous elastic pressure upon the hypertrophied internal condyle. The retention braces, which are to be worn during the day, are most effective. They consist of two steel uprights carried well up the thigh, attached by a stirrup to the shoe and to a band above and below the knee, jointed at the knee and ankle and having at these two points pressure-pads which make direct pressure against a long rectangular pad attached upon the outer upright about the middle of the leg. They must be fitted to the deformity with bending irons, and as the deformity diminishes they are to be straightened and the pressure increased. In severe cases these braces can be rendered still more efficient by fixing the knee-joint.

Manipulation of the limb as described, the retention day brace, and the elastic-traction night brace furnish together a plan of treatment that has proved very curative in the hands of the writer.

Operative treatment. The surgical procedures now employed for the relief of this deformity may be included under the following heads :

    1. Tenotomy.
    2. Forcible manual straightening.
    3. Osteotomy.
    4. Osteoclasis.

1. Tenotomy of the biceps and division of the ilio-tibial band and the external lateral ligament have been performed for the relief of

---

[1] Medical News, February 4, 1888.

this affection. These operations are now only employed as adjuncts to other methods in paralytic and arthritic cases.

2. Forcible manual straightening, or *rédressement brusque*, proposed by Delore and performed by Delore, Ormsby, Fochier, Paoli, and others in over 300 cases, has not been favorably received in this country. Delore has recorded one death from scarlatina, and Tillaux one from pyæmia. The operation consists in forcible manual reduction of the deformity, the limb being bent laterally or over-extended until it snaps and gives way; the lesions being rupture of the external lateral ligament, separation of the external condyle, and fracture into the joint. The bad results of the operation are said to be arthritis with effusion, severe periostitis, and necrosis, and a weak and lax condition of the joint. The operation is rough and unsurgical, is not entirely safe, and should be abandoned for other methods. In some cases of severe knock-knee, in the stage of softening, rapid manual reduction may with benefit be performed without the occurrence of any of the lesions before referred to.

3. Osteotomy for the relief of badly-united fracture has been performed since the time of Hippocrates. Although Rhea Barton, of this city, performed osteotomy for hip ankylosis as early as 1826, and Malgaigne suggested subcutaneous osteotomy in 1847, it was not until 1851 that Meyer, of Würzburg, first performed osteotomy for genu valgum. Operations upon other articulations became more numerous, and were performed in this country by Pancoast, Sayre, and Brainard.

Encouraged by the antiseptic method, Annandale in 1875 excised a portion of the condyles for genu valgum, and in the same year antiseptic osteotomy was introduced by Volkmann, of Halle, for ankylosis of the knee.

In 1876 Ogston, Jr., of Aberdeen, operated for genu varum through a small wound, using a saw, and was followed the same year by Schede, of Berlin.

In 1878 Macewen introduced antiseptic supra-condyloid osteotomy, which removed the dangers attendant upon opening the knee-joint, which had been the objection in all previous operations. In 1879 Reeves introduced his extra-articular condylotomy, a modification of Ogston's operation, the object of which was to minimize the danger of opening the joint and to loosen and properly replace the displaced condyle. The lines of the different operations are given in the cut. (Fig. 170.) The superiority of the Macewen operation is admitted by all, and except in rare cases it is now performed in this country to the exclusion of all others.

The operation of osteotomy consists of a subcutaneous division of part of a bone with a chisel or saw, and fracture of the undivided portion. Of the different operations the writer's personal preference is for that of Macewen, and this and its modifications is the only one that will be described.

FIG. 170.   FIG. 171.

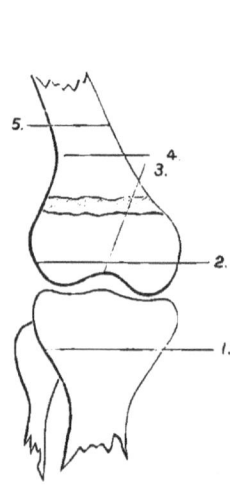

1. Mayer, Billroth, Schede; 2. Annandale; 3. Ogston,   Macewen's osteotomes.
Reeves, Chiene; 4. Macewen; 5. Taylor.

The osteotome, or bone-chisel, is a bone-knife and should be employed as such. (Fig. 171.) Different widths should be used, the temper should be midway between the temper of a wood and cold chisel, the point should gradually taper so as to avoid a shoulder, and the side should be marked in half or quarter inches to show how deeply the cutting edge has penetrated. The instrument should be firmly grasped with the left hand with the thumb against the head to steady it. (Fig. 172.) An ordinary carpenter's mallet answers every purpose as well as the more expensive instruments.

The Macewen operation is performed as follows: The limb having

been sterilized at the point of election for the introduction of the osteotome, the limb is flexed as suggested by Poore and placed upon a sand-
bag, and a longitudinal incision is made two fingerbreadths above
the internal condyle of the femur upon the inner aspect of the thigh.
The osteotome is introduced down to the bone, turned at right angles
to the long axis of the shaft, and with a few blows of the mallet the
bone is nearly divided. After each blow the osteotome must be
moved from side to side to prevent wedging, and the deeper layers
of bone should be cut in a fan-like manner. When the section is
nearly completed, as indicated by the depth to which the chisel has

FIG. 172.

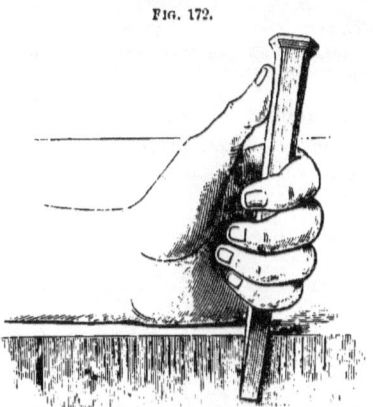

Position of hand in holding osteotome. (JONES.)

disappeared, the instrument is withdrawn and the fracture is completed with the hands. The limb is to be at once set in a slightly
over-corrected position, an antiseptic dressing is applied, and the
limb is incased in a plaster-of-Paris bandage and held in the corrected position until it hardens. The dressing need not be removed for three weeks, unless there be elevation of temperature,
pain, or the dressing becomes fœtid. In six weeks the patient may
be permitted to stand with the casts on, or light retention braces may
then be fitted. (See Fig. 173.)

Under full antiseptic precautions the dangers attending the operation are slight.

Hemorrhage is seldom of any moment, but the anastomotica magna
has been wounded (Marsh, Gibney) and the popliteal has required

ligature (McGill). Fatal hemorrhage has occurred,[1] and amputation for gangrene has been performed by Langton.[2]

FIG. 173.

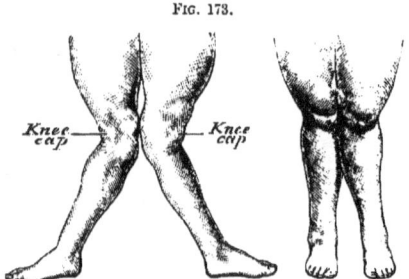

Appearance of limbs before and after Macewen's operation in a patient aged thirty-three. The legs and feet rotated outward and abducted. (BRYANT.)

As compared with other operations of the same character and magnitude, the number of hemorrhages have been very few. Thus, in 525 operations by Ogston's operation there were thirteen severe hemorrhages, while in 580 by Macewen's operation there were only two.

FIG. 174.   FIG. 175.

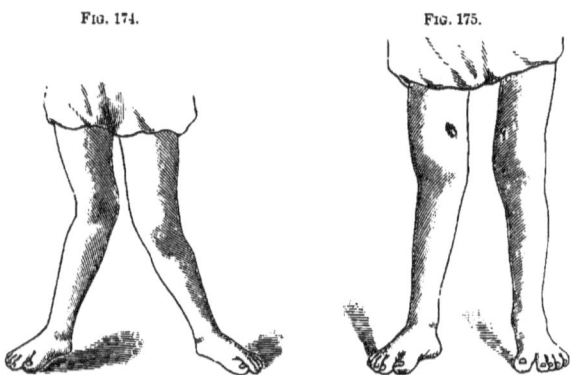

Result of Macewen's operation for knock-knee. (ASHHURST.)

According to Macewen these accidents occur from allowing the instrument to slip, by using too broad an osteotome, and by not cutting the posterior part of the bone with the chisel pointed forward and outward, but allowing the chisel to point backward. Some

[1] Medical News, November 1, 1884.   [2] Lancet, March 29, 1884.

operators prefer to perform the operation upon the outer side (Mac-Cormac, Brenner), and Hahn advocates its performance on both the outer and inner side.

FIG. 176.

Unilateral knock-knee before osteotomy. (HOFFA.)

FIG. 177.

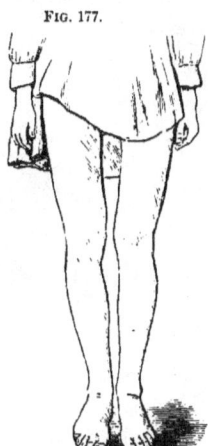

Unilateral knock-knee after osteotomy. (HOFFA.)

FIG. 178.

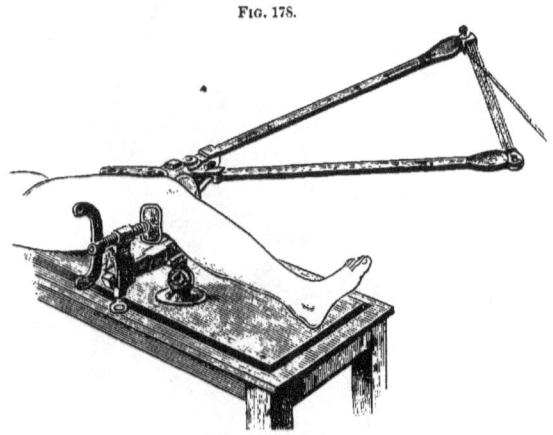

Collins' osteoclast.

When the deformity lies chiefly in the head of the tibia, osteotomy of this bone is a rational and necessary operation, and may be performed after the method of Billroth. A transverse incision one-half inch long is made three-quarters of an inch below the spine of the tibia, extending through the skin and periosteum. The outer compact portion of the tibia is then cut transversely and the bone fractured. Wedge-shaped sections of the tibia and femur are seldom necessary.

FIG. 179.

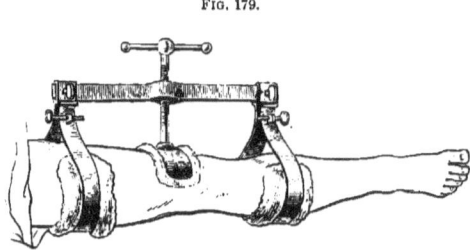

Rizzoli's osteoclast.

Osteoclasis, or forcible fracture of bones by instrumental means, is said to have been practised by the fathers of medicine, and has again recently been advocated, especially by the French, for the relief of

FIG. 180.

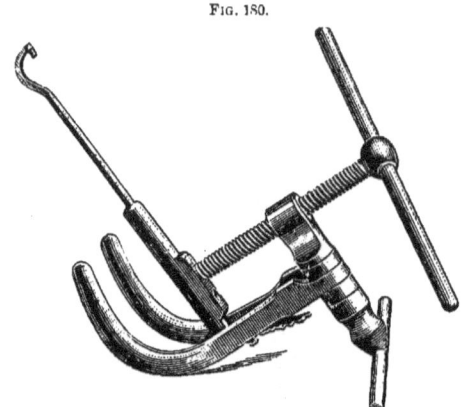

Grattan's osteoclast.

genu valgum. Excellent osteoclasts have been invented by Rizzoli, Collin, Robin, Bruns, Volkmann, Beely, and Grattan.

The objections to the method are the lack of precision, splintering of the fragments, rupture of the ligaments, and separation of the epiphysis. The more recent instruments of Collin, Rollin, and Molière are said to have overcome these objections and to be capable of breaking the femur within two fingerbreadths of the joint without affecting the articulation.[1]

Osteoclasis is more applicable to the shaft of the bone, as in the correction of bow-legs, and is slightly more safe in this locality than osteotomy, whereas osteotomy is considered superior for the correction of genu valgum by most American surgeons.

[1] Bull. et Mém. de la Soc. de Chir., Paris, 1883, ix. 885.

CHAPTER XXI.

BOW-LEGS.

Bow-legs, or genu varum, is that deformity in which the knee is thrown outside a perpendicular line drawn from the head of the femur to midway between the malleoli. It is the opposite condition to knock-knee, an adduction contracture.

Synonyms. English, *Bandy-legs.* German, *Säbel-bein; Sichel-bein; O-bein.* French, *Genou en Dehors.* Latin, *Genu Extrorsum.* Greek, *Exogonyancon.*

Varieties. It is generally double, but may be single, or may accompany knock-knee of the opposite side. It may be due to a gradual curvature outward of the shafts of the femur and tibia, to bowing outward of the lower third of the tibia, the femur remaining normal, or both bones may be bent forward and outward. The knee is very seldom primarily affected.

Etiology. It occurs in children about the time of walking, and is almost without exception rhachitic in origin. The chief factor in producing the deformity is the body weight in standing and walking. Muscular action is a factor in some children who have never walked, but it cannot be accepted as the chief cause, as Kassowitz has asserted. The direction in which the bones yield to the superincumbent weight of the body will depend upon the direction in which the pressure is transmitted, the weak part of the bone normally, and the location of the rhachitic softening.

Children affected with rickets stand with the thighs flexed, the feet wide apart, and the lumbar spine arched forward, this position being assumed on account of muscular weakness, or more probably on account of the pendulous and weighty abdomen. In this position the line of gravity falls outside the knee-joint, the shafts of the long bones yield to the pressure, and in time the internal condyle atrophies and the outer one hypertrophies. In some cases the bowing is extreme and the outline of the legs is almost circular.

Symptoms. These are principally the deformity of the legs, the diminished height, and the peculiar waddling gait in walking.

In cases where bow-legs and knock-knee coexist on opposite sides the bow-leg is always the more secure. The feet are usually inverted, in an instinctive effort to contract the base of support, and favor progression.

Diagnosis. The condition is clear upon inspection, but it is always necessary to determine also the location and degree of bowing, and the condition of the bones.

The only affection which it resembles is congenital dislocation of the hip, and the differential diagnosis from this has already been pointed out.

Prognosis. The prospects of spontaneous outgrowth of this deformity are more favorable than in knock-knee, but the fact that bow-legs are met with among adults proves that some cases never entirely recover. When the bones harden in this deformed position recovery without mechanical treatment or operation is not to be expected.

Treatment. The treatment of knock-knee may be included under three heads: 1. Hygienic; 2. Mechanical; and 3. Operative.

The hygienic plan of treatment includes the correction of the rhachitic diathesis by proper medication and improved hygiene, daily rubbing of the limbs, and manipulation of the deformity with the hands, grasping the limb with the hands upon the outer side and applying pressure in such a manner as to gradually reduce the deformity.

Fig. 181.

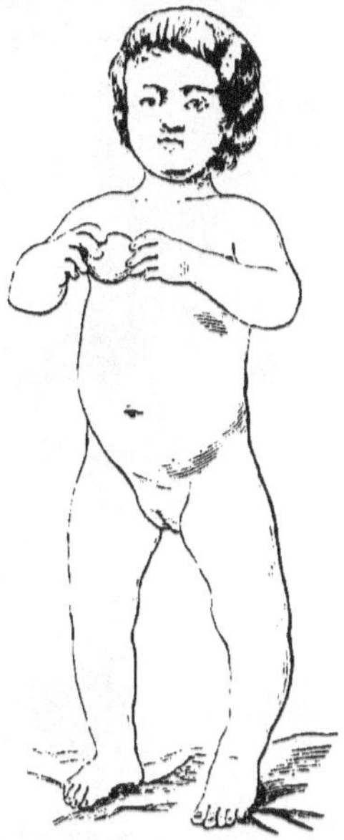

Characteristic attitude in a case of moderate bow-legs. (BRADFORD and LOVETT.)

The mechanical treatment is to be conducted upon the same general principles as knock-knee. Traction may be applied from an inner elastic or rigid upright applied to the inner side of the limb, and making pressure at the upper part of the thigh and at the ankle. In cases where the curve is principally in the tibia the elastic brace of Roberts is of service as a night appliance. (Fig. 182.) In most cases the leg-brace before described, but with the pressure-pads differently located, answers well as a retention and walking brace. The pads are applied to the inner ankle and outer side of the knee, and the uprights are to be bent as the curves improve. Mechanical treatment is not of service after the bones have hardened, but may be employed for three or four months to ascertain if any benefit accrue before resorting to operation. Careful tracings from time to time will indicate the progress of the case.

Fig. 182.

Fig. 183.

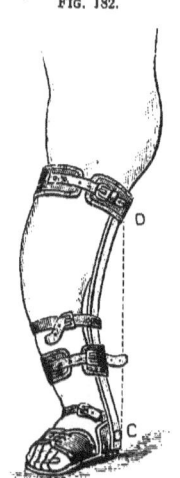

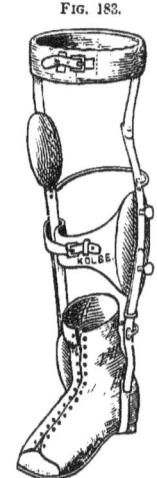

Roberts' elastic bow-leg brace.

Apparatus for bow-legs. (HOLMES.)

The rubber bandage has been employed with great benefit by Dr. G. G. Davis, of this city. His method of application is illustrated in Fig. 184.

The operative treatment of bow-legs includes: 1. Osteotomy, and 2. Osteoclasis.

Osteotomy is the operation usually performed by American surgeons, although the results of osteoclasis are more favorable in bow-legs than in knock-knee, and the same dangers are not to be feared. Osteotomy for bow-legs is performed in a similar manner to that for knock-knee, and several sections of the bone may be required, as high as six and even ten osteotomies having been performed at one time, these sections being made wherever they appear most necessary. The same rules and technique should be observed as in operations for knock-knee. As a rule, in osteotomy for bow-legs, simple linear osteotomy will be all that is required, and removal of a wedge of bone will be required only in the most exceptional cases.

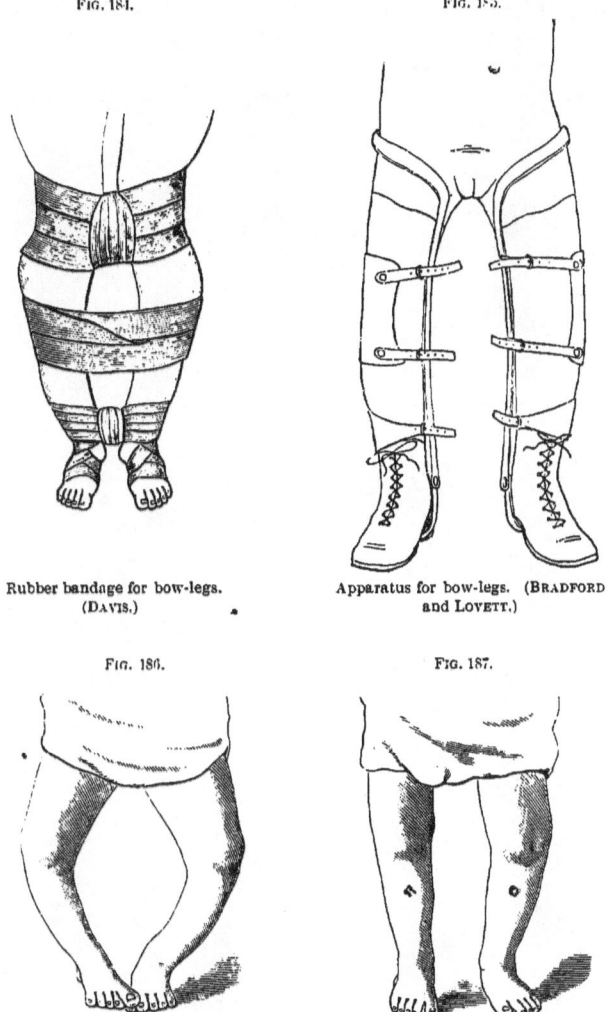

FIG. 184.

FIG. 185.

Rubber bandage for bow-legs. (DAVIS.)

Apparatus for bow-legs. (BRADFORD and LOVETT.)

FIG. 186.

FIG. 187.

Result of osteotomy for bow-legs. (ASHHURST.)

Osteoclasis is most efficient in outward bowing of the femur and tibia, since the sharp edge of the tibia offers a great disadvantage its application to anterior bowing of the leg. It is not applicable in adults nor where the bones are very strong. The fracture with the osteoclast is usually a transverse one without splintering, and will usually be found to occur opposite to the screw-pad plate.

After correction the deformity should be slightly over-corrected.

CHAPTER XXII

CURVATURES OF THE DIAPHYSES.

CURVATURES of the diaphyses occurring in knock-knee and bow-legs have already been described, but anterior bowing of the tibia or anterior tibial curves requires special notice.

Anterior bow-legs may consist of a long gradual curve from the knee to the ankle, or an abrupt angular curve in any portion of the bone, but usually in the lower third.

The course, symptoms, and diagnosis are the same as in other rhachitic curves; but the prognosis is not so favorable as in knock-knee and bow-legs, the mechanical conditions being entirely different. The treatment may be mechanical or operative. In the early stages, before the third year, much may be accomplished by manual straightening and the use of pressure apparatus, but Gibney,[1] Willard,[2] and most orthopedic writers agree that after once the bone has hardened anterior curves of any magnitude cannot be reduced by apparatus.

The form of apparatus employed during the early stages consists of lateral steel uprights attached by a stirrup to the shoe below, secured to the leg above by a leather band, and making pressure over the curve in front by means of a broad leather strap connected with the steel uprights. (Fig. 188.)

The operative procedures include forcible fractures, osteoclasis, and osteotomy.

Forcible fracture in children under three years of age is safe and efficient, both bones being fractured by the hands, a roller bandage, sand-bag, or the knee of the operator being used as a fulcrum. After correction the limb should be secured in a plaster-of-Paris dressing.

Osteoclasis offers the advantage that pressure can be applied to a definite point, but in this locality is open to the objections already detailed.

Osteotomy gives uniformly excellent and speedy results. In deformities of moderate degrees simple linear osteotomy with a chisel,

---

[1] Trans. Acad. Med., N. Y., 1886, vol. xiv. [2] Trans. Amer. Orthop. Assoc., vol. i.

or subcutaneous osteotomy with a saw are efficient, but in the severer grades cuneiform osteotomy, or removal of a wedge-shape piece of bone, is necessary. (Figs. 189 and 190.)

FIG 188.

FIG. 189.

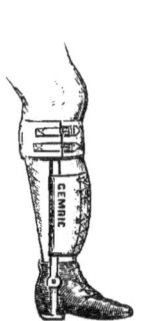

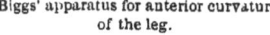

Biggs' apparatus for anterior curvature of the leg.

Anterior curvature of bones of legs. (ASHHURST.)

Simple linear osteotomy is performed as already described. In performing cuneiform osteotomy the most rigid antisepsis must be enforced. In addition to sterilizing the region of operatiom, instruments, etc., the hands of the operator should be sterilized.

FIG. 190.

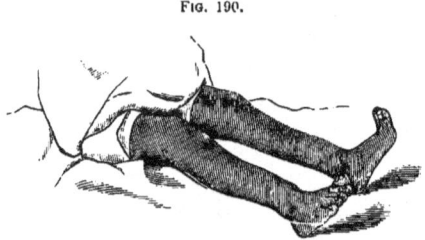

Result of osteotomy for anterior curvature of boues of leg. (ASHHURST.)

To determine the amount of wedge to be removed, the sphenometer[1] may be used, or, what most surgeons prefer, a paper outline of the anterior aspect of the leg may be drawn and cut out. This, when cut through the point of curvature and straightened, will give the size of the wedge.

[1] Therapeutic Gazette, 1887, p. 154.

The fibula, if rigid and strong, should first be divided. The section with the chisel should extend two-thirds through, and the wedge should be taken away by a series of chippings, and the bone finally fractured by manual force. This is necessary on account of the close proximity of the posterior tibial artery and vein, Dandridge[1] having lost a patient from pyæmia following an injury of these vessels.

Tenotomy of the tendo Achillis is required is almost all cases. Drainage with sterilized catgut or horsehair is necessary, and full antiseptic dressing and plaster-of-Paris cast should be employed.[2]

[1] Boston Med. and Surg. Journ., 1885, vol. cxiii. p. 25.
[2] For full details *vide* article of Dr. Willard : Trans. Orthop. Assoc., vol. i.

CHAPTER XXIII.

TARDY HEREDITARY SYPHILIS OF THE BONES.

THE deformities resulting from hereditary syphilis were entirely overlooked until Hutchinson and Fournier called particular attention to them. Since then Hoffa[1] has referred to syphilitic deforming osteitis of the lower extremity, and Dr. G. G. Davis has reported[2] four cases of tardy hereditary syphilis of the bones which have come under his observation. To the last-named I am indebted for the following data.

This is a disease of the bones producing at times considerable deformity and accompanied with constitutional disturbance, but which is often suffered to pass unrecognized. It cannot be a very rare affection, for in the past four years at least eight cases have come under his notice.

Synonyms. French, *Syphilis Hereditaire Tardif* (Fournier). German, *Osteitis Deformans Syphilitica.*

It is well known that hereditary syphilis is most apt to manifest itself in infancy, say within the first three months. If, however, the child escapes any evidences of the disease at that period they may appear at various times thereafter. Usually there is a period of three or more years in which the child is comparatively healthy. When tardy hereditary syphilis affects the bones it is most apt to do so between the ages of six and ten years, although cases do occur even in adults. In its commencement the affection is apt to be regarded as rheumatic because of the pains; these may be located either in the shaft of the bones or in the neighborhood of the articulations. They may be very severe and worse at night and after use of the limb, preventing the patient from sleeping. In some cases they vary in severity with the state of the weather. In cases in which the disease is active marked changes of temperature are observed, while, when it is chronic or subacute, it may be only slightly elevated or not at all. In very acute cases the temperature may rise

---

[1] Lehrbuch der orthop. Chir., 1891, p. 631.   [2] N. Y. Med. Journ., Jan. 23, 1892.

very high, in one of his cases reaching 104.8°. It only occasionally shows the marked rise and fall seen in septic diseases, and then only for a day or two at a time. Usually it resembles more that of typhoid fever, having a daily variation of from one to three degrees. Chills are not present. After the disease has existed a variable period of time enlargement of the bones is observed. This may occur either regularly around the bone or localized more to one side. In the tibia it is usually in a forward direction, causing it to project in a peculiar manner and forming the sabre-bladed deformity of

FIG. 191.

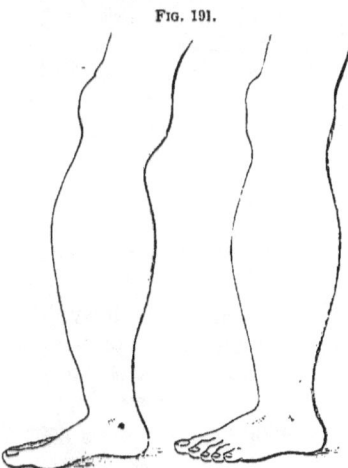

Syphilis of the bones showing "sabre-blade" deformity. (DAVIS.)

Fournier. The surface of the bone is also irregularly enlarged in the form of nodes, showing the presence of periostitis as well as osteitis.

When the affected bony surface is subcutaneous and the process active, the skin assumes a dusky red color, fluctuation and perforation occurs and a scale of bone is cast off. This exfoliation may consist of a few calcareous granules or of a large scale, or, if the process is exceptionally severe, may involve a considerable portion of the shaft itself. The tibia is the bone most frequently affected, but I have seen the fibula, radius, and ulna also attacked. (See Figs. 191 and 192, from a single patient.)' Pains have also been complained of by the patients in other parts of the skeleton, which would indicate that they too were affected. In examining into the history of these cases sometimes other manifestations of the syphilitic disease are discovered, such as imperfections of the teeth ; these are

FIG. 192.

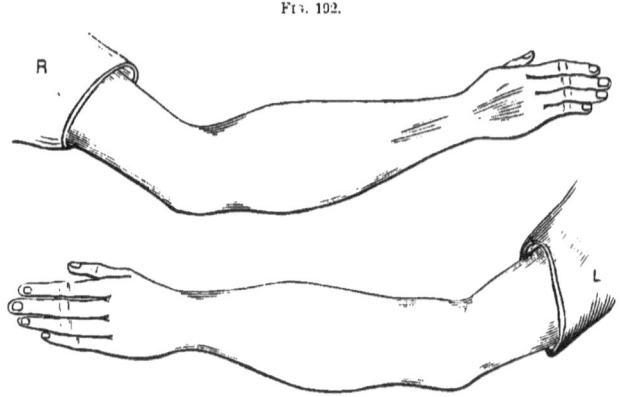

Tardy syphilis of the bones. (DAVIS.)

not so apt to be notched as pegged and decayed. Evidences of old eye-trouble, keratitis, etc., deafness, snuffles, and eruptions in infancy, and general malnutrition may also at one time have been present. The history of the parents and brothers and sisters usually yields confirmatory facts, the infant mortality being great, and miscarriages not uncommon ; hydrocephalus may also occur. The course of the disease is essentially chronic. Remissions occur only to be succeeded by a reappearance of the symptoms. Suppuration and exfoliation of bone are followed by healing with disfiguring scars. The swellings not infrequently disappear without leaving any trace of their presence.

The disease does not tend to a fatal issue, but exists in a more or less active state for four to eight or more years. It usually begins during childhood or youth, but Fournier gives a case occurring in a man aged twenty-six years, and I have seen one patient fifty-five years old with sabre-blade deformity of both tibiæ and other symptoms, precisely similar to the cases occurring in younger people, which I can only ascribe to the disease under discussion. He was of a family of seven children, one of whom besides himself lived, while the other five died before the age of six weeks. When the tibia is affected near its epiphyses its growth is accelerated and it becomes longer than the fibula, thus throwing the foot outward in the position of the valgus, as illustrated in Fig. 198, taken from a boy aged thirteen years. The total length of the limb is also considerably increased, thus impairing walking. In order to enable such cases to walk better, the shoe of the opposite or healthy limb should be elevated and the one of the affected side evened up by means of an inclined insole, and supported by a brace if necessary. By these measures locomotion may be considerably improved.

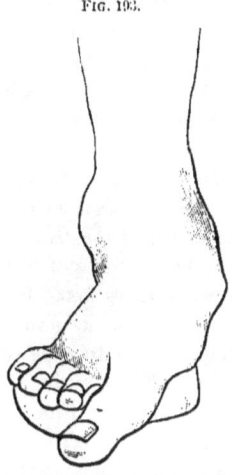

FIG. 198.

Extreme valgus from tardy syphilis of the bones. (DAVIS.)

As regards internal medication, iodide of potassium in as large doses as the patient will bear for long periods of time should be tried, to be alternated with mercurials, tincture of iron, syrup of the hypophosphites, syrup of the iodide of iron, cod-liver oil, and other alterative and tonic remedies.

CHAPTER XXIV.

ARTHRITIS DEFORMANS.

A CHRONIC disease of the joints (in later adult life) characterized by great deformity following changes in the cartilage and synovial membrane, with peri-articular formation of new bone.

Synonyms. *Rheumatoid Arthritis; Nodular Rheumatism; Chronic Articular Rheumatism; Chronic Rheumatic Arthritis; Rheumatic Gout; Dry or Proliferating Arthritis; Nodosity of the Joints; Malum Senile;* when present in the hip-joint, *Morbus Coxæ Senilis; Malum Coxæ Senilis; Senile Coxitis; Chronic Rheumatoid Arthritis of the Hip.*

The existence of these affections among the ancients[1] is attested by the evidences found by Chiase in the bones unearthed at Pompeii, in an Egyptian skeleton of the Ptolemaic period, in the Roman skeleton found in a sarcophagus at Smithfield, England,[2] and in the skeleton of the Norse viking which was found entombed in his warship, in the Christiania Fjord.

Etiology. The true nature of this disease is very obscure. It has been ascribed to rheumatism or gout; to exposure or traumatism; to mental worry or shock; to the tubercular diathesis; but of late the theory that it is due to some primary lesion in the cord, or to changes the result of peripheral irritation, seems to be gaining ground. This disease affects the knee, shoulder, hip, wrist, and elbow, either separately or together in the above order of frequency, and is most common above the age of forty-five.

Pathology. The changes begin with thickening of the synovial membrane, with hypertrophy of its fringes, the branching appearance of which has given rise to the name of lipoma arborescens.

The cartilage disintegrates, presenting a yellow shreddy appearance, and finally is absorbed or gradually thinned by attrition, thus laying bare the ends of the bone, which become smooth, polished, and eburnated. The head of the bone may be locked in a splint of osteo-

---

[1] Lebert: Handbuch d. prakt. Med., 1859, ii. p. 874. Eve: Brit. Med. Journ., 1890, i. 423.
[2] Norman Moore: Path. Soc. Trans., 1883, p. 226. Virchow's Archiv, 1869, xlvii. 298.

phytes which form from the proliferated cells; or, denuded of its cartilage, may undergo an ivory-like condensation, with final atrophy, exposing the Haversian canals and presenting a worm-eaten appearance. The muscles around the joint become wasted, the ligaments greatly thickened, which, together with the osteophytes, often form a false ankylosis of the joint.

Symptoms. In its acute form this disease resembles articular rheumatism. (Figs. 194 and 195.) If, as is more usual, it begins

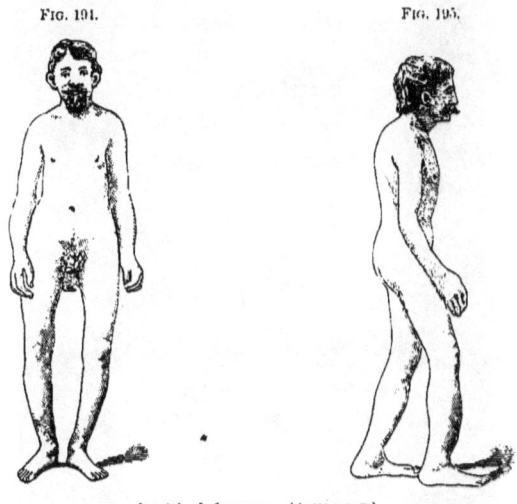

Fig. 194.　　　Fig. 195.

Osteitis deformans. (Ashhurst.)

in its chronic form, the first symptom is pain on motion, stiffness and swelling of the joint. From involvement of one joint the disease may spread until nearly all of the articulations of the body are affected. The amount of pain and of the effusion into the joint vary greatly. The shape of the joint is permanently altered by the deposition of new bone, by the thickening of ligaments, and atrophy and retraction of the muscles. A peculiar crepitation produced by the friction of roughened synovial membrane is recognizable. Motion may be entirely lost, either from muscular spasm or from the splint of osteophytes.

In the femur, from atrophy of the neck, the trochanter may be found above Nélaton's line. Heberden's nodosities is a variety of this disease in which little nodules develop upon the sides of the distal phalanges.

Spondylitis Deformans, or Rheumatism of the Spine.

A primary ankylosing arthritis of the vertebral column is a complication of arthritis deformans as it occasionally is of gonorrhœal

FIG. 196.

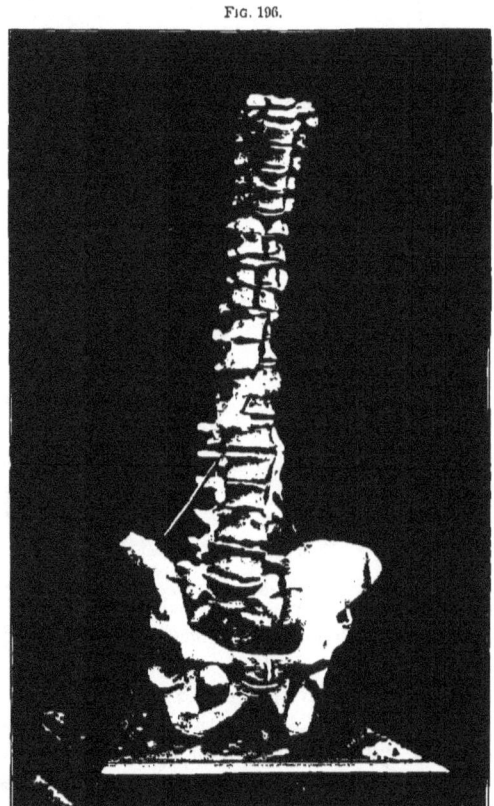

Photograph of specimen of spondylitis deformans, showing deposits.
(From the Museum of the Philadelphia Hospital.)

rheumatism. It presents, as its characteristic symptom, marked stiffness of the spine, indeed, the whole vertebral column may be as rigid as an iron rod. (Fig. 196.) Paroxysmal pain aggravated by every jar is present. The normal curves are exaggerated, especially the lumbar and dorsal ones, so that the patient may walk as though bent with Pott's disease. The cervical vertebræ are the last to become involved. In that case the up-and-down movements of the head are lost, though rotation is still possible. In severe cases the ankylosis of the ribs with the spine is so complete that chest expansion is almost abolished, the breathing being entirely abdominal. Its course is chronic in the extreme; the bone inflammation, having no destructive tendency, accomplishes nothing more than stiffening of the column.

Diagnosis. The general form must be distinguished from rheumatism; the mono-articular, when the hip-joint is involved, from impacted fracture of the neck of the femur; when the shoulder is the one affected, from the local arthritis of that joint—a curable disease.

Treatment. The pain and irritation of the acute attacks are relieved by rest in bed, with more perfect rest of the joint by the use of a traction splint.[1] Counter irritation by means of the Paquelin cautery, blisters, or iodine is of value in meeting this indication, as is the use of hot or cold douching. In spondylitis deformans a posterior spinal splint is of service, and when the cervical vertebræ are involved an antero-posterior head support with a chin-rest.

The disease is an incurable one, but its progress may be arrested by a visit to some of the numerous foreign or domestic mineral springs: Carlsbad, Baden, or Vichy; Buxton, Harrogate, or Bath in England, or Richfield, Sharon, or Arkansas Hot Springs in our own country. The general health must be maintained by tonics, good food and fresh air. Massage may help to promote the absorption of the effusion and to restore mobility and maintain the nutrition of the muscles, but passive movements are to be avoided, as increasing the amount and degree of ankylosis. The medical treatment should be the same as for chronic rheumatism, and potash iodide or nitrate, lithia or natural waters, as the Londonderry or Farmville, colchicum in the form of the old and reliable wine of the root, or one of the more modern alkaloids, will be found of the greatest service.

In senile coxitis a protection traction hip-splint will be of service, and crutches and canes are often useful.

Taylor: "Senile Coxitis," N. Y. Med. Journ., December 15, 1858.

CHAPTER XXV.

DUPUYTREN'S CONTRACTION.

DUPUYTREN'S contraction is a name applied to a contraction of the palmar fascia or its digital prolongation producing permanent flexion of one or more of the fingers.

Synonyms. French, *La Contraction des Doigts* (Dupuytren); *La Maladie de Dupuytren.* German, *Die Dupuytren'sche Contraction der Finger.*

History. This condition had been described prior to Dupuytren (by Plater, Chomel, Boyer, Astley Cooper, and others), but he was the first to demonstrate clearly its pathology by careful dissection of a hand he was fortunate enough to possess.

Etiology. The cause of this disease has never been satisfactorily explained. Two theories account for the major part of the cases— the influence of long-continued slight traumatism, and of a rheumatic or gouty diathesis. Syphilis is undoubtedly an occasional factor in the causation, and recently Abbe has advanced a theory of central nervous irritation producing nutritive changes in the affected hand.

In Keen's recorded seventy-two cases, only eighteen were amongst the laboring classes. This, together with the fact that the left hand is as frequently involved as the right, and that the ring and little fingers are most commonly affected, while certainly they are not the most exposed, sufficiently explodes the theory of traumatism. On the other hand, there are few American families in whom it is not possible to find a rheumatic member.

Pathology. The deformity is caused by a scar-like contraction of the palmar fascia and the areolar network overlying it, by means of which the digital prolongations of the fascia are retracted and the fingers flexed into the palm. The tendons lying below the fascia remain free and uninvolved. Along the course of the contraction small hard bodies are found, which, upon examination prove to be small fibromas. The skin, if affected at all, is only so late in the disease.

Symptoms. The first sign is a small body, the size of a shot, appearing in the metacarpal phalangeal crease, with some stiffness of the ring or little finger. This is entirely painless. After some months or years a cord is noticed running to these fingers, which are contracting into the palm.

Diagnosis. The forced flexion of the fingers, the fact that ankylosis of the joints does not exist, the absence of pain, the advanced age of the patient, the non-existence of cerebral or spinal disease, or of injury to the part, with loss of substance followed by a scar like contraction, present so clear a clinical picture that the diagnosis cannot easily be mistaken.

Prognosis. An absolutely favorable prognosis can be given if an operation is allowed. The condition is not easily corrected without it.

Treatment. The medicinal treatment is *nil* except where there is a syphilitic taint, when iodides should be freely used with prospects

FIG. 197.   FIG. 198.

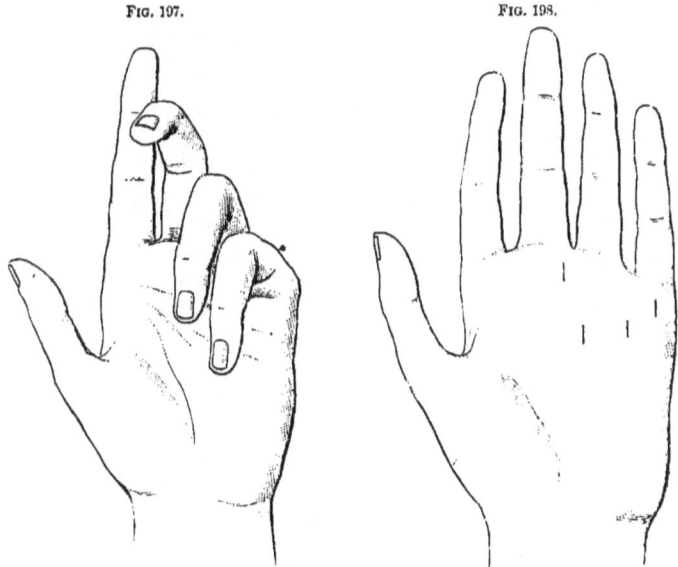

Dupuytren's finger contraction. (ASHHURST.)   The same hand after operation. (ASHHURST.)

of cure, as in the successful cases recorded by Richet and Ricord. The mechanical treatment has been abandoned for the surgical, which consists of either a subcutaneous incision of the fibrous bands, or an open one, with or without removal of part of the contracting tissue. (Figs. 197 and 198.) Mr. Adams enters a small pointed tenotome a little to one side of the cord and cuts down upon the fascia. This is repeated in several places until the fingers can be partially straightened. Dupuytren's original operation of open transverse incision through the band in its two or three most prominent places is still practised, or the skin may be reflected in a V-shaped flap and the contracting material dissected out. A curved splint should be applied for the first few days, after which time it may be replaced by a straight one, to be worn for two or three weeks. Ether need not be used, cocaine anæsthesia (2 per cent. solution) being sufficient.

CHAPTER XXVI.

TALIPES, OR CLUB-FOOT.

THE generic term talipes, or club-foot, as suggested by Little, signifies an abnormal position of the foot, whether in antero-posterior or transverse plane, in its anatomical relations to the leg, or of the foot to itself.

This abnormal position may consist of a flexion, extension, inversion, or eversion, or combinations of these, but is popularly applied to that deformity in which the foot is twisted inward so that the weight of the body is borne upon the outer side or dorsum of the foot.

Synonyms. English, *Reel-foot; Stump-foot.* Latin, *Pes Contortus.* Greek, *Kyllosis.* French, *Pied-bot.* German, *Klumpfuss.* Italian, *Piede Torto.* Spanish, *Pié Truncado.*

Anatomy of the Foot. It may be instructive here to review briefly the anatomical construction and mechanism of the natural foot. The foot includes all that portion of the inferior extremity below the tibio-tarsal articulation, consisting of the tarsus, metatarsus, and phalanges, and in the adult has the form of two arches, an antero-posterior and a transverse, each with its convexity or dorsal surface above, and its concavity or plantar surface below.

The antero-posterior, the most important, is supported upon two piers or pillars, and has its summit at the astragalus and ankle-joint. This has been still further divided into two arches, an outer and an inner, by an imaginary line drawn posteriorly between the third and fourth metatarsal bones. The inner portion of the antero-posterior arch is much more curved than the outer, and forms the instep.

The posterior pier, formed by the posterior parts of the astragalus and os calcis, is shorter, more curved, has but one joint, and is more solid, receiving the greater part of the weight of the body. The anterior pier, composed of the scaphoid, three cuneiform, and three inner metatarsal bones, is longer, less curved, has many joints, and is more elastic, serving to diminish the force of shocks transmitted to the arch. The head of the astragalus, fitting into the concave surface of the scaphoid, and its postero-inferior surface, articulating with the anterior surface of the os calcis, may be regarded as the keystone, though differing in many respects from such bodies as usually employed. The weak part of the arch is strengthened by the interosseous ligaments, particularly the inferior calcaneo-scaphoid, which supports it from below, while beneath, the inner portion of the plantar fascia adds additional strength. The outer portion of the antero-posterior arch consists of the outer portion of the os calcis, cuboid, and the two outer metatarsal bones. It is strengthened by the calcaneo-cuboid ligaments and the outer portion of the plantar fascia. Both arches are still further maintained by the tibialis posticus and peronei muscles, particularly the P. longus. The transverse arch, formed on the inner and outer sides by the bones entering into the inner and outer antero-posterior arches respectively, varies in degree of curvature in different portions of the foot, being most marked across the cuneiform bones. It affords protection to the soft parts of the sole, and adds to the elasticity.

To appreciate more thoroughly the deformities under consideration, the foot is best divided anatomically into an anterior portion, the "pes" or foot proper, and the posterior portion, the "talus" or ankle. These articulate at the *medio-tarsal* or Chopart's joint, formed between the os calcis and astragalus behind, and the cuboid and scaphoid in front, a joint which admits in a limited degree of every variety of operation—flexion, extension, abduction, adduction, and rotation.

The amount of lateral deviation of which the normal foot is capable is well shown in the outline tracings of the thirty-two normal feet measured by Dr. Roberts,[1] after the method advocated by Rohmer (*Les Variations de Forme normales et pathologiques de la Plante du Pied*, Thèse, Nancy, 1879), consisting of first covering the plantar surface of the foot with lampblack, which leaves a correct impression of the sole upon the white paper upon which the patients are then requested to walk. To obtain a correct basis of measurement, and still further to carry out Rohmer's researches, he selected the medio-tarsal joint as a base-line of measurement, erecting upon it a perpendicular corresponding to the long axis of the os calcis, thereby determining by a comparison of the degrees of variation the position and character of the deviation or deformity. In this manner the angle of deflection was found to range between 26 and 27 degrees (*average*, 20 males, 34.8 degrees ; 12 females, 31.5 degrees), typical examples of which are seen in the accompanying figures.

¹ Lectures on Club-foot, p. 18.

The tibio-tarsal or ankle joint admits of flexion and extension, and, according to Gray, in extreme extension of a slight amount of adduction also. This is in consequence of the articular surface of the astragalus being wider in front than behind, so that in complete

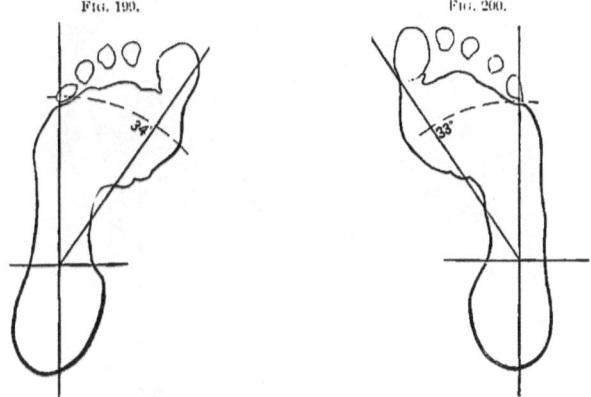

Angle of deflection, normal male feet.   (ROBERTS.)

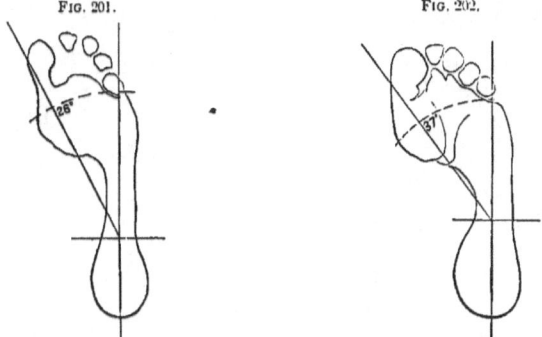

Angle of deflection, normal female feet.   (ROBERTS.)

extension the narrowest part of the astragalus is lodged in the widest part of the tibio-fibular arch, admitting of lateral motion. In the flexed position, however, *no lateral motion is possible.* (Fig. 207.)

The normal amount of flexion and extension has been summarized by Shaffer[1] as follows : Extension of the foot in the adult is limited

[1] Non-deforming Club-foot, 1885.

at about 135°, or 45° more than a right angle, using the long axis of the tibia as the plane of measurement. Flexion stops at about 70°, or 20° less than a right angle (see Fig. 207). The position of the foot in standing upon an even surface, with the knee in full extension, is

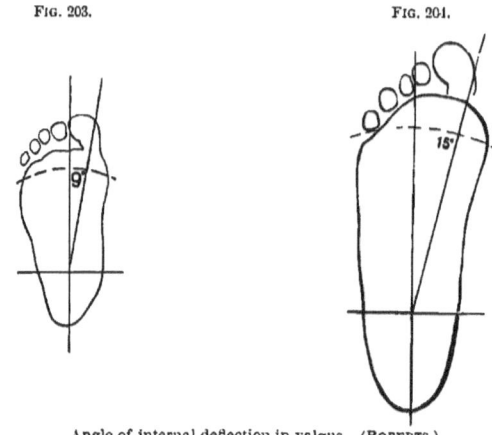

FIG. 203.   FIG. 204.

Angle of internal deflection in valgus. (ROBERTS.)

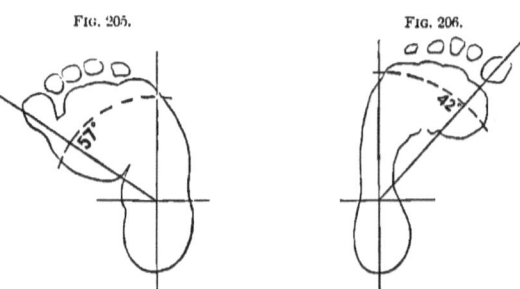

FIG. 205.   FIG. 206.

Angle of adduction of varus. (ROBERTS.)

about 90°. The amount of flexion and extension varies in different individuals, but these figures, based upon actual experiment and measurement, represent, he thinks, the average of normal movement in the living adult subject.

In the normal condition of the foot these movements are accomplished, and a correct anatomical relation of the parts is preserved, by the muscles of the leg, which may be conveniently divided into three groups, moving the foot in four directions.

Thus, the anterior group of muscles—the tibialis anticus, extensor longus digitorum, and peroneus tertius—act upon the foot as flexors; the superficial set of the posterior group—the gastrocnemius, soleus, and plantaris, assisted by the peroneus longus—act as extensors; the deep set of the posterior group—the flexor longus digitorum and tibialis posticus, assisted by the tibialis anticus—act as adductors;

FIG. 207.

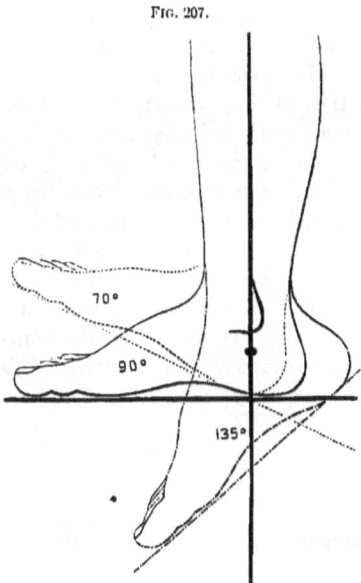

Diagram illustrating range of motion in normal foot.

and the external or fibular group, the three peronei (longus, brevis, and tertius) act as abductors.

The weight of the body is received by the astragalus, the highest part of the arch, and transmitted to the ground through the two piers of the antero-posterior arches. The foot in extension rests normally upon the heel, the tips of the metatarsal bones, and the outer side of the sole, the weight of the body in standing, walking, running, or dancing being transmitted through the heel, the ball of the great toe, and that of the little toe—the natural tripod of the foot —in the order named.

Thus the direction of the weight upon the arches is constantly changing, and it is only through the action of these muscles that the normal arches are conserved. Thus in flexion the antero-posterior arch is increased by the action of the tibialis anticus, peroneus tertius, and extensor longus digitorum; and in extension, by the action of the gastrocnemius, soleus, plantaris, and peroneus longus, both the curves are diminished, and the foot flattened. Then, also, when the foot is markedly flexed the foot is adducted, and when markedly extended it is abducted, in which positions the arches are respectively increased and diminished. In this connection, Mr. Le Gros Clark says: "In reviewing the action of the various muscles around the foot it is obvious that their attachment is designed to preserve the plantar arch, and that such healthy condition must depend in great measure on the evenly-balanced action of those muscles upon their several attachments. Thus the peronei and tibial muscles antagonize each other, and the expanded insertion of two of them into the tarsal bones is very instrumental in preserving the transverse as well as the antero-posterior arch."

If the equilibrium be disturbed the action of a single group will produce one of the simple varieties of the deformity, or, two groups acting together, one of the compound varieties. This is well exhibited in the following table:

Extension (equinus) .	Gastrocnemius. Soleus. Plantaris. Peroneus longus.
Flexion (calcaneus) .	Tibialis anticus. Peroneus tertius. Extensor longus digitorum.
Adduction (varus) .	Tibialis anticus. Tibialis posticus. Flexor longus digitorum.
Abduction (valgus) .	Peroneus longus. Peroneus brevis. Peroneus tertius.

Varieties. Systematic writers usually classify varieties of club-foot under congenital and acquired, but they are most conveniently divided into two classes, the simple and compound. Of the former there are four varieties, two on a lateral plane, talipes varus and talipes valgus, and two on the antero-posterior plane, talipes equinus and talipes calcaneus.

1. In talipes varus the inner side of the foot is raised, the sole turned inward, and the anterior portion adducted.

2. In talipes valgus, its opposite, the outer side of the foot is elevated, and the sole everted.

3. In talipes equinus the heel is elevated, the foot extended, and the patient walks on the balls of the toes.

4. In talipes calcaneus the toes are raised, the foot flexed, and the patient walks on the heel.

To these simple forms some authors have added others, and three additional simple varieties are now recognized, talipes cavus, talipes planus, and non-deforming club-foot. In talipes cavus the arch of the foot is increased; in talipes planus, its opposite, the arch is diminished and the sole rests on the ground; and in the non-deforming club-foot of Shaffer—an incomplete equinus—there is little or no deformity, but normal flexion is limited and mobility modified. Combinations of these simple forms will present the compound varieties, as equino-varus, equino-valgus, calcaneo-varus, calcaneo-valgus, etc. These may be conveniently arranged as follows :

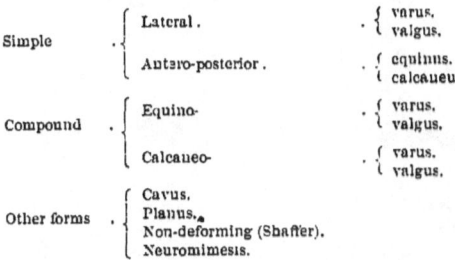

Simple	Lateral.	{ varus. valgus.
	Antero-posterior.	{ equinus. calcaneus.
Compound	Equino-	{ varus. valgus.
	Calcaneo-	{ varus. valgus.
Other forms	Cavus. Planus. Non-deforming (Shaffer). Neuromimesis.	

Relative frequency. In estimating the relative frequency of the different varieties some difficulty is experienced, owing chiefly to the different nomenclature employed by those collecting statistics. Especially is this true of varus, which, by Lonsdale and his followers, is equivalent to equino-varus, the former being rare, whilst the latter, congenital equino-varus, is the most common of all varieties. Allowing for these discrepancies, the published statistics of Tamplin, Chaussier, Lannelongue, Duval, Reeves, Roberts, and others furnish sufficient material from which to estimate the relative frequency of these affections.

Tamplin[1] tabulated 1780 cases of club-foot, of which 764 were congenital and 1006 acquired, by which it is shown that the greater

---

[1] Lond. Med. Gaz., Oct. 1851.

number were of the acquired variety, the proportion being, as given
by Adams, about as 3 to 2 :

Congenital.	Cases.	Acquired.	Cases.
Talipes varus	688	Talipes equinus	401
" valgus	42	" valgus	181
" calcaneus	19	" equino-varus	162
" varus of one foot and valgus of		" calc. and calc.-valgus	110
the other	15	" equino-valgus	80
	—	" varus	00
Total	764	" varus of one foot and valgus of	
		the other	5
		Total	999
		Unclassified	7
			1006

Out of 688 cases of varus, which included the equino-varus, 182
occurred on the right side only ; 138 on the left side only ; 363 were
double, and 5 were complicated with other deformities.

The relative frequency of the congenital and acquired forms is also
well shown in the 746 cases recorded by Dr. Roberts[1] from records
of the New York Orthopedic Hospital and the Orthopedic Dispensary of the University of Pennsylvania, of which 213 were congenital and 533 acquired. They were as follows :

	Congenital.	Acquired.
Equinus	5	87
Calcaneus	3	31
Varus	73	66
Valgus	29	236
Equino-varus	95	68
Equino-valgus	3	9
Calcaneo-varus	0	2
Calcaneo-valgus	5	34
Totals	212	533

Duval has recorded 1000 cases of club-foot, of which 574 were
congenital, and of these 364 were in males, 210 in females. Though
including a greater number of congenital cases, these figures are
interesting as exhibiting the relative frequency of the varieties and
greater liability of the male sex.

	Cases.	Boys.	Girls.
Equinus and equino-varus	417	215	202
Varus	532	302	230
Valgus	22	14	8
Calcaneus	9	6	3
Extreme calcaneus	20	13	7
Totals	1000	550	450

Out of 167 cases of club-foot reported by Detmold, 93 were double.
The primitive congenital equinus is exceedingly rare, Little and

[1] Medical News, March 13, 1886.

Brodhurst having met with but 2 cases, and Tamplin discredits the congenital origin entirely, not having met a case.

Chaussier, out of 23,923 newly-born infants, found 132 affected with various deformities, 37 of them having club-foot; and Lannelongue, out of 15,229 births at the Paris Maternity Hospital, covering a period of ten years, from 1858 to 1867, collected 108 deformed infants, of which 8 had club-foot, or a proportion of 1 case in 1903 births.[1]

K. Roser found among 100 deformed fœtuses 36 with club-foot, of which 8 were double talipes varus, 11 double calcaneus, 9 calcaneovarus, and 9 with unilateral club-foot.

F. Busch, Adams, Reeves, Roberts and Ketch, and most modern writers, give equino-varus as the most frequent variety.

These statistics taken together are sufficient to show the frequent association of club-foot with other deformity ; its greater frequency on the right side, and in the male ; the relative greater frequency of double club-foot ; the preponderance of equino-varus as a congenital variety, and the variety of primitive forms.

Etiology. The etiology of talipes may be divided into two great classes—the congenital and the acquired.

The etiology of congenital club-foot. The study of the cause of congenital club-foot involves much that is mysterious and unexplained, from the fact that there is no direct scientific means of investigation except such data as embryological research, comparative physiology, and post-natal life and disease furnish. Hence there is much that is purely speculative, and numerous views have been advanced and defended from time to time. These may be considered under the following heads :

1. Theory of heredity.
2. Theory of mechanical pressure, or intra-uterine pressure theory.
3. Theory of pre-natal disease, or musculo-nervous theory.
4. Theory of arrest of development, or osseous theory.
5. Theory of retarded rotation or non-rotation of extremities.

Maternal impressions. In exceptional instances the striking resemblance of deformity to some object seen by the mother during her pregnancy has led the laity to consider maternal impressions as in some occult manner producing these deformities. Upon closer investigation, however, the fright or observation is found to correspond

---

[1] Beiträge zur Lehre vom Klumpfuss und vom Platfuss. Leipzig, 1885.

to a period of gestation when it could have had no influence upon the production of the deformity. They can be readily accounted for under existing physical causes, or are simply striking coincidences. Dabney,[1] in ninety collected cases of maternal impressions apparently producing deformity, did not find a single case; and, according to a recent authority,[2] there is no recorded case of the development of club-foot produced by maternal impression.

1. *Heredity*, exerting its influence through one or both parents, is undoubtedly a factor in the etiology of these cases. Particularly is this true of consanguineous marriages, which are so productive of deformities generally.

Devay and Boudin report 1 case in 164 births from marriages of kin, against 1 case in 1903 of other marriages. Reeves[3] and Brodhurst[4] have known several striking illustrations, in which some form of club-foot, and generally equino-varus, ran through various members of the same family. Adams gives a very interesting history of a club-footed family.[5] The writer has recently operated upon a child whose father, in his infancy, was operated upon in Russia, by Pirogoff, for a similar deformity.

2. *The theory of mechanical or intra-uterine pressure* on the fœtus, through deficient liquor amnii, compression of the uterus by another part of the fœtus, or the abnormal position and action of the umbilical cord, or of amniotic bands, is as old as Hippocrates. Ambroise Paré supported it, and ascribed club-foot to the circumstance that "the mother, during her pregnancy, had been sitting too much with her legs crossed." Cruveilhier supported the pressure view, but added traumatism as an elementary factor. Malgaigne, Volkmann, Kocher, Banga, Parker, Vogt, Cocher, and others supported this mechanical theory of pressure and malposition *in utero*. Against this theory many objections have been raised. If uterine pressure were an influential factor, other organs would exhibit pressure evidences, and club-hands, legs, and thighs would be common, instead of the rarest of deformities. Then, again, no appreciable difference is observed in the quantity of liquor amnii over previous births of healthy children, and Duval has asserted the reverse. Furthermore, the deformity is observed before the fourth or fifth month of intra-uterine life, when

---

[1] Cyclopædia Dis. of Children, vol. i. p. 207.
[2] Bradford and Lovett: Orthopedic Surgery, 1890, p. 454.
[3] Practical Orthopedics, Philadelphia, 1885, p. 149.
[4] Club-foot, August, 1893.
[5] Adams: Club-foot. Jacksonian Prize Essay, 1866, p. 200.

the amniotic fluid is abundant and no pressure is possible. Cases of twin births have been recorded, notably one of Roberts and Ketch, of "double equino-varus in a twin, the other child showing no deformity whatsoever."

3. *The musculo-nervous theory*, the alteration of the muscles, with or without a central nervous lesion, has been supported by a galaxy of celebrities. Morgagni, Benjamin Bell, Duverney, Rudolphe, Béclard, Jules Guérin, and Delpech, have all given it the weight of their indorsement, the last-named being among the first to consider the influence of the malformation of the tarsal bones.

The majority of these considered a contraction of certain muscles the cause of the deformity; Rudolphe, intra-uterine convulsions, Chance and Little also inclining to this view, while Béclard and Barwell attributed it to weakness or paralysis of other muscles. Guérin believed this convulsive muscular retraction to be secondary and consecutive to a central nervous lesion, in some cases demonstrable, in others not. The microscope has not revealed changes in the fœtal brain or cord, nor do the electrical reactions correspond to those met with in acquired cases. Chaussier, in 37 cases of club-foot, and Duval, in 574 cases of the same, found no other deformity associated. The frequent association of club-foot with spina bifida hydrocephalus, and anencephalus, is confirmatory of this association; contrariwise, many fœtuses are met with having extensive nervous lesions but devoid of club-foot. Thus, Lannelongue found only four fœtuses with club-foot among the 32 cases of spina bifida and encephalocele.

4. *The osseous theory*, or theory of arrest of development, the permanence of the feet in the physiological position of the sixth or seventh fœtal week, with the sole turned inward, has been maintained by Geoffroy Saint-Hilaire, Meckel, Breschet, Adams, Hueter, Eschricht, and others. This anatomical fact has been denied by Cruveilhier, but has received the support of Martin and others. While the coexistence of club-foot with such deformities as spina bifida, harelip, and cleft palate would seem to be confirmatory of this theory of arrest of development, yet, in such cases, the feet themselves exhibit no arrest of development, but only such changes as occur in cases unassociated with other deformity.

A modification of this theory of arrest of development, ascribing the cause to primary changes in the tarsal bones, principally the astragalus and os calcis, has been supported by Scarpa, Broca, Bouvier, Brocher, Robin, Lannelongue, Thorens, and others.

These osseous changes are by no means constant, and much difference of opinion exists as to whether they are primary and causative, or secondary, as the result of pressure. A. Luecke favors the former, but recent investigations tend to support the latter view.

5. *The theory of non- or retarded rotation* is the latest, and probably presents the best explanation that has yet been advanced—founded as it is upon embryological research. It was proposed by Berg, of New York (*Archives of Medicine*, New York, December 1, 1882), announced independently by Parker and Shattuck (*British Medical Journal*, 1886, vol. ii. p. 10), confirmed by Scudder ("Boylston Prize Essay," *Boston Medical and Surgical Journal*, October 27, 1887), and favored by Roberts and Ketch, Bradford and Lovett, the writer, and others.

Berg, from a series of embryological investigations, studied the changes in the position of the lower extremities at different periods of fœtal life. From the flexed and crossed position of the lower extremities, all the intra-uterine pressure was brought to bear directly upon the outer side of the thigh and leg corresponding to the fibular border of the leg and upon the dorsum of the foot, resulting in the foot being placed in a position of equino-varus, which he believes to be a stage in the normal development of every healthy fœtus. To provide against the permanence of this, Nature provides an inward rotation of the extremity, carrying the leg away from its position against the abdomen of the fœtus, bringing the soles of the feet against the uterine walls in extreme flexion, and the force of the intra-uterine pressure directly upon them. Upon the completeness of this internal rotation depends the rectification of the early varus, and upon its failure or incompleteness depends the deformity.

In conclusion, while the subject of etiology of club-foot is still unsettled, these recent investigations seem to point to a failure of or non-rotation as a demonstrable theory, and one which is eminently scientific and feasible.

Etiology of acquired forms. As before pointed out in speaking of relative frequency, by far the greater number of cases are of the acquired variety, and the majority of these are paralytic.

Those which are not paralytic are usually mechanical in origin. Thus an equinus may result from the pressure of the bedclothes in long-continued decubitus, as in continued fevers, as in a case recorded by Volkmann following typhoid. Osteitis of the tarsus, ankle-joint, or the lower end of the tibia not involving the ankle-joint, as in a case of equinus in a young woman seen by the author, are frequent causes of acquired talipes. In addition to equinus, varus and valgus positions may also be produced.

Occupations necessitating long-continued standing, as cooks, bakers, printers, and other trades, frequently give rise to inflammatory forms of the affection, especially valgus.

A somewhat peculiar form of valgus—valgus adolescentium—is met with about puberty, depending upon rapid growth and increase of weight without a corresponding increase of strength in the plantar arch.

Rhachitic changes, particularly knock-knee, occurring about the seventh year, are another cause of acquired club-foot of the valgus variety.

Injuries, particularly Pott's fracture, luxations of the ankle-joint, burns, and extensive cicatrices about the foot or ankle-joint, are causes of the acquired varieties.

That deformities of the feet may depend upon the neurotic diathesis, as pointed out by Shaffer (?), is now admitted, but that they may result from reflex paralysis, as a muscular spasm from functional disturbance of the nervous system, has been doubted by many authorities. Of the former—the neuromimeses of club-foot—numerous interesting cases have been recorded, notably those by Weir Mitchell, Little, Paget, Skey, and others.

A few cases of acquired club-foot have been recorded from pseudo-hypertrophic paralysis, post-hemiplegic contractions, a large number resulting from tetanoid paraplegia, but by far the greater number occur from infantile paralysis or poliomyelitis anterior. These latter will be considered in detail in their appropriate places, but it may here be stated that the talipes resulting from the tetanoid paraplegia is of the equinus variety, and that resulting from infantile paralysis is usually of the calcaneus and valgus varieties.

In these paralytic forms, as pointed out by Volkmann, the club-foot does not result from tonic contraction, but faulty positions of the feet assumed to better support the body become permanent deformities through growth of the limb.

There is also a temporary form of equinus, spastic in nature, accompanying the paraplegia of Pott's disease.

Symptoms. In addition to the unsightly appearance, club-foot necessitates a peculiar and characteristic method of progression, depending upon the variety of the affection. Thus, in double equino-

22

varus the feet are lifted over one another, in valgus the foot is swung outward and forward, flail-like, as each step is taken, in equinus the locomotion partakes of a springing gait, etc. In severe cases, bursæ, callosities, and severe inflammation and ulceration result from pressure upon unprotected parts, which diminishes the activity and capabilities of the patient, necessitating sitting occupations, and at times becoming so aggravated as to demand amputation. Those afflicted, likewise, are liable to become a burden to the State, for, as Dieffenbach[1] suggested, it is impossible to employ the club-footed as cavalry.

Aside from the physical discomfort and suffering, club-foot when severe is a source of great psychical suffering. Thus Talleyrand is said to have entered the Church, and Lord Byron is said to have suffered greatly, on account of deformities of this kind.

Varieties—Specially Considered. As pointed out before, the purely primitive forms are so rare that writers have usually considered the compound varieties under the simple forms, but to avoid confusion, which otherwise occurs where this plan is followed, each variety will be separately considered, especially concerning its morbid anatomy and pathology.

Simple Forms of Club-foot. Talipes Varus.

Synonyms. *Supination Contracture; Adduction Contracture.* German, *Klumpfuss.* French, *Pied-bot Varus.*

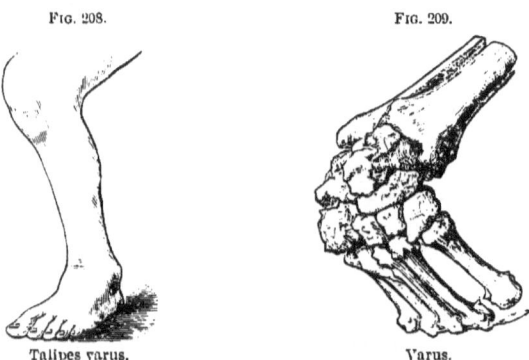

FIG. 208.   FIG. 209.

Talipes varus.   Varus.

As a pure pes varus this variety of club-foot is one of the rarest, but as a part of the congenital equino-varus it is exceedingly

---

[1] Ueber die Durchschneidung der Sehnen und Muskeln. Berlin, 1841, p. 76.

common. As a purely primitive type the deviation inward of the anterior portion of the foot, and its supination and adduction, occur at the tibio-tarsal articulation, the deformity being entirely upon a lateral plane. (Fig. 208.) In severe cases the internal border is elevated, the sole curved inward, and the external border depressed. (Fig. 209.) It may here be remarked that in most infants at birth there exists a tendency to varus, which disappears as the child attempts to walk.

The morbid anatomy of this rare affection will be described under the compound form—Equino-varus.

Talipes Valgus.

Synonyms. *Flat-foot; Splay-foot.* French, *Pied-bot Valgus.* German, *Plattfuss.*

This variety occurs both as a congenital and acquired affection, the former being rare and usually of the compound variety—equino-valgus; the latter being very common, occurring also as a compound form—calcaneo-valgus. The frequency of this form is well shown in the interesting statistics collected by Hoffa [1] in the Münchener Poliklinik. Of the 17,619 surgical affections, 338, or 0.49 per cent., were flat-foot; of 1444 deformities, 338, or 23.41 per cent., were flat-foot. Out of 235 cases, he found that 10, or 4.5 per cent., were congenital, and 225, or 95.7 per cent., were acquired. Of these 225 cases, 11, or 4.9 per cent., were traumatic; 7, or 3.1 per cent., were paralytic; 7, or 3.1 per cent., were rhachitic; 200, or 88.9 per cent., were static.

Flat-foot often coexists with lateral curvature as cause and effect (*vide* Scoliosis.)

Congenital Valgus.

As a congenital affection, in its mildest form it is often found at birth, but seldom in such a degree as to be considered pathological. It is, moreover, a well-established fact that all infants on commencing to walk are flat-footed, and do not acquire a perfect plantar arch until they have exercised some time and the leg muscles have become developed.

Of well-marked congenital valgus, Adams found 42 cases in 764

---

[1] Lehrbuch der orthop. Chir., p. 697.

patients afflicted with congenital deformity of the feet, and Küstner[1] found 13 cases of well-marked valgus in 150 newborn infants taken consecutively.

The statement that most infants have flat-foot, with a tendency to varus, may seem paradoxical, but when it is understood that the varus relates to the inverted position of the anterior portion of the foot, and the valgus, or flat-foot, to the sole of the foot, the correctness of the statement will be evident.

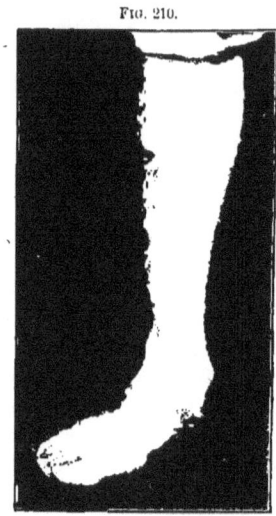

FIG. 210.

Medium degree of flat-foot.

Some recent observations of Dane[2] seem to show that at birth most infants have a distinct arch, which persists for eighteen months, after which a breaking down of the arch occurs. (Figs. 211 and 212.)

As a persistent congenital deformity three degrees of severity have been established, according to the amount of outward deviation of the anterior part of the foot—slight, intermediate, and severe. In the severer forms the entire sole of the foot rests upon the ground, the internal border is lengthened, and the external border shortened and creased over the medio-tarsal articulation, or the foot may even be rotated outward, the sole being everted and turned backward.

But few opportunities have been offered to examine the dissected specimens of congenital valgus. Adams, Ringnetta, and Lacour have each recorded the results of their dissections. According to these, the astragalus is tilted downward and forward, the head depressed on its outer side, its upper part projecting from the rotation of the scaphoid, and the changed direction of the astragalo-scaphoid articulation. The tuberosity of the os calcis is short and thrown forward. The cuboid and scaphoid are rotated inward. The malleoli are depressed, the inner one resting in severe cases upon the ground.

---

[1] Archiv f. klin. Chir., 1880, 25, p. 397.
[2] Boston Med. and Surg. Journ., October and November, 1892.

FIG. 211.

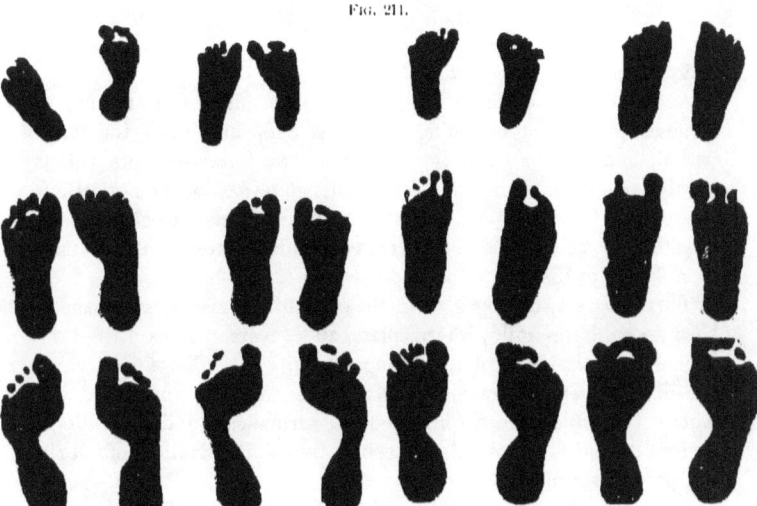

Tracings of male feet. Ages: Four weeks, three months, one year two months, one year ten months; two years, two years six months, three years, three years six months; four years, five years seven months, six years three months, seven years. (DANE.)

FIG. 212.

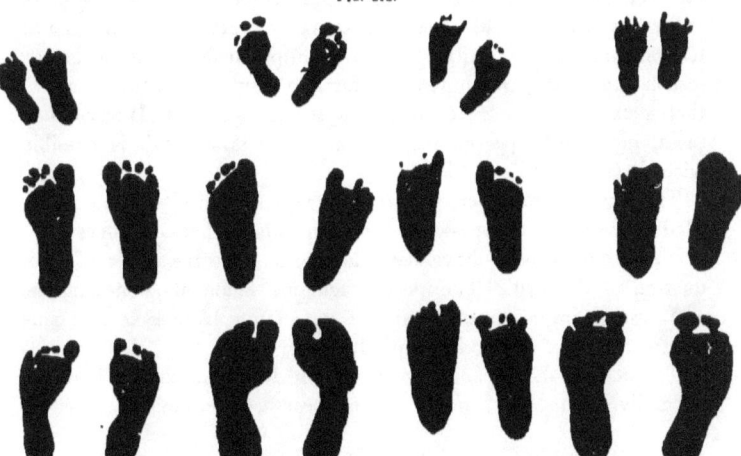

Tracings of female feet. Ages: Four weeks, three months, one year two months, one year ten months; two years, two years six months, three years, three years six months; four years, five years seven months, six years three months, seven years. (DANE.)

Acquired Valgus.

Synonyms. *Splay-foot; Flat-foot.*

As an acquired affection flat-foot is most common from infancy to adolescence, seldom becoming a serious deformity until the latter period, and always occurring in one of two forms—as pes valgus paralytica, resulting from poliomyelitis anterior, or as pes valgus acquisitus, from simple muscular relaxation, as associated with rhachitis or as a sequela of burns, traumatism, osteitis of the tarsus, or ankle-joint disease.

Three degrees, or stages, as in the congenital variety, have usually been recognized—mild, intermediate, and severe. Thus Little[1] describes three varieties, of increasing intensity : 1. Those " in which the displacement is transient during the erect position only." 2. Those " in which the displacement is permanent, and *deformity* is present ;" and 3. Those characterized by an aggravation of all the symptoms to a marked degree.

A very good division, also, is that of Roth,[2] who describes three varieties : " 1. Cases in which it is possible to restore the foot completely to the normal shape by passive manipulation, without any force exerted by the surgeon, or by making the patient stand with the heels raised. 2. Cases where the tarsal bones have become more or less fixed in their displaced positions by shortened ligaments and tendons, osseous deformity of the articulation surfaces, and fibrous or osseous ankylosis, which requires forcible manipulation under anæsthetics to restore more or less of a normal arch. 3. Intermediate cases, in which a partial restoration of the tarsal arch is possible without *brisement forcé.*"

The division, however, which I prefer is that into—1. Slight ; 2. Medium ; and 3. Severe—which corresponds to that of Reeves.[3]

Numerous theories have been advanced as to the mode of production of flat-foot. The muscles, tendons, ligaments, bones, fasciæ, and even an improper mechanical construction, have each been considered the chief factors in its production.

Thomas, of Liverpool, and Roth,[4] of London, believe that the normally constructed human foot is incorrect anatomically, because

---

[1] Little : On Deformities, 2d edition, p. 233. London, 1853.
[2] Roth : New York Medical Record, March, 1888, p. 51.
[3] Practical Orthopedics, 1885, p. 185.
[4] Lewis, H. K. : New York Medical Record, March, 1888, p. 53. Also, Lateral Curvature, etc., London, 1889.

" the lower end of the tibia is placed too much toward the inner border of the foot," and that this is the primary cause. Stromeyer[1] believed that atony of the plantar fascia and the tarsal interosseous ligaments were responsible.

H. von Meyer[2] considered it due to a rotation of the astragalus inward and pressure atrophy of the bones; while Reisman[3] considered it due to muscular contraction, first of the extensors and then of the pronators; but Volkmann and Lorenz[4] have disproved this, the latter supporting the view of Henke,[5] who announced that muscular insufficiency and body pressure were the chief factors; an explanation first proposed by Duchenne (of Boulogne), who considered the peroneus longus particularly at fault.

Of these theories, the last has always appeared to me the most satisfactory: that it is a muscular affection from the first. Thus muscular relaxation, either from debility or paralysis of the peroneus longus, leads to an improper distribution of the body weight upon the tarsal arches, which, by undue pressure upon the interosseous ligaments, especially the inferior calcaneo-scaphoid and calcaneo-cuboid ligaments and plantar fascia, induce their subsequent relaxation, stretching, and obliteration of the normal arch.

The symptoms differ in the two varieties—pes valgus acquisitus and pes paralytica—but in each are characteristic.

a. *Pes Valgus Acquisitus.*

Synonyms. *Talipes Valgus Spurius; Inflammatory or Acute Flat-foot; Tarsalgia of Adolescents; Contractured Static Flat-foot.*

Patients suffering from pes valgus acquisitus have an attitude and walk that is characteristic. They stand with the feet everted, resting on the inner side of the sole, with the knees in a flexed valgoid position, and walk with a difficult, heavy gait, the knees flexed and the feet placed everted flat upon the ground in an uncertain, careful manner. Such patients have an expression of anxiety, tire easily, and complain of great pain of a dull, aching character, or sharp, excruciating in nature, about the inner malleolus, instep, or ball of the great toe, accompanied in some cases with swelling. On inspec-

---

[1] Beiträge zur operativen Orthopädik, 1838.
[2] Ursache und Mechahismus der Enstelung des erworbenen Plattfusses, 1883.
[3] Der erworbene Plattfuss, Langenbeck's Archiv, 1869, Bd. 2, Heft 3.
[4] Die Lehre vom erworbenen Plattfusse, 1883.
[5] Zeitschrift für rationelle Medicin, 1859, 3d series, vol. v.

tion the external appearance varies with the degree. In the mild variety the plantar arch is depressed, the inner malleolus lowered and prominent; the foot is cold, bluish, and perspires freely, and an outline taken to ascertain the angle of deflection [1] is found reduced from thirty-four degrees normal to about twelve degrees in mild, to eight degrees in severe, cases.

The patient cannot raise himself on his toes, but the deformity can be readily restored by manual pressure. In the medium variety all these conditions are exaggerated, the foot is completely flattened, and somewhat everted, the inner side presents two prominences below and anterior to the internal malleolus, the head of the astragalus, and the tubercle of the scaphoid; and on taking an outline the angle of deflection is found reduced to probably eight degrees. The pain is common about the inner lower portion of the astragalus, and may be so severe as to render exercise intolerable. The arch can still be restored, and the deformity is not prominent.

In the severe variety the foot is more everted, making an angle of deflection in severe cases as low as five degrees. This is due to functional paresis of the peroneus longus, and increased reflex contraction of the peroneus brevis, tertius, and extensor communis digitorum. The pain is increased by the unbalanced action of muscles, giving a painful spot at the base of the first metatarsal bone; by the direct pressure upon the ligaments, giving a point over the calcaneo-cuboid and astragalo-scaphoid articulations, and from direct pressure upon the plantar nerves referred to their distribution. Fibrous and osseous changes have occurred, and the arch cannot be restored without some operative procedure.

The pathology of the affection was for some time obscure, especially of this the non-paralytic valgus, it being supposed at one time to be inflammatory. Such cases have frequently been mistaken and actively treated for inflammatory lesions of the bones or ligaments; complete inflammation, with effusion of lymph, serum, or pus, never, however, occurs. All the inflammatory symptoms are simply the result of pressure.

The extensors, particularly the peroneus longus, from some cause, become relaxed. This is followed by stretching and elongation of the plantar muscles and tendons, the long and short plantar ligaments, the calcaneo-cuboid and astragalo-scaphoid ligaments, and the

---

[1] Rohmer: Les Variations de Forme normales et pathologiques de la Plante de Pied. Thèse, Nancy, 1879.

plantar fascia, and a giving way of the inner antero-posterior arch. The astragalus is pushed downward, forward, and inward; the scaphoid is rotated, the internal surface downward and the external surface upward; the ligaments, generally on the inner and plantar surfaces of the bones, are stretched, the joint surfaces separated on the plantar and crowded together on the dorsal aspect. This may become so great that the bones entering into the medio-tarsal (Chopart's) joint divide, the anterior portion resting upon the ground and the posterior portion drawn up by the extensor so that the heel does not touch the floor, making the so-called "canoe-shaped sole."

FIG. 213.

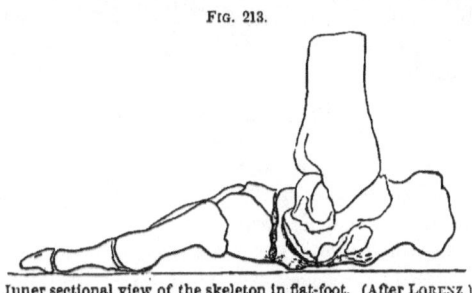

Inner sectional view of the skeleton in flat-foot. (After LORENZ.)

The articular surfaces become eroded from the continued pressure, and osseous ankylosis renders the deformity permanent. The diagnosis in many cases can readily be made, but in others, especially the inflammatory variety, with which we are most concerned, very careful attention is required to avoid confounding it with other conditions.

From the congenital forms of valgus it can be distinguished by the latter always being bilateral and the peroneus responding to electrical stimulus. (The slight normal tendency to flat-foot observed in infants has already been referred to.) Its occurrence about the period of adolescence, the previous history of rheumatism, knock-knee, bow-legs, lateral curvature, traumatism, bone or joint disease, should be all carefully inquired into.

It is most commonly mistaken for rheumatism, of which it may be a sequela, and for inflammatory conditions of the tarsal and metatarsal joints or ligaments. Its angle of deflection determined from the outline will be the best single symptom to distinguish it from the former, and the same with reflex muscular contraction, or a collection of inflammatory fluid, from the latter. The lesions from which it would be most difficult to distinguish it are the early stages of perforating ulcer, and such neuropathics as Charcot's joint disease. From these in the earlier stages a differential diagnosis cannot be made, but the subsequent course would soon decide the true nature of the affection.

The prognosis will depend on the cause and general condition and surrounding of the patient and degree of the deformity. If advice is sought early, and the disease is still in the first or even second degree, the prospects, under more appropriate treatment, of great improvement or even complete cure are good. Certainly entire relief of pain may be expected. If, however, ligamentous or osseous ankylosis has occurred, some relief may yet be hoped for, but resort must be had to *brisement forcé*, or tarsectomy, before the deformity can be overcome.

b. Pes Valgus Paralytica.

As a result of infantile paralysis (poliomyelitis anterior) paralytic flat-foot is favored by the flail-like condition of the foot upon the ankle-joint, the relaxed ligaments, the effect of gravity, and is aggravated by the bearing the body weight upon the foot in its weakened everted position. Neglected cases grow progressively worse until the severest grades are reached. This tendency is increased by the so called antagonistic contractions, which, as pointed out by Hueter, are not at all muscular actions, but are due to atrophy and lack of growth.

Volkmann has shown that these deformities gradually become permanent, are not dependent upon tonic retraction; nor is the shortening of the muscles due to contraction, but to the growth of the limb, the foot remaining in its deformed position. This is particularly true of the peronei group of tendons, which often stand out prominently behind the external malleolus. The walk of patients suffering with paralytic valgus is peculiar and characteristic. The everted foot is thrown outward and forward, describing the arc of a circle, the centre of which corresponds to the position of the other foot. The foot rolls inward and the body sinks downward on that side as the superincumbent weight is thrown upon it. The sound foot is brought forward and the act is repeated, giving to the patient an oscillating progression. Aside from the general paralysis of the muscles the morbid anatomy of this variety of flat-foot is identical with that of the pes valgus acquisitis, and need not be here repeated.

The prognosis is influenced by the degree of the deformity and the amount of paralysis. The paralyzed muscles cannot be restored, but the position and usefulness of the foot can be much improved by appropriate measures. The electrical reactions of the muscles establish the diagnosis of the deformity.

Talipes Equinus.

Synonyms. French, *Pied-bot Equine.* German, *Spitzfuss; Pferdefuss.*

As a primitive form talipes equinus may be either acquired or congenital. The former as a paralytic deformity is common, but the latter is considered by Adams and Little very infrequent, and Tamplin, according to Adams, discredits its existence.

The most common causes of acquired equinus are infantile spinal paralysis, spastic paralysis, post-hemiplegic contractions, protracted decubitus, as typhoid, typhus, etc.; inflammation of the ankle-joint; wounds, burns, cicatrices, abscesses, etc., of the calf of the leg; hysteria, or neuromimesis. Compensatory equinus occurs from a shortening of the extremity from any cause, as hip disease, fracture, etc. The foot is extended upon the leg varying in degree from a condition in which the foot is a trifle over 90 degrees in relation to the leg, to one in which the foot is in a continuous line with the leg or even beyond 180 degrees. ᐧ According to the degree of deformity the external appearance, the walk of the patient, and the structural changes vary greatly. In the mildest form the general appearance of the foot is but little changed. In the severer grades the arch is increased, the plantar surface diminished, the plantar fascia contracted, the toes are extended, the calf muscles, particularly in the paralytic forms, are shrunken. When the foot is retroverted the skin of the dorsum becomes callous, thickened, and inflamed from walking upon it.

In mild cases locomotion is but little affected, the increased length of the extremity being overcome by flexion of the knee. The individual walks upon the ball of the foot and toes alone, dragging the limb or swinging it forward by circular movements. When the foot is retroverted it becomes practically useless.

The morbid anatomy of equinus consists essentially of a contraction of the superficial extensor muscles[1] and a relaxation or paralysis

[1] Pancoast asserted that the soleus, particularly its internal head, was at fault.

of the flexors of the leg. The bones of the foot may be unaltered, as in a case observed by Adams, or they may be and usually are much deformed in severe cases. The astragalus projects prominently on the dorsum of the foot, the head projecting free above the astragalo-scaphoid articulation, and the scaphoid being subluxated. The articular surface of the astragalus extends further posteriorly than normal.

FIG. 214.   FIG. 215.

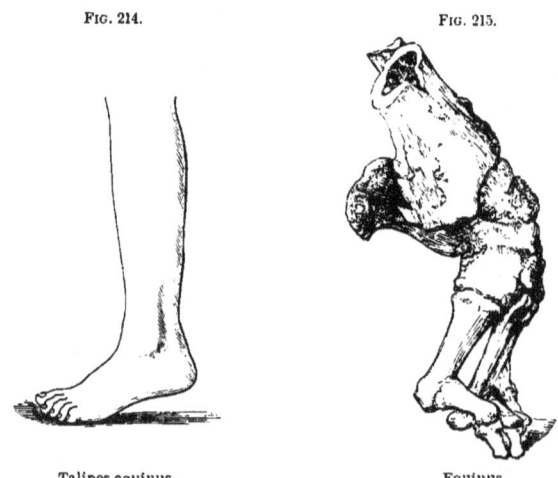

Talipes equinus.   Equinus.

The calcaneo-cuboid articulation is likewise affected, the anterior portion of the os calcis being subluxated. The ligaments on the plantar surface and the plantar fascia become contracted ; those on the dorsal surface become elongated.

The diagnosis should always be made with the knee extended, as this increases the deformity. The previous history is important in determining the cause.

The prognosis in general is good, particularly if osseous changes have not occurred.

Talipes Calcaneus.

Synonyms. French, *Pied-bot Talus ; Pied-bot Calcanien.* German, *Hakenfuss.*

This deformity, the direct opposite of the equinus, occurs both as a congenital and acquired affection, but as a purely primitive form it

is the rarest of all the varieties. Associated with valgus—as calcaneo-valgus—it is quite common. The deformity consists of a flexion of
the foot upon the leg, at the tibio-tarsal
articulation, upon an antero-posterior
plane. (Fig. 216.)

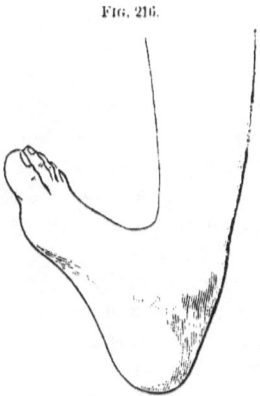

FIG. 216.

As a congenital affection it may be
the remains of the normal intra-uterine
position of flexion, or may depend upon
intra-uterine disturbances. According
to Adams, it is common in breech cases
with extended limbs. In congenital
cases the foot is observed to be flexed
upon the leg, and on attempting to walk,
only the heel comes in contact with the
ground. The dorsum of the foot is
shortened and wrinkled. The plantar
surface later becomes somewhat concave (pes cavus) from the action of

Talipes calcaneus.

the flexors of the foot and of the deep extensors of the leg.

Reeves has described three grades of congenital calcaneus, according to the degree of deformity: "1. When the foot is at a right
angle to the leg, and extension cannot be carried beyond that point.
2. The foot is at an acute angle to the leg. 3. The dorsal aspect of
the foot nearly, or quite, touches the anterior surface of the leg"
(p. 218).

Considering the amount of deformity, the tarsal bones are not
much altered, the changes being principally in the articular surfaces.
The astragalus seems to be displaced backward toward the posterior
part of the os calcis, and rotated backward, its neck in contact with
the inferior tibio-fibular surfaces and the anterior part of the upper
articular surface behind the tibia and uncovered. The posterior part
of the os calcis was bent downward, and the tubercle lifted slightly
forward in Nicoladoni's case.

Lannelongue considered the astragalus displaced backward toward
the posterior part of the calcaneum. The superficial extensors—gastrocnemius, soleus, and plantaris—are usually atrophied; the deep
muscles are usually normal.

As an acquired affection it frequently results from infantile paralysis, caries of the ankle-joint, wounds, burns, cicatrices, rupture of
the tendo Achillis, rapid stretching of the tendo Achillis after

tenotomy, etc. Cases are also recorded from pathological displacement of the epiphyses, following osteomyelitis,[1] and from spontaneous separation of the epiphysis.[2]

In the acquired form, particularly the paralytic variety, the relation of the foot to the leg is much the same; but there is usually a greater tendency to cavus, from the retraction of the plantar muscles and fascia. The extensor proprius pollicis, extensor longus digitorum, tibialis anticus, and sometimes the peroneus tertius, are all involved. The leg and foot are much atrophied, and the skin cold and purplish. The bones are more displaced, and the ligaments are lengthened posterior to the ankle.

There is no difficulty about the diagnosis of this deformity, and the prognosis in the congenital cases is excellent. The acquired variety can be much improved by suitable mechanical means, but the paralytic nature of the majority of these precludes the possibility of absolute cure.

Artificial Calcaneus.

The deformed feet of Chinese women bear a close resemblance, if not an exact identity, to this deformity, especially in its severer forms, and have been described by Mr. Adams and others as *artificial calcaneo-cavus*.

This artificial deformity is effected early in life, about the fifth year, by an ingenious method of foot-binding, peculiar to the Chinese proper and not accepted by the Manchu Tartars, or reigning power. The process of production has been well described by Dr. Robert P. Harris,[3] to whom we are indebted for the following:

"There are," he says, "three points to be gained by the binding, which collectively, and under the influence of an atrophic action, the result of defective nutrition, make the feet small, and give the skin of both feet and legs a shrivelled appearance. The first step, and that to which the parts make the least resistance, is the bending of the four smaller toes under the sole of the foot, and the narrowing of the parts supported by the metatarsal bones. The second step is best accomplished in early childhood, when the tarsal bones appear to move with a joint-like flexibility upon each other, and consists in a forcing together of the plantar portions of these bones, whereby they are subjected to a continuous pressure, and in an opening, or

---

[1] Kundrat.　　　　　　[2] Meusel: Centralbl. f. Chirurg., 1877, No. 50, p. 603.
[3] "Foot-binding in China," Trans. College of Phys. of Phila., December 3, 1879.

rather attempt at opening, of the articular crevices of the instep, keeping their ligaments constantly tense at the same time. The third step is the last to be accomplished, and can only be completed in the young subject. The os calcis, with the astragalus, is forced downward until the heel is vertical and its bone on a line with those of the leg. The calcis is rounded in form, losing its processes, which are so prominent in the normal bone; and its anterior articulating face is brought up near to the high, instep-like arch just in front of the ankle-joint, while at the same time the point of attachment of the tendo Achillis is made the base upon which the girl stands when erect, and steps in walking."

The anatomical changes are well shown both externally and by dissection. The characteristics are the bending under of the four smaller toes, especially the disproportionate size and length of the fifth toe, which reach to or even beyond the central line of the sole; the deep indentation of the sole at the medio-tarsal articulation, which, when perfected, amounts to a mere fissure an inch and a half deep; the vertical direction of the os calcis, and its square base; the height and prominence of the instep; the posterior position of the external ankle; and the diminutive circumference of the ankle.

Dissections have been made by Mr. Bransby B. Cooper and others, and almost every museum possesses dried specimens and casts of this interesting deformity. The Mütter Museum of the College of Physicians of this city is the fortunate possessor of a very fine collection.

Little did not entertain the slightest doubt that it might be entirely cured, even after thirty or forty years' duration, by means similar to those employed in its production—that the assiduous labor of years wilfully occupied in distorting Nature might be undone in a few weeks. Adams, however, doubted this, and did not believe a complete restoration possible, because "the general adaptation of all the ligaments of the foot and of the tibio-tarsal articulation would present an insurmountable obstacle to the restoration of the form of the foot in this severe grade of distortion—at least at the adult period of life."

Pes Cavus.

Synonyms. German, *Hohlfuss.* French, *Pied Creux; Pied-bot Talus.*

This consists of an elevation of the arch and an excavation of the sole of the foot. It may be either congenital or acquired, though the former is rare.   As an acquired affection it results usually from paralysis of the superficial calf muscles, the gastrocnemii, and soleus.

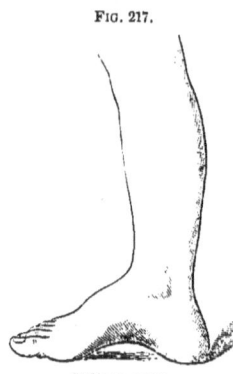

FIG. 217.

Talipes cavus.

It may also result from a contracted condition of the tibialis anticus and peroneus longus muscles, and the plantar fascia. The dorsum of the foot is increased and prominent, the sole of the foot is shortened, and in severe cases the heel and balls of the little and great toes only rest upon the ground in walking. (Fig. 217.) As a primary affection it is rare, though the writer has met a case, it being usually associated with calcaneus; as such it usually depends upon contraction of the plantar fascia.   As a complication of congenital equino-varus, it sometimes persists after the complete correction of the equinus and varus by mechanical extension.   The walk in congenital cases is not much affected, but may become painful and even impossible from callosities and ulcerations the result of pressure.   In others, described as "painful cavus," tarsalgia may result from over-use, independently of the pressure symptoms.   The anatomical changes, except in severe cases, are not, as a rule, marked.   The plantar fascia is contracted, the dorsal ligaments are elongated.

The diagnosis is easy, and it has only to be distinguished from the compound form, calcaneo-cavus.   The prognosis in the congenital cases is excellent, and even in the acquired forms the deformity may usually be much relieved by appropriate measures.

Pes Planus.

Synonyms.   *Splay-foot; Spurious Valgus.*   German, *Plattfuss.* French, *Pied-plat.*

Splay-foot, the direct opposite of pes cavus, consists essentially of a flattening of the plantar arch without abduction of the anterior part of the sole.   The latter element differentiates it from talipes valgus.

It may be either congenital or acquired.   It is hereditary in, and characteristic of, certain races, particularly the Jews and negroes.

The anatomical changes consist in relaxation of the ligaments which support the plantar arch, allowing the tarsal bones to rest upon the ground.

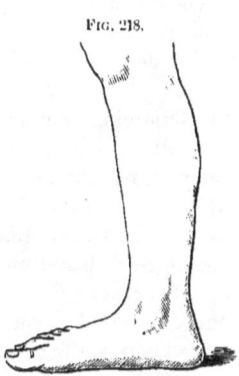

FIG. 218.

Talipes or pes planus.

The entire sole is applied to the ground in walking, and progression loses its normal elasticity. Tarsalgia from abnormal pressure is a frequent symptom. It never reaches a degree beyond the first stage of valgus. The normal condition of the bones and joints, especially the astragalo-scaphoid joint, serves to distinguish it from talipes valgus.

Non-deforming Club-foot.

Under this head Dr. Shaffer,[1] of New York, described a variety of equinus which hitherto had escaped the observation of surgeons and writers on deformity. (Fig. 219.)

"In non-deforming club-foot," he writes, "all the conditions found in certain forms of talipes exist, with the exception of the exaggerated deformity. That is, there is a loss of normal relation between the articulation at the ankle and the muscles which act upon it, involving also in many instances the tarsus, producing a condition which prevents flexion at the ankle-joint, and modified mobility, with slight deformity at the tarsal, metatarsal, and phalangeal articulations. With this state of affairs we find as a result, varying with the conditions present, actual disability, pain, sometimes very severe, in various parts of the foot, ankle, leg, and even reflected to the lumbar region, and tender and inflamed articular surfaces, especially at the junction of the first metatarsal bone with its phalanx." It may occur at any age, but occurs most frequently among young and rapidly growing females,

FIG. 219.

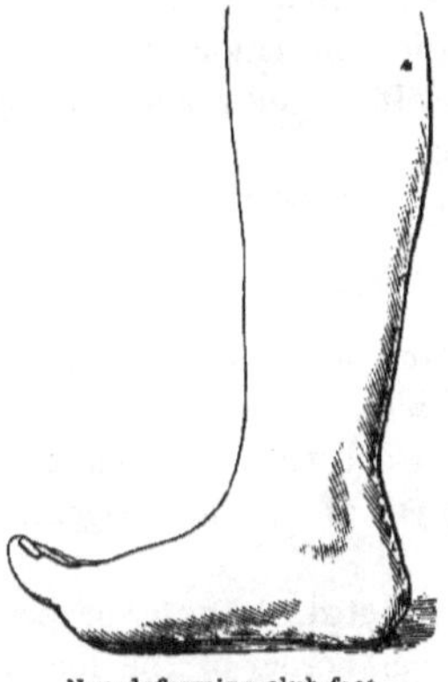

Non-deforming club-foot.
(SHAFFER.)

[1] N. Y. Med. Record, May 23, 1885, p. 561.

23

and is often associated with true rotary lateral curvature of the spine.

The diagnosis is simplified by a previous hysterical history, but is in all cases difficult and demands unusual caution.

Non-deforming club-foot may result from five different causes: "1, non-deforming club-foot seen after poliomyelitis anterior; 2, non-deforming club-foot which follows simple and uncomplicated malposition, habit, etc.; 3, non-deforming club-foot produced by traumatisms, sprains, etc.; 4, non-deforming club-foot found after infectious diseases of childhood, especially diphtheria and scarlet fever; 5, non-deforming club-foot due, as I believe, to some remote trophic disturbance and seen quite frequently coexisting with true lateral curvature." The diagnosis is made by finding that the foot cannot be flexed beyond a right angle. The prognosis under appropriate treatment is excellent and recovery prompt.

Neuromimesis of Club-foot.

The mimicry of club-foot, the so-called *hysterical club-foot,* is more familiar to neurologists than to orthopedic surgeons, but the present article would be incomplete without some reference to this interesting deformity. It occurs most frequently in young females about puberty, and consists of a chronic spasmodic contraction of some of the muscles of the leg, particularly the calf muscles, and usually associated with some other hysterical element.

Talipes Equino-varus.

Synonyms. *Club-foot; Reel-foot; Stump-foot; Pes Contortus.* German, *Klumpfuss.* French, *Pied-bot Varus.*

This, the most common variety of club-foot, may be either congenital or acquired. It is the most frequent of all the congenital deformities.

The deformity is threefold: the heel is elevated, the anterior portion of the foot is adducted, and the internal border of the foot is raised upward.

It has been customary to subdivide the deformity into five grades or degrees; more recently three forms have been distinguished. Thus Schreiber gives:

1. Where the foot can still be brought into the normal position

and the angle between the foot and the lower extremity is greater than 90 degrees.

2. Where correction is not possible to such a degree, the foot being drawn more internally.

3. Where it is impossible to correct the deformity manually, the foot being at an acute angle with the inner surface of the lower extremity, all the tissues on the inner side being shortened.

The former division, according to the etiology, into congenital or acquired is preferable, and the deformity can best be described from well-marked cases in the infant and neglected varus in the adult. As pointed out before in the etiology, the theory of non- or retarded rotation is believed best to explain the congenital production of this deformity, though there are many eminent authorities who still favor the mechanical theory. The acquired forms may result from one of three causes: 1, nervous, either spastic, as from tetanoid paraplegia, spinal sclerosis, pseudo-hypertrophic paralysis; or, paralytic, especially

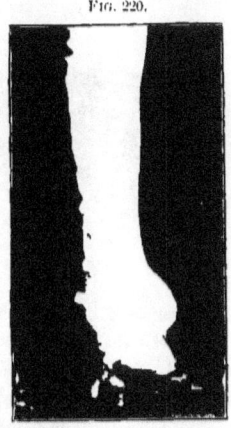

FIG. 220.

Talipes equino-varus.

infantile or progressive muscular atrophy; 2, traumatic, such as fractures, dislocations, sprains, etc., in the neighborhood of the tibio-tarsal or mid-tarsal articulations; 3, articular, from chronic joint affections of the same articulations.

Congenital Equino-varus.

In severe congenital equino-varus the anterior portion of the foot is turned inward, the adduction ranging from 40 degrees to 63 degrees, with an average of about 51 degrees; the sole is directed backward, and its inner border directly upward; the heel is small and misshapen; the internal malleolus less prominent, the external more prominent than normal and located more downward and backward. The dorsum of the foot is irregular, through the prominence of the head of the astragalus and anterior extremity of the os calcis, and the obliquity of the neck of the astragalus.

In cases of greater severity the inner edge of the foot may approximate the inner side of the leg, and the great toe is separated from the second by the contraction of the extensor pollicis muscle.

The *morbid anatomy* in congenital equino-varus consists in marked changes of all the structures except the muscles, which at birth are almost always healthy, but soon atrophy from disuse.

The *os calcis* is altered both in form and position. Its form in severe cases is slightly concave on the inner side. Its position in

FIG. 221.

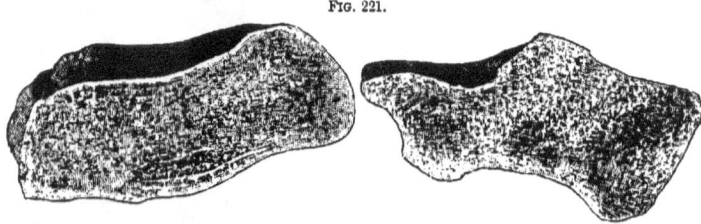

Section of calcaneum from congenital foot and from well-developed normal foot. (HOFFA.)

severe cases is more oblique, almost vertical, from the contraction of the superficial calf muscles, and the anterior extremity is directed obliquely forward and inward. Its tuberosity is directed outward and in severe cases in contact with the fibula.

The *astragalus* is altered in position and form and rotated on an antero-posterior axis. It is tilted obliquely forward and downward

FIG. 222.

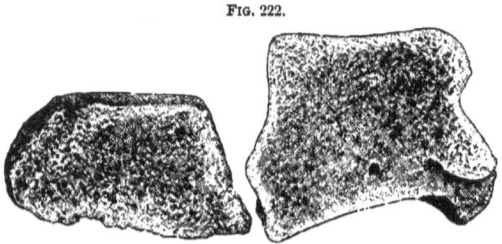

Section of astragalus from congenital foot and from well-developed normal foot. (HOFFA.)

and the superior articular surface somewhat displaced, the anterior portion being prominent on the dorsum of the foot. The lateral facets of this articulating surface are extended from the ankle-joint forward. The neck of the bone is deflected obliquely inward, forming an obtuse angle with the body of the bone, bearing, according to Parker, a strong resemblance to the os calcis of the ape.

While the normal angle (found by a line drawn parallel with its external border and the sagittal axis of the body) amounts at birth to an average of 38 degrees, and in the adult healthy foot to 26½ degrees, in congenital equino-varus it varies from 64 degrees to 31 degrees, the average being 49.5 degrees.

In consequence of this obliquity of the neck of the astragalus the articular head does not project directly forward, but in an antero-lateral direction, and in severe cases the surface has two articular facets instead of one, the larger of which articulates with the displaced scaphoid.

The *scaphoid* is not altered in form, but is drawn upward, inward, and backward by the tibialis posticus, assisted by the tibialis anticus and extensor proprius pollicis muscles.

The *cuboid*, as Adams remarks, presents no alteration either in form or position. Later, as the effect of increased lateral growth, its shape, as seen from above, is quadrilateral rather than triangular.

The cuneiforms and metatarsals are inverted, but not otherwise altered in relation to one another or to the scaphoid and cuboid bones.

The *malleoli* in recent congenital cases are normal, but from the inward twisting of the lower portion of the tibia, there is an apparent deficiency of the internal malleolus, and an exaggerated projection of the external malleolus, (Figs. 223, 224.)

The *ligaments* are only markedly affected in the severe grades. In these the internal (lateral, especially the deltoid portion of it), the ligaments on the plantar surface, the calcaneo-cuboid, calcaneo-scaphoid, and the plantar fascia are all contracted. Adams considered that the severity of the case might be judged by the ligamentous rigidity rather than by the external form, and Parker considers the ligaments as the only constant anatomical hindrance to rectification. The dorsal ligaments, the astragalo-scaphoid, and the anterior portion of the capsular ligament of the ankle-joint are elongated.

The *muscles* are at birth almost always healthy and retain their electrical reactions. They are subject later, in uncorrected cases, to the atrophy of disuse. In rare cases hypertrophy of one or more muscles has been found, as in the case recorded by Mr. Adams.

The *tendons* are always much displaced and adapted to the altered position of the bones. A knowledge of the altered relations of the tibialis anticus, tibialis posticus, and tendo Achillis is of direct practical importance in tenotomy. The tendon of the tibialis anticus, as it crosses the ankle-joint, is much displaced to the inner side, passing in severe cases obliquely downward across the internal malleolus.

The tendon of the tibialis posticus at the point usually selected for its division—just above the malleolus—is placed relatively more forward than in the normal foot, being exactly midway between the anterior and posterior borders of the leg on its internal aspect.

Fig. 223.

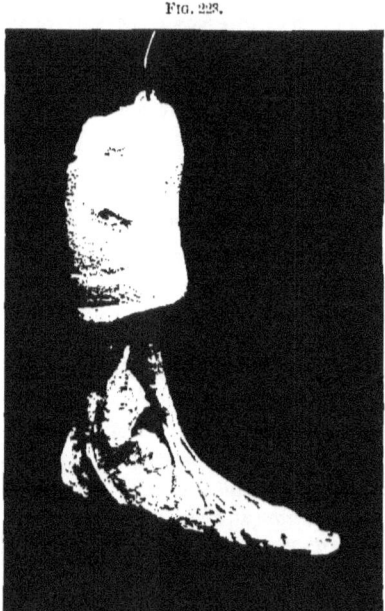

Normal foot (right).   (BURRELL.)

The tendo Achillis, in consequence of the lateral obliquity of the os calcis, is inclined toward the external malleolus, and consequently further removed from the posterior tibial artery than normal. The relations of the other tendons, though altered, have no practical bearing. The vessels and nerves are contracted on the internal aspect. This contraction of the vessels, particularly the posterior tibial, should be remembered in forcible manual rectification of the foot, lest the circulation be arrested or the artery be ruptured.

In *neglected varus* and in uncorrected adult cases as the result of pressure in walking, the deformity is very greatly exaggerated. The inversion of the foot brings the part at right angles to the leg, the sole is directed upward and backward, the dorsum downward and forward, the latter callous from walking; the heel is drawn up, and the foot is shorter than its opposite, from atrophy. A deep transverse groove marks the tibio-tarsal joint, and a deep longitudinal groove divides the sole. The calf muscles are much atrophied. There is

Fig. 224.

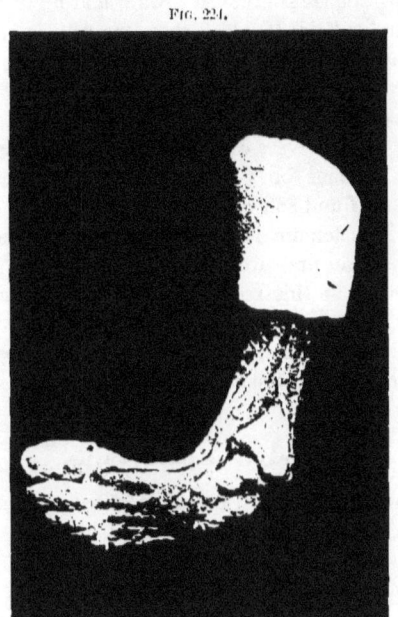

Talipes equino-varus (left). (BURRELL.)

slight motion at the ankle-joint, the feet in walking being lifted one over the other, a circumstance which probably gave origin to the term "reel-foot."

The anatomical changes are great. The astragalus is completely vertical in position, the neck is much elongated and misshapen, and the head has two articular facets at right angles to each other. The os calcis occupies an oblique, almost vertical position, its tuberosity being deflected toward the fibula. The body of the bone is curved, the convexity being outward.

The scaphoid is atrophied, compressed in its inner half, and drawn completely under the internal malleolus, its long axis being vertical and parallel to that of the astragalus instead of being at right angles to it.

The cuboid bone is displaced inward, exposing two-thirds or more of the anterior articular facet of the os calcis. In its altered position it sustains the greater part of the superincumbent weight, and becomes from pressure somewhat wedge-shaped from below upward.

The cuneiform bones are moved inward and upward together with the scaphoid and cuboid, to which they retain their normal relations.

The greatest deviation from the normal occurs in the metatarsal and phalangeal bones, which give to the deformity its characteristic appearance. These are placed at right angles with the inner side of the leg, or, in severe cases, are even inclined backward. The ligaments and tendons on the plantar surface are contracted, the muscles of the calf and sole of the foot are atrophied, and bursæ are developed on the dorsum and outer side of the foot. The skin becomes callous from pressure, but otherwise retains its natural appearance, differing in this respect from that in the paralytic variety. The deformity in severe cases is so great that the foot no longer resembles the normal member, but its appearance suggests the extremity of an animal ; hence the popular names of "devil's hoof," "cow-foot," etc.

Paralytic Equino-varus.

This, the most common variety of acquired equino-varus, is produced by the healthy muscles predominating over the paralyzed. The external appearances are not so severe nor so marked as in the congenital form. The foot is not rigidly fixed. The dorsal projections and plantar depressions are absent. The skin is purplish and cold, and the leg is atrophied and its outline obscured by a thick pad of fat. The bones of the tarsus are also usually atrophied, and the foot and limb are smaller and shorter than its fellow. Fatty degeneration of the paralyzed muscles commences early, differing in this respect from the muscular changes occurring in the congenital affection. Marked changes also occur in the tendons, these being often but one-half the size of those in the opposite leg.

Talipes Equino-valgus.

This deformity is characterized by an elevation of the heel, associated with an inversion of the anterior portion of the foot, combining the conditions present in the two simple varieties—equinus and valgus.

It may be either congenital or acquired, the latter variety being much the most common. As an acquired affection it includes many of the severer grades of valgus, so that it would seem advisable, as suggested by Adams, in the nomenclature of club-foot either to do away with equino-valgus as a separate variety, or to add to its importance by classifying under this term a large number of severe

FIG. 225.

FIG. 226.

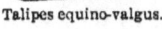

Talipes equino-valgus.

Talipes calcaneo-varus.

deformities described hitherto as simple valgus. The external appearances, symptoms, and morbid anatomy have already been described under the two simple forms, and need not here be repeated.

The diagnosis depends upon the association of the two simple forms, and the prognosis will be much influenced by the amount of equinus present.

Talipes Calcaneo-varus.

In this variety the heel is without elevation or is even depressed, and the inner margin of the foot is raised, the sole contracted, with the dorsum prominent toward the outer side. It occurs as a congenital affection, but as an acquired deformity from paralysis or spasm it is exceedingly rare, and Little, among others, has doubted its existence.

As a congenital affection it results from a preponderance of action of the tibialis anticus, tibialis posticus, and contraction of plantar fascia. The degree of deformity is usually slight, and always unilateral. The external appearance, morbid anatomy, and symptoms combine what has already been said of the two simple forms. A

deformity resembling the one under consideration sometimes results
from a partial correction of equino-varus, from omission to remedy
adduction and plantar contraction, after improper section of the
tendo Achillis. For this reason the modern method of first correcting the varus should always be adopted. The prognosis in the
congenital variety is good.

Talipes Calcaneo-valgus.

This is characterized by a depression of the heel, associated with
an elevation and abduction of the anterior part of the foot. It may

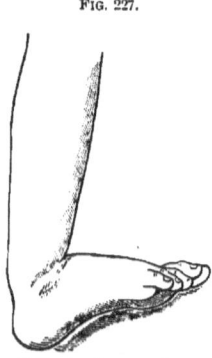

FIG. 227.

Talipes calcaneo-valgus.

be either congenital or acquired, being in
either case due to contraction of the peronei
and extensor longus digitorum muscles combined with contraction of the anterior tibial
and extensor proprius pollicis muscles.

Being a compound variety it combines all
the peculiarities of these two simple forms.
As an acquired affection it usually results
from paralysis or articular disease. In
infantile paralysis, the most common cause,
the foot is abducted by the stronger action
of the external extensors of the toes and
the peroneus brevis. This, as pointed out
under valgus, is increased by the growth of
the limb, the foot remaining in its deformed
position. The diagnosis is easy, and the prognosis in the congenital
cases is excellent; in the acquired cases it will depend upon the
amount of paralysis present, and will be much improved by operative
and mechanical measures.

Diagnosis. Little difficulty is experienced in recognizing club-
foot or in differentiating its individual varieties. In estimating the
degree of the deformity, due attention must be given to the range of
mobility of the foot as given under its normal anatomy; and in
establishing the severity of the apparent distortion, due attention
should be paid to the age of the individual or length of time the
affection has existed, the extent to which it can be corrected by
manual means, and the existence of spasm or paralysis. Authors
have attempted to divide the congenital equino-varus into three
stages, according to the degree of severity, but these divisions being

more arbitrary than real, have but little practical value, and a division into (1) infantile, (2) neglected, and (3) relapsed, would probably be of more service. The acquired forms, however, particularly pes valgus acquisitus, admit of such a division—three forms, (1) slight, (2) medium, and (3) severe, being the one preferred (*vide* Flatfoot).

Prognosis. These deformities do not correct themselves, but if uncorrected grow progressively worse, but two cases of spontaneous recovery from congenital club-foot ever having been recorded, one by Baron Dupuytren [1] and one by the writer.[2] The prognosis, therefore, will depend much upon the form of treatment employed. Under the older mechanical methods failure in severe cases was frequently the result even after prolonged treatment. Adams,[3] in this connection, refers to the well-known case of the distinguished poet before alluded to, who during his childhood was placed under the care of the late Mr. Sheldrake, as well as other mechanicians, without deriving any permanent benefit. Under present methods club-foot, certainly congenital equino-varus, and all but the paralytic varieties, there is a certainty of curing club-foot never before obtained, and, as Bradford [4] remarks, "In no branch of surgery can a cure be more confidently promised than in the treatment of club-foot." The time required to accomplish a cure varies with the method adopted, from a few weeks to two or three months, and this will depend upon the merits of the individual case. If the cases have been slightly over-corrected, and this should be the rule, relapses will be rare occurrences. To avoid this, however, retentive or walking apparatus will be required, particularly in the younger cases.

The author has met with congenital cases in adults who refused treatment because they considered the deformity a dispensation of Providence, and dared not incur Divine displeasure. Such cases must remain deformed. The mental effect upon the individual after a correction is ofttimes marked, and, as suggested by Tamplin, patients are enabled to pursue occupations and responsibilities from which they were previously debarred. According to Dieffenbach, of all the women treated by him only one was married, indicating the asceticism of humanity in matrimonial affairs, and the impediment of these deformities to marriage ; and the writer is also acquainted with one instance where a female, beautiful in every other respect,

---

[1] Lectures on Clinical Surgery, p. 201.   [2] Medical News, May, 1894.
[3] Adams: Loc. cit., p. 30.   [4] Trans. Amer. Orthop. Assoc., vol. i. p. 112.

contracted a matrimonial alliance notwithstanding the presence of this horrible deformity.

Treatment. The successful treatment of club-foot varies according to the variety, age of patient, duration, and nature of the deformity ; but in the words of Mr. Adams, it demands in most cases "a judicious combination of operative, mechanical, and physiological means." The physiological and mechanical means include manipulations, *massage* and electricity, and the application of splints. The operative means include tenotomy, division of ligaments, myotomy, tarsotomy and tarsectomy, *brisement forcé*, multiple tenotomy and open incision, and amputation. All these measures aim to slightly over-correct the deformity, and to retain it in this position until the tendency to return has been overcome.

Manipulations, preferably by the hand, are as old as Hippocrates, and have from his time till the present been successfully employed in the correction of many congenital cases. In fact, in all mild congenital cases an attempt should be made to correct the deformity by manipulations before resorting to more severe measures. In the compound varieties—for example equino-varus—the correction should be divided into two stages : first, the eversion of the anterior portion of the foot; second, the flexion of the foot upon the leg. (Fig. 228.)

Fig. 228.

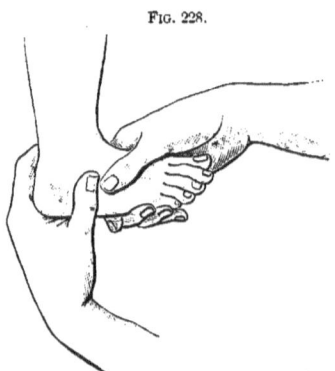

Manual correction of club-foot. (HOFFA.)

The foot must be firmly grasped with one hand while the other gently but firmly forces the foot into the correct position. This should be daily performed, and in the intervals the feet should be retained in their improved position. These manipulations must not be omitted until the foot is forced into an over-corrected position or the manipulations are abandoned for some method more radical. These forcible methods may also be employed under an anæsthetic, and in this manner it is often astonishing how much can be accomplished by manual measures alone.

Massage and *electricity* are particularly useful in the paralytic varieties, but they are almost as important in congenital cases as a part of the after-treatment; indeed, the writer is inclined to look upon the after-treatment of tenotomy, which includes these two measures, as almost as important as the operation itself.

The value of these measures in the acquired varieties cannot be overestimated. Especially is this true of infantile paralysis, in which massage not only keeps up the nutrition of the muscles and reduces the wasting, but assists in relaxing the shortened and contracted muscles and tendons. It should, if possible, be given by a person skilled in its use, and should not be continued too long at each application. Ten minutes is sufficient if but one leg be paralyzed. As a substitute for massage, Reubsam's rubber muscle-beaters are of service in these cases. Electricity may be applied in one week in infantile paralysis, provided no fever nor hyperæsthesia of the muscles be present, but for one month only the mildest currents may be employed. In deciding between the faradic and galvanic currents, the current which gives the greater amount of contraction with the weakest current, and the least amount of pain, should be the one selected. The treatment must be used daily for a short duration only; four or more contractions for each muscle are sufficient. One pole (the anode) is best applied over the nerve trunk, and the other (the cathode) applied firmly over the entire surface of the limb.

The *splints* employed in the treatment of club-foot may be divided into two classes: first, apparatus intended to correct the faulty position; second, apparatus employed to retain the foot in a corrected position. These can best be described under the varieties for which they are designed, and the principles upon which they act need only here be mentioned.

Almost all the corrective apparatus are but modifications of the ancient Scarpa shoe, by means of force applied by leverage, pushing or traction screw power, elastic traction, etc. The leverage brace is well illustrated in St. Germain's *appareil à plaquette*, which consists of a foot-piece and a lateral upright piece, more or less vertical according to the deformity. The foot is applied to the sole-piece and maintained in this position by strips of lead plaster applied to each

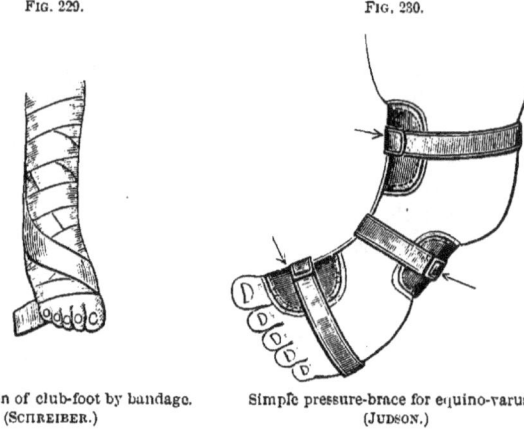

FIG. 229.

Correction of club-foot by bandage.
(SCHREIBER.)

FIG. 230.

Simple pressure-brace for equino-varus.
(JUDSON.)

side of the leg, passed through the openings in the foot-piece, and then brought over the foot. These vertical strips are retained by a

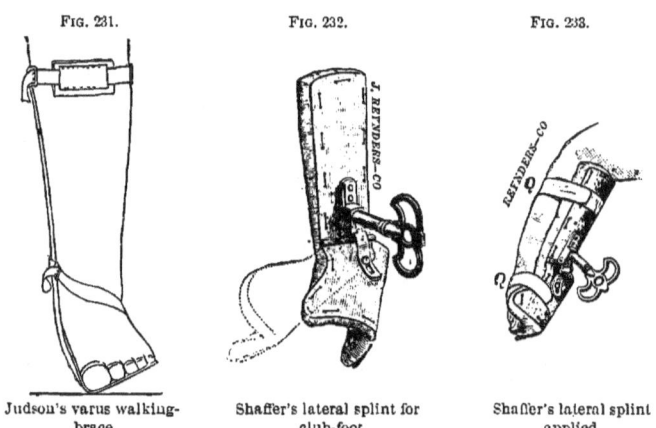

FIG. 231.

Judson's varus walking-brace.

FIG. 232.

Shaffer's lateral splint for club-foot.

FIG. 233.

Shaffer's lateral splint applied.

circular strip of plaster, and the vertical piece is drawn to the side of the leg by a bandage and the deformity corrected. As applied by the inventor after tenotomy, it is both a corrective and retentive splint. The pushing and traction force are the principles upon which Shaffer's and Roberts' modifications of Taylor's ankle support operate.

The internal lateral traction shoe of Shaffer is remarkable for its ingenuity. It consists of a steel trough fitted to the inner side of the leg, from the upper part of the tibia to the internal malleolus. From opposite the latter point a hinge, obliquely placed, connects this upright with the sole-plate. An endless screw acts upon this

Fig. 234.   Fig. 235.

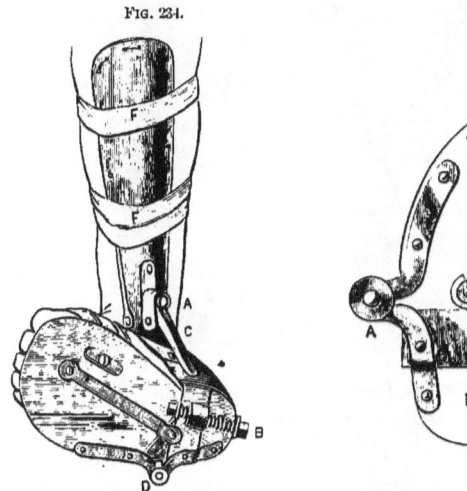
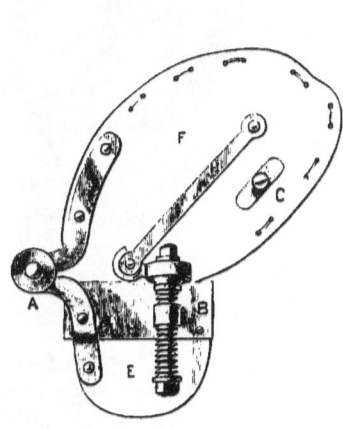

Shaffer's modification of Taylor's
ankle-brace.

Divided sole-plate of Shaffer's equino-varus
brace.

hinge in such a manner as to correct the deformity of the anterior portion of the foot. The sole-plate is divided opposite the medio-tarsal joint, and by means of a simple screw and lever, or the more powerful triple screw of Roberts, allows of extreme and powerful abduction of the anterior portion of the foot.

The apparatus is applied to the foot in its deformed position, and by means of keys the correction is gradually accomplished by means of this powerful screw force. In a somewhat similar but still more powerful manner external lateral traction is applied.

The use of elastic traction after the method of Barwell and Sayre has the advantage of not interfering with the free use of the joints and muscles, as well as simplicity of application to recommend it. It consists essentially of a zinc plate fastened to the front of the leg

FIG. 236.

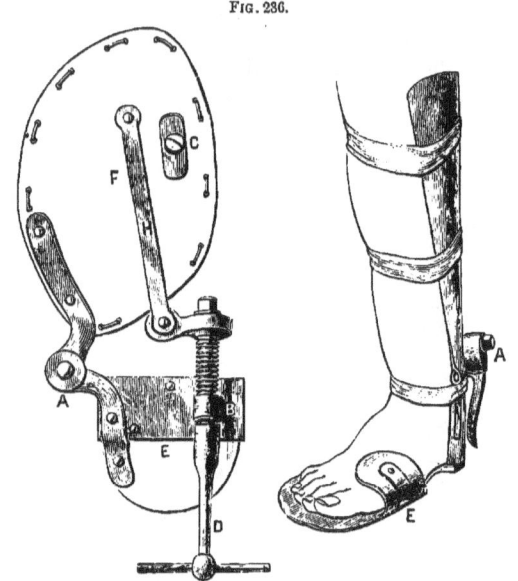

Sole-plate extended.   Brace adjusted to corrected position.

FIG. 237.   FIG. 238.

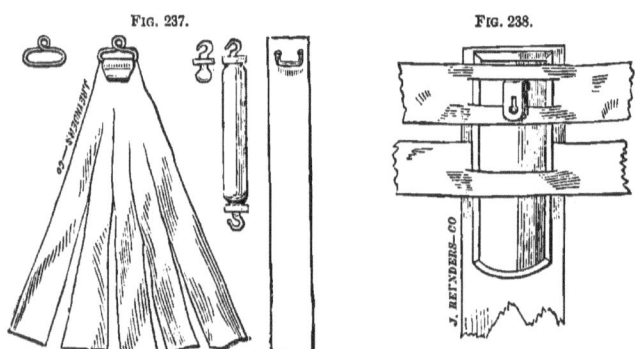

Burwell's appliance for club-foot.

by plaster, which serves as a fixed point of attachment for the elastic traction, and a fan-shaped piece of plaster cut into a number of strips and carrying a wire loop to serve for the second point of attachment. An elastic tube, with a hook at one end and a chain at the other, furnishes the traction force. Elastic traction is most valuable in paralytic deformities to supplement the action of the weaker muscles.

Fig. 239.

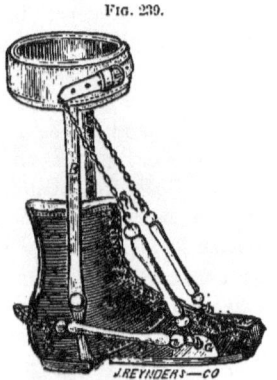

Sayre's club-foot shoe.

The retentive apparatuses are employed in the intervals between the manipulations to maintain what has been secured, or are applied after tenotomy and other cutting operations, to maintain the parts during the healing process. These may be constructed of metal (zinc, tin, lead, steel), rubber, felt, or prepared leather, and retained by an ordinary roller or plaster-of-Paris bandage. More elaborate apparatuses are sometimes employed, and of these preference is given to the universal talipes Scarpa shoe of Reeves. The plaster-of-Paris cast, well applied, furnishes the simplest, neatest, and best retentive apparatus. In small children, and when the dressing is carried high up the thigh, the plaster-of-Paris dressing should be reinforced with or substituted by a silicate of soda bandage to render it impervious to water. The dressing should be removed and reapplied every week or ten days, at which time the limb should be washed with soap and water, and vigorously rubbed with alcohol or soap-liniment.

Tenotomy. The history of subcutaneous tenotomy is the history of the advances in orthopedia as a scientific branch of surgery. Prior to 1784, orthopedic methods were limited to mechanical measures. In this year, Thilenius, a physician of Frankfort, proposed, and Lorenz performed for him, an open section of the tendo

24

Achillis for a case of acquired talipes equinus, which was entirely successful. In 1799, Petit, according to Handcock, subcutaneously divided the same tendon for the retraction of the heel after Chopart's operation. In 1806 Sartorius repeated the operation of Thilenius, and later Michaëlis (1809) and Delpech (1816) repeated the same with certain modifications. To Stromeyer (1831), however, belongs the credit of having first accomplished tenotomy as it is now performed. The success of the operation spread rapidly, and Dieffenbach, Bouvier, Pauli, Duval, Jules Guérin, Bonnet, and Scoutteten quickly adopted it. Into England it was introduced by Whipple (1830) and Little (1837), and later into this country by Rogers (1834), Dickson (1835), Detmold (1837), and Mütter (1844). It is interesting at this early period to observe the difference of opinion which existed as to the relative merits of immediate restoration of the foot as advised by Bouvier and sanctioned by Symes, and the plan of gradual mechanical extension as recommended by Stromeyer and indorsed by Little. Particularly is this true since a few authorities, even at the present time, do not in their works advocate immediate restoration of the parts. In this connection the exact pathological process by which divided tendons reunite becomes of great practical importance, for upon this devolves the solution of the problem of immediate or subsequent rectification of the deformed parts. Numerous experiments have been made by surgeons and physiologists to determine this. Von Ammon,[1] Adams,[2] Hunter,[3] Guérin,[4] Pirogoff,[5] Brodhurst,[6] Paget,[7] Mayo,[8] and others have all carefully studied the steps of this process, and the following is a *résumé* of their efforts.

Immediately following a well-performed subcutaneous operation a very little blood is effused into the space from which the upper part of the tendon has retracted. If much blood be effused it retards or defeats the formation of the proper exudate, this latter being not an organization of the blood-clot, but a specially formed reparative substance. The first result of this division is the effusion of a fluid or semi-fluid inflammatory lymph exudate. The adjacent bloodvessels enlarge and the tissues about the wound become infiltrated, yellow,

[1] De Physiol. Tenot., etc., Dresden, 1837.　　[2] Loc. cit., 1873, p. 16.
[3] The Works of Hunter, Lond., 1837, p. 34.
[4] Essais sur la Methode sous-cutanée, Paris, 1841.
[5] Ueber die Durchschneidung der Achillossehne, etc., 1840 and 1843.
[6] "On the Repair of Tendons, etc.," Proc. Roy. Soc., 1860, No. 37.
[7] Lectures on Surgical Pathology, 1863, p. 201.
[8] Outlines of Physiology, Lond., 1827.

and succulent. This exudate takes no part in the reparative process, and usually ceases in twenty-four hours. Somewhat later, about forty-eight hours in rabbits, the proper reparative material makes its appearance (the inflammatory exudate ceases or degenerates) and fills the entire space between the divided extremities of the tendon and unsheaths them both for a short distance. This gradually becomes firmer, stronger, and grayer, and about the fifth day forms a distinct bond of union between the ends of the tendon. The differentiation gradually advances, at first rapidly, later more slowly, in time becoming identical, even microscopically, with the original tendon.

Adams[1] in his dissected cases showed that the tendon and sheath at the point of section had almost entirely returned to its normal condition, and could scarcely be detected on the closest scrutiny; and Sir James Paget[2] says, in referring to the specimens of tendo Achillis, posterior and anterior tibial tendons deposited by Tamplin in the Museum of the Royal College of Surgeons, which had been divided four months before death, that "no trace of division of any of the tendons could be detected even with microscopic aid."

There is, however, in all instances a difference in the separability of the sheath of the divided tendon over the seat of the previous operation, if the dissection be carried from above downward or from below upward.

Of late years the method of immediate restoration, and in fact, of slight over-correction, after tenotomy is the one usually adopted by all practical surgeons. Under these circumstances it is very interesting to observe that this reparative exudate is always sufficient to fill the space between the severed ends, which in some instances is very great. In a recent case related to me by Prof. Willard, the ends were separated three inches and firm union occurred. The most interesting feature of the process, however, and the one bearing directly upon the subject of over-correction, is that brought out by some recent investigations of the writer upon rabbits, which demonstrated conclusively that if the ends of a divided tendon be separated a very short distance, but a small amount of exudate fills the space ; if they be separated a great distance the entire space is filled in ; and if they be separated at first but a short distance, or the ends remain in contact, and are subsequently separated, the resulting tendon will be a weakened spindle-shaped one, from the stretching-out of the plastic exudate.

[1] Club-foot, etc., 1873, p. 36.   [2] Loc. cit., p. 199.

In performing tenotomy much of the success depends on the technique. If the pointed tenotomes be employed, as, for example, the slightly sickle-shaped knife of Dieffenbach, no other instrument is needed; but if blunt, rounded tenotomes are used, another sharp-pointed bistoury must be employed for the puncture of the skin, and this is always a disadvantage, but one that cannot be overcome in certain localities.[1] (Fig. 240.) Anæsthesia is not absolutely essen-

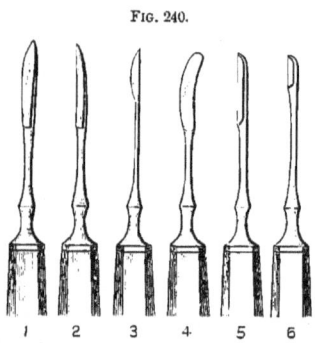

FIG. 240.

Author's tenotomes.

1. Straight, blunt-pointed. 2. Straight, sharp-pointed. 3. Convex, sharp-pointed. 4. Convex, probe-pointed. 5. Straight, probe-pointed, long cutting face. 6. Straight, probe-pointed, short cutting face.

tial, owing to the short time required for the operation, but is usually employed to overcome movements on the part of the patient which might interfere with the operation.

The skin having been previously rendered aseptic by cleansing with 1:1000 bichloride of mercury, the parts are so held as to render the shortened tendons prominent. The skin over the tendon is then punctured about its middle in such a manner that this incision and the one in the deeper parts shall not correspond when the parts are relaxed. The tenotome is carried flatwise beneath and close to the under surface of the tendon, the cutting edge is turned against it and the division accomplished by a slight rocking motion. The section is usually evidenced by an audible crackling and the extension

---

[1] Tenotomes are best made in sets of six or eight, the largest of which should not exceed three-quarters of an inch in length and one-eighth of an inch in breadth. They should be made of one piece of metal, nickel-plated, strong in the shank to avoid the danger of breaking off in the wound; and a set of this kind should include one or two pointed knives with rounded bellies for cutting small superficial tendons, and one pointed, flat-bellied tenotome with long cutting face for cutting the plantar fascia.

of the contracted part. The assistant should at once relax the part. The tenotome is then turned flat and withdrawn, the operator compressing the wound to prevent any ingress of air. The puncture is then dressed with a small wet bichloride compress, and by a superficial dressing of gauze and cotton, and a prepared or flannel roller. The parts should then be placed in a slightly over-corrected position and retained in position by a metal retention-shoe which allows inspection, or a plaster dressing. The wound need not be inspected for ten days unless pain or elevation of temperature calls attention to the part.

Owing to the prominence of the contracted tendons there is not the same difficulty there would be in the normal state, and not the same danger of wounding adjacent structures, hence the complications following tenotomy are therefore few and usually slight. There may be:

1. Too great wound in the tissues or puncture of the opposite side from movements of the patient. With an antiseptic dressing such accidents are of no importance.

2. Profuse hemorrhage from wounding a large artery, as the anterior or posterior tibial or plantar fascia, is sometimes alarming, but calls for an antiseptic compress, a ligature being seldom necessary.

Such an accident can usually by avoided by inserting the tenotome in such a manner as to avoid the artery, and then cutting away from it. For this reason, in section of the tendo Achillis, if a sharp-pointed tenotome be employed the puncture should be made on the inner side to avoid the risk of wounding the posterior tibial artery with its point.

3. A traumatic aneurism may be treated with a firm compress, or the vessel may be ligatured.

4. The section of a nerve, which, although a disagreeable complication, calls for no especial treatment, as the ends will probably unite in a short time.

5. Incomplete section of a tendon, which should be guarded against by observing that the tendon is completely severed before the tenotome is withdrawn. Otherwise the operator must forcibly rupture the undivided fibres or reinsert the knife.

6. Suppuration, which very rarely occurs, but should be met by free drainage and antiseptic dressings.

7. The non-union of a divided tendon is an exceedingly rare occurrence, but one which may happen. Adams reports having

witnessed it once, and this where the tendon had been divided in a situation not usually selected by orthopedic surgeons. The possibility of such an occurrence should be avoided by selecting in all cases the proper locality for section.

While experience confirms the truth of the statement that the accidents of tenotomy are few and usually slight, these operations are not, however, entirely free from dangerous complications. Thus, Agnew[1] records the fact that he has seen death from erysipelas follow the division of the adductor longus tendon at its origin ; has known a child to perish from concealed hemorrhage after an operation for club-foot, and has seen a leg rendered useless in a great measure from section of the peroneal nerve during tenotomy of the outer hamstring.

Surgeons differ much as to the proper time at which to perform tenotomy in club-foot, some operating very early, within the first month—Stromeyer operated upon one within twenty-four hours of birth—and others delaying until the child shows an inclination to walk. Since in all instances it is proper to first endeavor to correct the deformity by manual and mechanical means, and failing in this, to resort to operative procedures, a later period appears to me preferable. Much will depend upon the individual case, but from eight to twelve months would appear to be a good time to operate upon infantile cases. The later period has also this advantage, that failing by subcutaneous tenotomy to entirely rectify the deformity, resort can at once be made to other surgical procedures, to be discussed presently.

The individual tendons most frequently requiring division are, in the order named, the tendo Achillis, tibialis anticus, tibialis posticus, plantar fascia, and peronei. (Fig. 241.)

Division of the tendo Achillis. In dividing this tendon the patient, etherized, is placed upon the breast or side ; an assistant renders the tendon moderately tense by flexing the foot. The puncture is then made with the sharp-pointed tenotome, about one inch above its insertion, a short distance from the tendon to be divided, and preferably to the outer or fibular side, to avoid injury of the posterior tibial artery. The blunt-pointed tenotome is then inserted into the puncture flatwise, and carried behind the tendon as close as possible to its posterior surface ; the edge is then turned against the tendon, and by a slightly rocking, not a swinging, motion the tendon is divided, care being

---

[1] Surgery, 2d edition, vol. iii. p. 336.

taken not to divide the skin nor to enlarge the puncture wound. The division of the tendon is evidenced by a sensible and audible snap, the sinking-in of the soft parts and the extension of the heel. The assistant should at once relax the parts immediately the section is

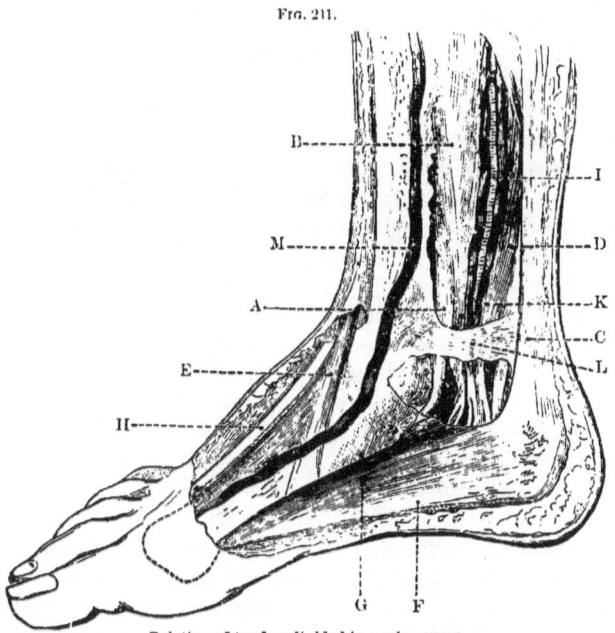

Fig. 211.

Relation of tendon divided in equino-varus.

A. Tibialis posticus—point at which cut. B. Flexor longus digitorum. C. Tendo Achillis—point at which cut. D. Flexor longus pollicis. E. Tibialis anticus—point at which cut. F. Plantar fascia. G. Abductor pollicis. H. Extensor longus pollicis. I. Posterior tibial artery. K. Posterior tibial nerve. L. Part of annular ligament. M. Saphena vein.

accomplished. The knife is then turned flat and withdrawn, pressure being made with the finger to prevent the ingress of air. The parts are then dressed as before given.

The writer has twice met with quite a severe hemorrhage in dividing the tendo Achillis, once in an infantile case and once in a young adult who had been previously operated upon. It appeared to come from a small unnamed artery which supplies the sheath of the tendon, and was in each case readily controlled by a firm compress. An attempt should always be made to control even the slightest amount of bleeding from tenotomy, lest the blood-clot in the sheath interfere with the normal healing process.

Division of the tibialis anticus tendon. This tendon is divided about two inches above its insertion, the points for its division being found from the prominence of the tendon on the inner aspect of the foot a short distance above the scaphoid tubercle. In many cases the puncture used for dividing this tendon will also answer for the division of the plantar fascia. For the division of this tendon a sharp-pointed tenotome alone will suffice. The knife is inserted beneath the tendon, its edge turned upward and its division easily accomplished, care being taken to avoid wounding the skin.

Division of the tibialis posticus tendon. The division of this tendon requires both practical anatomical knowledge and precision in the use of the knife. As pointed out by Little, the position for the division of this tendon is exactly midway between the anterior and posterior borders of the leg on its internal aspect. The puncture is best made about two centimetres above the tip of the internal malleolus in a line drawn vertically midway between the posterior border of the malleolus and the corresponding border of the tendo Achillis. The tendon itself must always be divided by means of a blunt tenotome for fear of wounding the posterior tibial artery. It must be recollected that at this position the posterior tibial artery is located between the tendons of the flexor longus digitorum and flexor longus pollicis, to the inner side of the former being the tendon of the tibialis posticus, and to the outer side of the latter the posterior tibial vein and the posterior tibial nerve, the artery being, therefore, the middle of these five structures. The internal saphena vein and nerve are superficial and anterior to the puncture. It is also evident from these anatomical relations, as pointed out by Little, "that an incision made a little in front of this line might wound the internal saphena vein, and if made behind it would run the risk of dividing the flexor communis digitorum instead of the tibialis posticus, or the knife might even pass posterior to the former tendon, and if carried deep enough might wound the artery and nerve without touching any tendon whatever." The blunt-pointed tenotome should be carried perpendicularly downward to the depth of one or one-and-a-half centimetres, the handle made to describe the arc of a circle, so as to turn the cutting surface of the blade forward and against the tendon, and the latter be divided vertically inward. Its division will be evidenced by a slight snap and a slight change in the position of the foot.

The tendon may also be divided by inserting the knife between the bone and the tendon and cutting directly backward, but it is probably safer to perform the operation as first described. The danger of wounding the posterior tibial artery is probably exaggerated, Adams having met it but once in a very large experience, and Bonnet having wounded it more than once, but only a few times and without subsequent serious injury. In Mr. Adams' case, and in a similar one recorded by Mr. Tamplin, the injury resulted in a false aneurism, but both patients subsequently recovered. The injury to this vessel is evidenced by a florid jet of blood and the blanching of the foot. In such an event a firm compress should be accurately applied and maintained in position by a roller for some days. This may be sufficient, as in a case recorded by Agnew. In the majority of cases puncture will be followed by aneurism, which will require ligation of both ends of the sac, as in the cases recorded by Tamplin, Bradford and Lovett, and others. The rule laid down for the section of this tendon is the one to be employed in infantile cases, in whom the difficulty of locating the position of the internal malleolus is enhanced by the small size of this bony prominence and the great quantity of fat which usually covers it. In adult cases the tendon can be located with greater ease and precision, and may then in some cases be divided by simply locating it with the finger and cutting forward against the bone.

Division of the plantar fascia. This fascia is divided at its most prominent point. A sharp tenotome alone may be employed, or if the division is to be extensive a blunt tenotome will be preferable. The puncture used for the section of the anterior tibial tendon may be employed, or a new puncture may be made in the inner side of the sole of the foot nearly midway between the os calcis and the ball of the great toe, but a little nearer the os calcis. The knife is carried close to the fascia flatwise between it and the skin, the blade is turned toward the sole of the foot, and the section accomplished. In infantile cases it is advisable to make but small sections at one time, repeating these if necessary. Should the internal plantar artery be wounded in division of the plantar fascia, a few days should elapse before instituting restitution and extension, lest aneurism result, as in the case reported by Walsham.[1]

Division of the peroneal tendons. If the slightly sickle-shaped

---

[1] London Lancet, Jan. 28, 1888.

knife be employed, no other instrument is needed, but if a probe-pointed tenotome be used, another sharp-pointed instrument must be employed to puncture the skin. Anæsthesia is not absolutely necessary, and in dividing this tendon the writer usually employs local freezing with ice and salt. The parts, rendered aseptic and insensible, are held in a position which renders the affected tendons prominent. A puncture having been made a short distance behind the external malleolus, so that it will not correspond with the one in the deeper tissues when the parts are relaxed, the blunt-pointed tenotome is passed either over the tendons, between them and the skin, and the division of the constricted tendons accomplished by cutting directly down upon the external malleolus, or, as the writer prefers, the tenotome is passed beneath the tendons and the cut is made directly outward, care being taken not to cut the skin nor increase the size of the puncture. The proper division of the tendons is evidenced by an audible crackling and the fact that the foot now remains in its normal position. The tenotome is then turned flatwise and withdrawn ; the operator, compressing the wound to prevent the ingress of air, places an antiseptic dressing upon the part, and secures it in position with a plaster-of-Paris bandage. Over-correction is not necessary in these cases, as relapse is not so likely to occur as after the section of contracted tendons.

After-treatment of tenotomy. The after-treatment requires as much if not more attention to detail than the operation itself, and on this depends much of the success. Indeed, many authorities believe treatment only to have commenced when the foot has been straightened by forcible means and tenotomy.

Speaking of this after-treatment, Reeves[1] aptly remarks : "I wish it thoroughly to be understood, that though correction by tenotomy and splinting are absolutely essential in the majority of cases, it is to the subsequent proper adjustment of splints, frictions, manipulations, massage, etc., of the foot and leg that very much of the success of these cases depends."

At the expiration of ten days the foot should be inspected, and if in good condition it should be placed in a well-constructed mechanical walking-shoe. At least once a day, and preferably twice a day, the foot and leg should be rubbed with linimentum saponis, bathing whiskey, or extract of witch-hazel, and manipulated, particularly

---

[1] Bodily Deformities, Philadelphia, 1885, p. 165.

over the muscles and tendons which were formerly contracted. This is very important and tends to free and strengthen the tendons. The skin over bony joints subjected to pressure should be hardened by the application of a solution of alum in alcohol. At night for some months the foot should be placed in a retention night-shoe, by which any tendency to relapse may be corrected.

Electricity should not be omitted. It should be applied to the whole leg to improve the nutrition of the skin and deeper structures, but it should be more particularly applied to the contracted or para-lyzed muscles.

Division of ligaments—syndesmotomy. In some instances the con-tractures of the ligaments, as the astragalo-scaphoid and calcaneo-cuboid ligaments in talipes equino-varus, offer the greatest resistance to the reduction of the deformity, and the division of these has, therefore, been advised by many surgeons. As early as 1819, Thomas McKeever[1] having divided the deltoid ligament found he could reduce the foot into its natural shape with the greatest facility. Later Guérin, Little, Lonsdale, and Adams referred to the part played by the ligaments, but so far as known none of these advocated their division as a routine practice. Stuckeisen, of Bâle, recorded cases and advocated section of ligaments, and more recently Parker and Shattuck, Phelps and others have advised and performed syn-desmotomy for the relief of severe club-foot. The division of these ligaments may be performed by the percutaneous and subcutaneous methods. Phelps, in his open incision, has advised the incision of the internal lateral or deltoid ligament and all its branches by an extensive subcutaneous curvilinear incision, the tenotome being in-troduced through the open incision. The astragalo-scaphoid and calcaneo-cuboid ligaments, the ones most frequently requiring divi-sion, can both be divided subcutaneously. To divide the astragalo-scaphoid ligament the tenotome held vertically, edge forward, is entered immediately in front of the anterior border of the internal malleolus, and passed between the skin and ligaments. The blade is turned toward the surface of the ligament, and by a gently sawing motion the division is accomplished. By directing the point of the knife to the plantar aspect of the foot and keeping close to the bone, the calcaneo-scaphoid ligament may also be easily divided. To divide the long and short plantar (calcaneo-cuboid) ligaments, the tenotome

---

[1] Jahrb. f. Kinderheilkunde, 1869, p. 49.

must be entered behind the head of the fifth metatarsal bone, as nearly as possible over the calcaneo-cuboid articulation on the outer border of the sole of the foot. The blade is passed close to the bone in the direction of the articulation, and the section of the two ligaments may be accomplished simultaneously.

Myotomy. The division of the muscles is but seldom resorted to at the present time. Pancoast advised and practised division of the internal head of the soleus muscle for equinus, but it soon fell into disuse. In Phelps' operation the abductor pollicis and flexor brevis muscles are divided. Except in this operation myotomy is rarely performed, and tenotomy, when possible, is always preferable.

Tarsectomy and tarsotomy. The cutting operations upon the tarsal bones for the relief of talipes—about sixteen separate operations in all—are included under these two heads; the former, tarsectomy, consisting of the removal of a wedge-shaped piece of bone; the latter, tarsotomy, consisting of the division of the bony structure of the tarsus with an osteotome.

The following enumeration, slightly modified from Roberts' and Ketch's[1] valuable article, includes all to the present time:

1. Linear osteotomy of the scaphoid practised on the plantar surface. (Hahn.)

2. Linear osteotomy of the tibia above the malleolus. (Hahn.)

3. Enucleation of the cuboid. (Solly.)

4. Enucleation of the astragalus alone. (Lund, Maron.)

5. Same, with resection of tip of external malleolus. (Maron, Reid.)

6. Excavation of the spongy portion of astragalus, leaving the articular surfaces. (Verebely.)

7. Enucleation of astragalus and excision of a wedge-shaped piece from anterior part of os calcis. (Hahn.)

8. Enucleation of astragalus and cuboid. (Albert, Hahn.) Enucleation of astragalus and scaphoid. (West.)

9. Enucleation of astragalus, cuboid, and scaphoid. (West.)

10. Enucleation of scaphoid and cuboid. (Bernet.)

11. Resection of head of astragalus. (Lücke, Albert.)

12. Excision of wedge from outer half of neck of astragalus. (Hueter.)

---

[1] Wood's Handbook of Med. Sciences, vol. ii. p. 196.

13. Excision of two wedges perpendicular to each other, with bases at Chopart's articulation and the astragalo-calcanean joint. (Rydygier.)

14. Resection of a wedge without regard to any individual bones. (O. Weber, Davies-Colley, R. Davy.)

15. Linear osteotomy of the lower end of the tibia and fibula. (Trendelenburg, Willy Meyer.)

16. Linear osteotomy of the neck of the astragalus. (Bradford.)

Of these sixteen operations three only, which are most employed, will be described, the others being either unsatisfactory, insufficient, or too mutilatory. These three are: (4) enucleation of the astragalus; (14) wedge-shaped tarsectomy; and (16) linear osteotomy of the neck of the astragalus.

Enucleation of the astragalus in inveterate cases of club-foot is best performed as suggested by Mr. Barker.[1] The foot, duly prepared and rendered evascular with an Esmarch bandage, is supported in a strongly

FIG. 242.

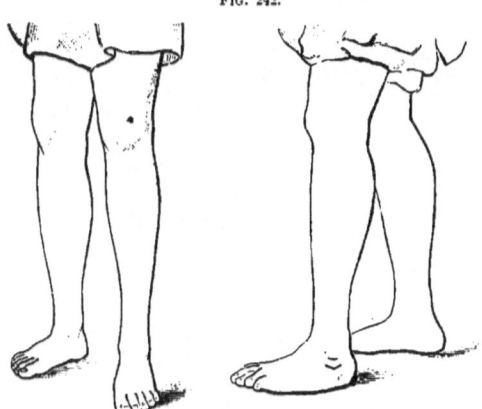

Result after removing astragalus of left foot. (BRADFORD.)

inverted position upon a sand-bag. An incision extending from just above the tip of the external malleolus is carried forward and a little inward, curving toward the dorsum of the foot. This crosses a space between the peronei tendons, in which no important structures are found,

[1] Manual of Surg. Oper., p. 175.

and being carried directly down to the bone, the latter may be readily exposed and removed with gouge chisel and forceps. Care should be

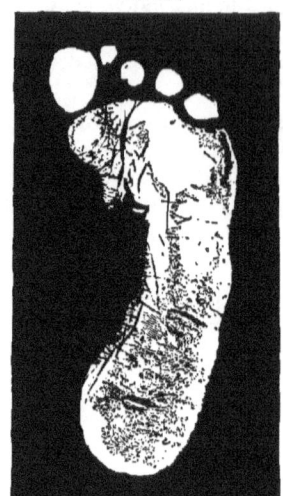

FIG. 243.

observed to avoid bruising the surrounding bony structures, and in dressing the strictest antiseptic precautions should be enforced. (Fig. 243.)

Resection of wedge-shaped piece from outer side is best performed after the methods of Mr. Davies-Colley and Mr. Davy.

The foot, cleansed and rendered evascular, is supported on a sandbag, and a T-shaped incision made upon the outer side of the foot. The straight incision is carried along the outer border of the foot, from the middle of the os calcis to the middle of the fifth metatarsal bone. The vertical incision at right angles to the centre of this enlarges the field of operation. The flaps are reflected, the tendons and vessels are held aside

Sole-imprint after removal of astragalus for club-foot. (BRADFORD.)

FIG. 244

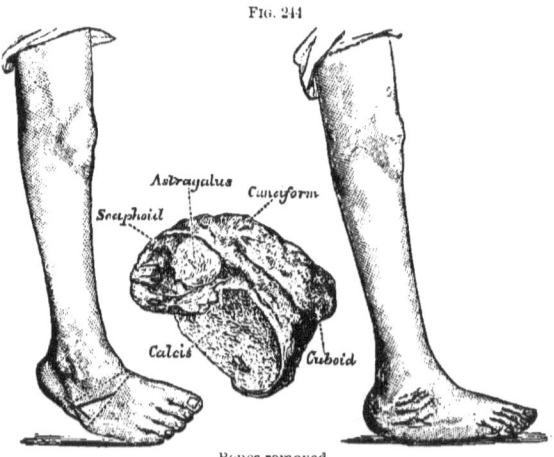

Astragalus   Cuneiform

Scaphoid

Calcis   Cuboid

Bones removed.

Foot before operation.   Foot after operation.

upon the dorsal and plantar surfaces with a periosteal elevator and retractor, and with a narrow-bladed saw or a chisel a wedge-shaped

Fig. 245.

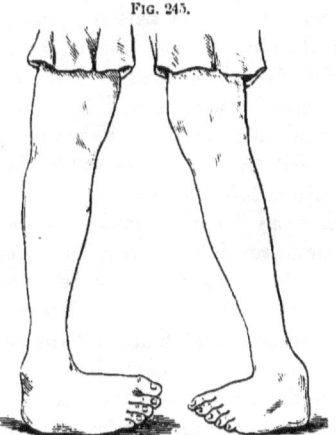

Drawn from photograph before operation in boy of twelve. (BRADFORD.)

Fig. 246.

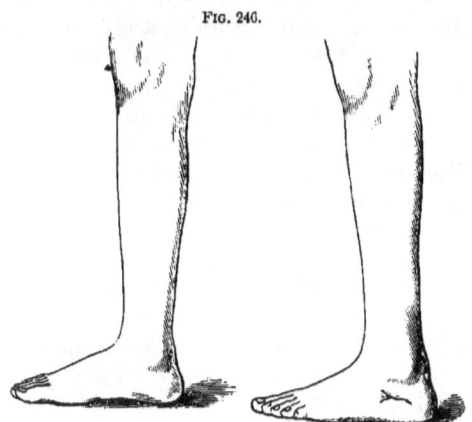

From photograph after operation in boy of twelve—osteotomy of neck of astragalus and os calcis. (BRADFORD.)

section is removed with a lion forceps from the tarsus, irrespective of the individual bones involved, sufficient in size to allow of the correction of the deformity. This section will include a portion of the os calcis and a portion of the cuboid—the entire cuboid or even a portion of the fifth metatarsal bone being removed in the severest cases. The tendons, if any have been divided, should be united with silk or chromicized gut, the vessels secured, a rubber drainage-tube is inserted, and the wound partially closed with wire sutures. A full antiseptic dressing is then applied before the Esmarch is removed. The foot is then secured in a plaster-of-Paris dressing or some suitable retentive apparatus. The strictest antiseptic precautions are essential to success and the safety of the patient, and during the subsequent treatment care should be taken to keep the wound aseptic. (Figs. 245 and 246.)

"*Astragaloid osteotomy*," as performed for severe cases of club-foot is best given in the words of its author, Dr. Bradford, of Boston :[1]

"The procedure," he says, "will not be found a difficult one. Tenotomy and division of the fasciæ and ligaments should be done, and the foot stretched and manipulated into as nearly a normal position as possible. An incision through the skin is made from the top of the malleolus to the inner side of the head of the first metatarsal, which will be found in severe cases close to the malleolus. The scaphoid will be seen before the astragalus is encountered, if the deformity is great, and it will be first within the reach of the knife in all cases. If the foot is still further stretched, the scaphoid begins to uncover the side of the astragalus, and the neck of the astragalus is seen ; a small osteotome is entered and placed upon the neck of the astragalus, to the distal side of the scaphoid articulation, and, as in the Macewen operation, the neck of the astragalus divided or nearly divided. The foot is then forcibly straightened, and the neck of the astragalus is chiselled or fractured. The result is similar to that in Macewen's operation for knock-knee, and the distortion at the neck of the astragalus is removed. It is manifest the line of section of the bone at the neck of the astragalus should be transverse to the axis of the bone, and at such a plane that when the equinus deformity is corrected, the resulting gap at the section should not be greater than necessary. Strict asepsis is essential. The foot should be fixed in the corrected position."

These three operative procedures, in addition to the tenotomes and open incision, are all that are required even in the most severe cases, and while in many cases they offer a means of quickly and brilliantly correcting very severe deformities, their field of application should

---

[1] Trans. Amer. Orthop. Assoc., vol. i. p. 109.

always be restricted, and in no instance should they be undertaken unless the surgeon is confident that antiseptic measures can be faithfully enforced. The relative merits of enucleation of the astragalus and wedge-shaped tarsectomy have given rise to much discussion, which has resulted in the substitution of enucleation of the astragalus and other methods for wedge-shaped tarsectomy. The objections to the latter may be briefly stated as follows:

It shortens the outer side of the foot.

It impairs the form of the foot, the stability, mobility, and usefulness of the osseous arch, and

It exposes to infection an extensive surface of cancellous tissue.

The objections to the cutting operations upon the bones in general (resections of the tarsus) may be briefly stated as follows:

1. Resection as an operation is not free from risk.

3. Resection removes all chance of future restoration by orthopedic treatment.

3. Resection is unjustifiable except in persistently painful club-foot in an old subject, where all orthopedic treatment has failed, and where it (resection) may be employed in preference to amputation.

In conclusion, excision of the tarsus should *never* be performed before the fifth year, and, preferably, from the twelfth to the fourteenth year, and where possible enucleation of the astragalus should have the preference over all other bone-cutting operations.

For additional information upon the operation of excision of the tarsus the reader is particularly referred to the excellent monographs of Dr. De Forest Willard [1] and Dr. H. A. Wilson.[2]

Brisement forcé. Under this title are included all operations which aim at the immediate forcible restoration of the foot, either by the hand or powerful instruments. It is necessary in all cases to employ an anæsthetic, to economize time and save the patient much suffering. It is remarkable how much can be accomplished by manual means alone, but when great force is to be applied, the club-foot wrenches of Thomas, Gibney, and Bradford; the club-foot stretchers of Morton or Bradford, or the club-foot machine of Phelps, should be employed. The objection to the wrenches is the limited application of force and the difficulty of rapid removal of the apparatus after the reduction is accomplished. The Bradford stretcher has this advantage over the Morton apparatus, that the force is applied by screw

[1] Trans. Med. Soc. Penna., vol. xvi., 1884, p. 381.
[2] Trans. Amer. Orthop. Assoc., vol. vi., 1893, p. 150.

power, and does not yield as do the leather straps of the former. The Phelps machine has the advantage of being able to apply any amount of force from a single pound to a ton in force. The chief objection urged against all these methods of forcible restoration are the supposed risks incurred—the sloughing of the skin, the rupture of the ligaments, and the breaking of bones. Experience in osteoclasy, however, proves that these fears are more theoretical than real. The pressure applied to the skin is so momentary that sloughs seldom occur, and the ruptured and stretched tendons readily heal in the fixed apparatus subsequently applied; fractures seldom or never occur.

FIG. 247.　　　　　　　　　　　FIG. 248.

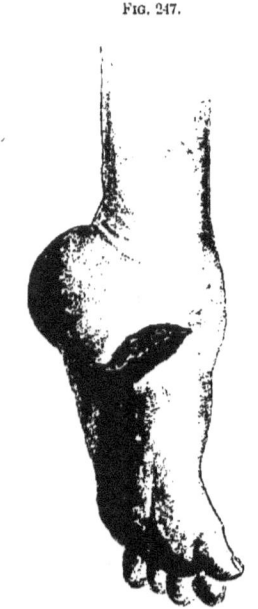 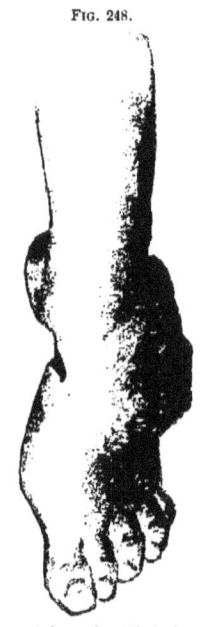

Line of incision and position of foot after Phelps' operation. (McKENZIE.)

Position of foot after Phelps' operation, showing large bursa on outer side of foot. (McKENZIE.)

Open incision. Since the first introduction of this method ten years ago by Dr. Phelps,[1] before the Eighth International Congress at Copenhagen, it has steadily grown in favor, until now it numbers

[1] New York Medical Record, November 29, 1890, p. 593.

among its advocates many of the foremost surgeons of Europe. This accumulation of experience has, moreover, enabled its author to more clearly define the proper place for the operation in surgery. As advised by him at the recent Tenth International Congress in Berlin, in 1890, after excluding all cases which by manipulation or force can immediately, or in a reasonable length of time, be restored,

FIG. 219.

Result after Phelps' operation.   (MCKENZIE.)

the following rules should be followed : "Cut the contracted parts as they first offer resistance, cutting in the order of those parts which first contracted when the deformity was produced. The operator will then proceed, after strong manipulation or force is applied with a club-foot machine or hands, to subcutaneously divide the tendo Achillis. *If the skin is not short*, subcutaneous tenotomy in the sole of the foot will usually suffice. *If the skin is short*, an open incision one-fourth the distance across the foot can be made, beginning directly in front of the inner malleolus and carried down to the inner side of the neck of the astragalus. Through this incision the following tissues can be cut, if they offer strong resistance, in the order given: (*a*) Tibialis posticus ; (*b*) division of abductor pollicis ; (*c*) division of plantar fascia through the wound; (d) division of flexor brevis muscle; (e) division of long flexors; (f) division of deltoid ligament or its branches."

The parts are thoroughly antisepticized before the operation, an Esmarch bandage is applied, the wound is thoroughly irrigated with bichloride solution, 1:2000, during the operation, and a full antiseptic dressing is applied before the bandage is removed,

FIG. 250.

September, 1890.   May 4, 1891.   May 9, 1891.

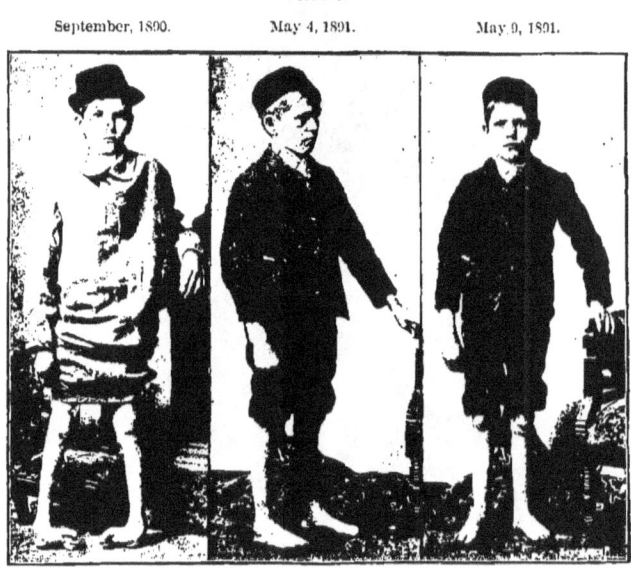

Showing result in Phelps' operation.   (McKENZIE.)

the foot being subsequently slung to a nearly perpendicular position for six hours or longer. In this manner Dr. Phelps has performed 161 operations in 93 cases, the average age being six and a half years, and the average time of the healing of the wound four weeks. In 140 of these cases 117 healed by blood-clot organization, 4 by catgut dressing in the wound, and 19 were failures. Four months after operating the feet were all straight, but out of 140 cases traced after one year, 10 cases were found relapsed or partially so, from neglect. In these cases there were in all 17 osteotomies performed.

The limits of the application of the operation are thus given: "1, eliminate all cases which by the hand can easily be placed in a normal position; 2, eliminate all of those cases which can by subcutaneous tenotomy be perfectly relieved with accompanying proper after-management. Then open incision will find its legitimate place in surgery."

The advantages of the operation are thus given: "1. Cutting parts as they offer resistance, in their respective order, prevents the operator from needlessly cutting tissue not deformed by contraction. 2. After the subcutaneous tenotomy of the tendo Achillis, the tibialis posticus tendon is easily cut through an open wound near its attachment to the scaphoid; ligamentous contraction at this point can also be divided. 3. Through this open wound contracted parts can be extensively cut without wounding the plantar arteries or nerves. 4. After all contracted soft parts have been divided, including the skin, ... the operator can ascertain the amount of the deformity of the bones, and if any considerable amount exists it can be easily remedied with a chisel. 5. It restores the foot to its natural length by lengthening the shortened side. 6. It makes the surgeon master of the situation; he advances step by step, in a proper order, and need not stop or retract until the deformity is overcome, beginning with manipulation and subcutaneous tenotomy and ending with osteotomy, if necessary."

Amputation. In the severer forms of club-foot, especially when complicated with bursitis and extensive ulceration, the individual is rendered practically helpless, and relief is sometimes sought in amputation of the affected parts. Operations of this kind at the present time are exceedingly rare, since with improved orthopedic measures even the most severe cases may now be rendered more or less useful.

Treatment of Special Varieties.

Treatment of Varus. The treatment of pes varus forming a part of the congenital variety, equino-varus, can best be considered under the compound variety. Then remains the treatment of the acquired and paralytic varus. The eversion of the anterior part of the foot in this deformity may be accomplished either by mechanical measures or by operative procedures. If the muscles are contracted this can best be done by means of Shaffer's varus shoe, a modification of Taylor's ankle support, the mechanical construction and action of which can best be given under the compound form. In the paralytic variety the elastic traction appliance recommended by Barwell and Sayre will be found most efficient. In some cases section of the plantar fascia and tenotomy of the tibialis anticus and posticus will be necessary. In dividing the plantar fascia in these paralytic cases it is best to make but a moderate division at first, and repeat the operation subsequently, lest the varus be converted into a valgus deformity. In severe cases it may be necessary to resort to *brisement forcé*, but tarsectomies and tarsotomies are seldom or never necessary in these cases. In this connection the treatment of varus, which remains after the correction of compound varieties in completely corrected cases, the so-called "residual varus," may be referred to. This may be corrected by the forcible correction of the deformity under an anæsthetic, with or without tenotomy as the case may require. Mild cases may be overcome by mechanical appliances alone, of which Stillman's modification of Gregory Doyle's apparatus will be found the most efficient. (Fig. 251.) This consists of a spiral spring attached to the outer part of the outer aspect of an ankle support and attached at its upper extremity to a pelvic band. By means of a key any amount of everting force may be given to the spring. As a therapeutic measure the use of skates both for out-of-doors and parlor use may be employed with great benefit.

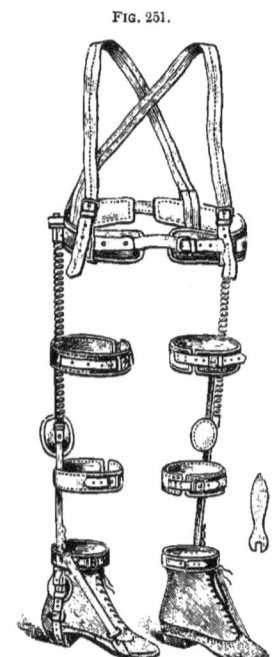

FIG. 251.

Gregory Doyle's apparatus for residual varus.

Treatment of Acquired Valgus.

The treatment of pes valgus acquisitus varies with its cause and degree. Any diathetical tendency must be corrected by appropriate measures. The mild varieties yield readily by electricity to the affected muscles, massage and removal from the exciting cause, with properly fitted metallic sole-plate. The massage should be especially applied to improving the condition of the peroneal group of muscles upon the outer aspect of the calf. These should be daily rubbed, preferably by a trained *masseur*, and electricity should be applied over the position of the peroneal muscles and to the musculo-cutaneous branch of the external popliteal nerve which supplies these muscles. Exercises directed to the development of the size and strength of these muscles are also advisable. Dancing, skipping rope, walking on the toes, and similar ,exercises are valuable. A very good exercise consists in standing with the heels together and feet turned out, then slowly rising on tiptoe, separating the heels, and again slowly lowering them. This should be repeated a number of times night and morning. To support the arch numerous devices have been recommended—pads of leather, felt, rubber, and other material attached to the inner or outer side of the sole, steel bars, springs, etc., added to specially constructed shoes, etc. The writer has attained the best results with the improved plate spring of Dr. A. Sidney Roberts, figured in the *Medical and Surgical Reporter* for April 6, 1889. It consists of a tempered steel plantar spring, so constructed as to supply an artificial arch. In describing it,[1] he writes:

Fig. 252. Fig. 253.

Plantar spring for flat-foot, under surface.   Plantar spring for flat-foot, upper surface.
(ROBERTS.)   (ROBERTS.)

"An outline of the patient's foot is first taken on stencil-board, the tracing being extended upward on the inner side of the foot. The elevated, corresponding to the depressed, arch of the foot can be tempered to the extent required by the particular case. The lateral pressure brought to bear by the elevated flanges is such that while giving support to the arch to a certain extent, the artificial arch also prevents further displacement of the astragalus and scaphoid. Again, in place of giving only a limited amount of support to the inner side of the foot, this appliance supports the foot as a whole."

[1] Med. and Surg. Rep , vol. xi., No. 14, p. 420.

After the use of this spring in a large number of cases in hospital and private practice, the writer is much gratified with the results obtained. The difficulty, however, has been in preserving the spring for any length of time. Steel—rough, polished, or galvanized—would rust from the saline perspiration and readily break. Silver, nickel, Japan coating, and bicycle varnish would peel off, and solid silver and aluminium were too expensive for general use, while the latter did not possess sufficient elasticity. Phosphor-bronze was objectionable for the same reason, and when employed for small springs for children it corroded and broke. A white leather covering riveted to the spring was found very durable, but the steel rusted beneath the leather, and soon became objectionable. Hard vulcanized rubber was then experimented with, with a view to obtaining a cleanly, elastic, and durable spring. It possessed the advantage of being easily bent and moulded when immersed in hot water, but, however, readily yielded to the heat of the foot and flattened. This led to the use of a steel spring coated with hard rubber in a similar manner to the coating

FIG. 254.

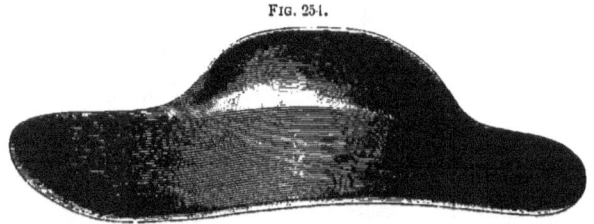

Author's hard-rubber-coated flat-foot support.

of trusses, and resulted in producing the "ideal plantar spring." (Fig. 254.) These springs are made either from a cast of the foot (modelled) or from a pattern. They are then highly tempered and finally coated with hard vulcanized rubber. This produces a highly polished, aseptic, elastic spring, which can on heating be altered from time to time by means of bending hooks, and has proved to be, after thorough trial, entirely satisfactory.

Dr. Whitman[1] has devised a very efficient form of sole-plate, and has demonstrated its practical use in a number of successful cases. It is necessary in employing this plate to make a plaster cast of the foot with the leg lying on its outer side and retained as nearly as possible

---

[1] Trans. Amer. Orthop. Assoc., vol. ii. p. 73.

in a corrected position. An iron counter-cast is then made and three points. A, beneath the ball of the great toe; B, beneath the inner tuberosity of the os calcis; and C, below and just in front of the internal malleolus opposite the head of the astragalus, are then marked as the boundaries, and upon this the plate is constructed. The writer has had no personal experience in the use of these plates, but doubts not they are very satisfactory.

In severe cases, after the restoration of the arch by operative measures, the spring may with advantage be supplemented by an ankle support—two lateral steel uprights connected with a band extending about the calf, and an internal oval ankle-pad over the mediotarsal articulation.

As a symptomatic condition in osteitis of the tarsus and ankle-joint disease, the valgus generally yields to the treatment employed for the primary osseous or articular lesion.

In severe cases, before resorting to extreme operative measures, all contracted tendons should be divided. Those most frequently requiring division will be the tendo Achillis, peronei, and extensor longus digitorum, and they are best divided in two stages—the peronei and extensor longus digitorum first, to correct the valgus, and the tendo Achillis subsequently, to correct the equinus, which in severe cases is present. Failing in the worse cases to restore the arch by such means, accompanied with manual force — *brisement forcé*, with either a Thomas or Bradford wrench, should be attempted, and, as a last resort, Ogston's operation should be resorted to, or, if much ankylosis be present, a regular tarsectomy, *i. e.*, removing a wedge-shaped piece of the tarsus from the inner side of the tarsus without reference to the bones or portions of bones removed (*vide* tarsectomy, *ante*) should be performed, and the restoration of the foot be completed. These severe operations should only be resorted to when milder measures, after due trial, have failed; for comparative relief will, in the great majority of cases, be obtained by the milder remedies in about six or eight weeks, whereas several months are required after any cutting operation upon the tarsal bones before the foot may with impunity be freely exercised.

Treatment of Paralytic Valgus.

The treatment of this variety requires the same attention to the employment of massage, electricity, and the use of plantar springs and other mechanical appliances as do the other forms of flat-foot.

The use of electricity in these cases is particularly serviceable, and the rules laid down for its application under the general head of treatment should be observed. It is of the greatest advantage in the infantile paralysis cases to prevent fatty degeneration of the muscles and to improve the condition of the skin. In these cases marked eversion of the foot upon the leg occurs from the growth of the foot, as pointed out before, and the tension upon the peronei tendons. This condition may be overcome in part or entirely by the division of these tendons after the rules previously given under the general head of tenotomy.

After an operation of this character over-correction is not so essential as in ordinary tenotomy performed for contracted tendons. Two weeks should elapse before the foot may with safety be used, and at first only in moderation. The greater number of these cases also require a high sole or patten to equalize the length of the limbs, since the paralysis has usually diminished the growth of the limb in length as well as in size. Elastic webbing-straps are often necessary to maintain the foot in a good position. These pass from the inner side of the front part of the shoe to the inner or anterior part of the band which encircles the leg below the knee. It is in these forms of club-foot that the elastic traction apparatuses of Sayre, Barwell, and others are most efficient.

Treatment of Equinus.

The treatment of the congenital equinus forming a part of the congenital equino-varus will be given under the compound variety. The contraction of the heel in talipes equinus may be overcome either by mechanical or operative means according to the degree of deformity present. In the milder cases, if the foot can be prevented by light mechanical appliances from turning to either side, the weight of the body in walking will overcome the deformity in many instances. A very useful apparatus is Shaffer's appliance for correcting equinus —one of the many modifications of Scarpa's shoe. (Figs. 255, 256, and 257.) This consists of two steel uprights extending from the upper part of the tibia to the ankle-joint, and attached to it a heel-cup and sole to hold the foot, the heel being held in its place by means of a strap of webbing, a bandage, or similar material passed over the instep. The efficiency of this apparatus has been increased by dividing the sole of the brace opposite Chopart's joint, this ante-

rior portion being worked by an extension-bar passing beneath the
heel-cup. The apparatus is applied extended to an angle corresponding
to the angle of deformity; the heel is secured by means of two straps,
one passing over the instep, as before described, and the other passing
around the heel and forward to be attached to either side of the sole-
plate. By means of a key at the ankle-joint the foot-piece is flexed
upon the upright, and by means of the extension-bar the anterior
portion of the sole is thrown forward, the os calcis is dragged upon by

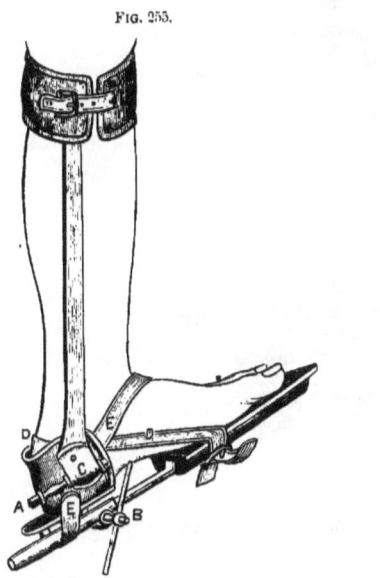

FIG. 255.

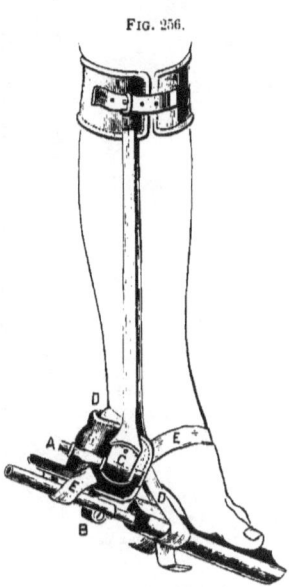

FIG. 256.

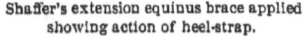

Shaffer's extension equinus brace applied,
showing action of heel-strap.

Shaffer's extension equinus brace applied
to foot in deformed position.

the strap passing over the heel, and the tendo Achillis is thus thoroughly stretched. This stretching is repeated several times at each
sitting, the pressure being not continuous, but a momentary over-
stretching followed by relaxation. By this means the tendo Achillis
may be extended until the deformity is slightly over-corrected, when
the apparatus is substituted by a retention shoe with a stop-joint at
the ankle to keep the foot in the corrected position. By these
mechanical measures the writer has corrected a number of cases of
severe equinus.

In the severer grades, in adults and where the time for treatment is limited, tenotomy of the tendo Achillis will be found a most satisfactory plan of treatment. The operation itself has already been given under the general subject of tenotomy. After the section of the tendon the foot should be placed in a slightly over-corrected

FIG. 257.

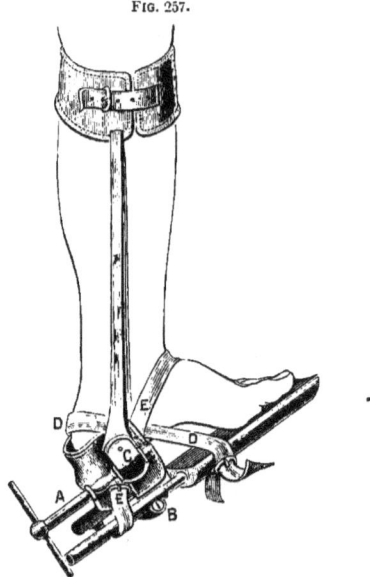

Shaffer's extension equinus brace applied, showing tendency of heel to slip away.

position, and retained by means of a plaster bandage. Two weeks after the operation the foot should be placed in a retention walking-shoe with a stop-joint at the ankle to keep the foot in the corrected position.

The after-treatment should be carefully carried out, and has already been given under the general subject of tenotomy. If the deformity has resulted from paralysis of the anterior group of muscles, in addition to these measures electricity and massage should be applied to the whole leg, and elastic straps should be added from the anterior portion of the shoe to the upper part of the brace to supplement the paralyzed or weakened muscles. In this connection the treatment of "compensatory equinus" deserves notice, since this usually results from an attempt to equalize the length of the shortened limb by an affected or acquired equinus. The simple use of a cork patten will in the majority of cases overcome the difficulty. If, however, the foot has remained in this position long enough for the bones, tendons, and ligaments to have become altered, the mechanical and operative measures before referred to will become necessary. In "non-deform-ing club-foot," which in many cases is simply an incomplete equinus, the stretching of the tendo Achillis by means of Shaffer's appliance for correcting equinus will readily overcome the difficulty. In many of these cases relief is obtained after one or two stretchings of the con-tracted tendon. If, however, these measures should fail to bring the foot into an over-corrected position, tenotomy should be resorted to.

Treatment of Calcaneus.

As a congenital affection the treatment of calcaneus seldom requires more than daily manipulation and extension of the foot to overcome the deformity. In the severest cases tenotomy of the tibialis anticus, peroneus tertius, and extensor longus digitorum may be required. The corrected foot may then be retained by a simple walking-shoe with a fixed joint to prevent flexion beyond a right angle.

Dr. Judson[1] has advocated a brace of this kind consisting of an upright and foot-piece, the joint between the two so constructed that the foot-piece falls but cannot be raised beyond a right angle with the upright. Attached to the foot and leg in standing and walking, the foot will remain at right angles with the leg.

The treatment of the paralytic cases from anterior poliomyelitis can best be accomplished by means of retention apparatus, with an elastic strap to limit flexion beyond a right angle. Operative inter-ference in these cases is seldom called for. Reeves[2] has described a method of shortening the tendo Achillis which offers an excellent surgical method of overcoming the deformity. An incision is made down to the tendon, the sheath is opened, and the tendon raised by a blunt hook or spatula, and folded or pinched between the fingers until the size of the piece required is ascertained. A silver suture is then passed through the tendon about a quarter of an inch above and below the points of the proposed section, to prevent a slipping away of the tendon within its sheath. The segment is removed, the

---

[1] Medical Record, May 16, 1885, p. 538.
[2] Bodily Deformities, Phila., 1888, p. 223.

extremities approximated, the ends of the wire twisted and buried in the tendon. In this manner one-half to two-thirds of an inch may be removed. Antiseptic measures should not be omitted. Walsham and Willett[1] have proposed a modification of this operation which

FIG. 258.

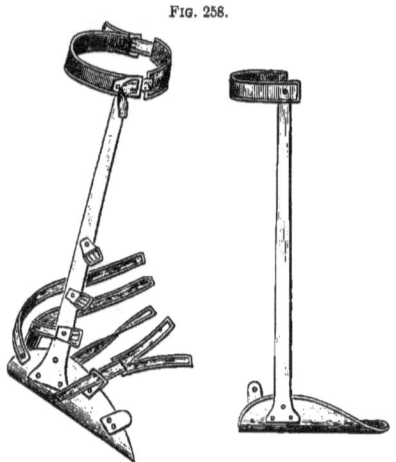

Judson's walking-brace for equino-varus.

consists of dividing the tendon obliquely, of sliding the cut ends past each other until the desired shortening is obtained, and stitching them very firmly together by chromicized catgut or kangaroo tendon, including the skin and tendon. The results of this modified operation are reported generally satisfactory, and the operation has the advantage of not removing any piece of the tendon.

Treatment of Cavus.

The treatment of the milder forms may be accomplished by extension with an equinus shoe, particularly the pattern of Dr. Roberts, by which powerful stretching force can be applied to the plantar fascia. The severe forms will demand aponeurotomy of the plantar fascia, and the division should at first be moderate and repeated, lest valgus result. The after-treatment will consist in the use of an extension night-brace to prevent relapse.

[1] British Medical Journal, June 14, 1884; May 31, 1884.

Treatment of Planus.

The plan of treatment advised for the milder forms of valgus acquisitus may with advantage be employed in the correction of pes planus. Friction, massage, and electricity to the peroneal muscles, and exercises calculated to improve the power and tension of these and the plantar muscles, should be employed. In addition to these measures a well-fitting plantar spring should be used, changing its height from time to time as the elevation of the arch demands. Dancing is to be recommended for this affection, particularly in young growing girls.

Treatment of Neuromimesis.

The treatment, as in other neuromimeses, should be conducted with the same care and attention to detail as if the patient had a genuine deformity, directing particular attention simultaneously to the nervous element present, and to the improvement of the general *morale* of the individual. Tenotomy is seldom required ; but after other means have failed it may become necessary, as in a case recorded by Dr. S. Weir Mitchell,[1] in which section of the tendo Achillis, performed by Dr. A. Sydney Roberts, completely restored the deformity and relieved the spasm.

Treatment of Equino-varus.

The treatment of this variety of club-foot can best be considered under two heads—the treatment of the congenital variety and that of the acquired form. The treatment will necessarily vary according to the patient's age and the degree of deformity present ; but the process of rectification is best divided into two stages—first correcting the varus and afterward correcting the equinus. The correction of the varus deformity will include manipulation, massage, and electricity, retentive dressings, extension and fixation, elastic extension, and tenotomy combined with extension and fixation. In the severer cases, to these must be added *brisement forcé*, tarsotomy, and tarsectomy.

Manipulation and massage. The milder cases, if treated from birth, will often yield to these simple measures alone, and the writer

---

[1] Lectures on Diseases of the Nervous System, 1885, p. 129.

can refer to several mild cases successfully treated in this manner. It is essential in treating cases in this manner to personally instruct the nurse or mother how to grasp the foot and how to apply the force. With one hand the heel should be firmly grasped, while the other firmly holds the anterior part of the child's foot in a correct position. Here it should be held for a few moments, and this should be repeated as many times as possible through the day. In addition to this correction the muscles of the calf and sole of the foot should be manipulated, and the electric current should be applied to the affected muscles and entire limb, according to the method already described.

Retentive dressings. The addition of some form of retentive apparatus enhances materially the prospect of a cure. For this purpose numerous materials have been applied, manufactured from tin, leather, pasteboard, stiffened felt, and other materials, held in position by bandaging. The best material, however, for this purpose is the plaster-of-Paris bandage, first introduced into orthopedic practice by Jules Guérin. In applying this dressing either for correction or retention after correction of the deformity, the technique is important. All the bony prominences must be protected by a padding of cotton, over which a flannel roller should be firmly and smoothly applied. In applying this, advantage may be taken of the direction of the turns in correcting the deformity. Fixing the bandage above the ankle by an oblique and circular turn, the roller should run diagonally across the instep to the ball of the great toe, across the sole of the foot to the little toe, to the outer side of the leg above the ankle, making firm traction before proceeding further, so as to evert and

FIG. 259.

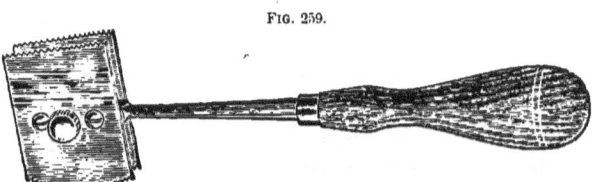

Steele's saw, double and single bladed, for dividing plaster splints.

elevate the outer border of the foot. This turn can then be repeated one-third higher up, until the entire foot and leg are covered, after which the knee, bent at a right angle, and the thigh, are covered in. The plaster bandage should be placed endwise in the water, until all bubbles have ceased to rise. The superabundant moisture should then be squeezed out and the roller applied to the foot, leg, and thigh in the same manner, taking advantage of the turns to correct the deformity, particular pains being taken to apply the first

FIG. 260.

Steele's skin-protector.

turns of the bandage smoothly. When sufficient turns have been applied, the foot and limb should be held firmly in the corrected position while the plaster is setting, applying the palm of the hand to the sole of the foot, and avoiding undue pressure of the individual fingers lest the indentations in the soft plaster produce sloughs. Hahn has devised a simple appliance for holding the foot in position while the plaster is being applied, without interfering with the operator. It consists of a T-shaped wooden splint, included in the bandage, traction upon the arms of which overcomes the inversion and equinus.

FIG. 261.

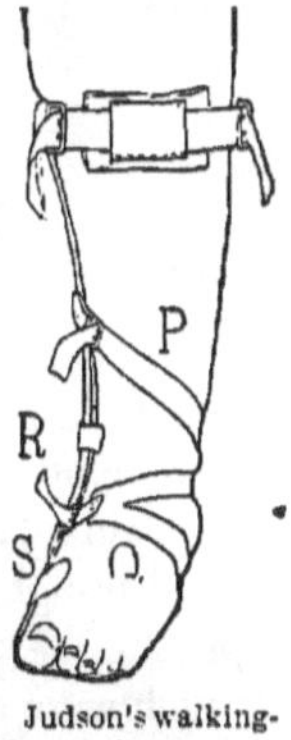

J. Wolff employs a silicate of soda bandage after first applying strips of adhesive plaster after Sayre's method. In very young children the upper part of the dressing may be protected by a coating of shellac or liquid glass to prevent the soiling of the dressing by discharges.

Extension and fixation. For the purpose of extension and fixation, one of the many modifications of Scarpa's shoe offers the best mechanical means. Of these, Stromeyer's, Little's, Adams', and Tamplin's appliances, and Langaard and Kolbe's shoe, may be mentioned. The best, however, at the present time is Taylor's ankle support and Shaffer's and Roberts' modification of this. Taylor's appliance consists of

Judson's walking-brace for equino-varus, applied in deformed position.

a flat steel sole-plate, made from the outline of the plantar surface of the foot, and a steel upright on the inner border extending at right angles to the sole-plate and jointed opposite the ankle. The foot is firmly fixed to the sole-plate by straps of webbing; the

26

foot is then firmly everted, and the upright is then brought into position at the side of the leg, and retained there by a buckle and strap. This appliance can be readily and easily constructed, and offers an efficient method for the correction of the varus. It has also been employed to correct the equinus, but for this purpose the additional application of adhesive plaster to the leg is necessary. Shaffer's appliance for the correction of varus is a decided improvement over this, and offers the best mechanical means for correcting this deformity. The instrument consists of a sole-plate

FIG. 262.

Equino-varus brace applied.

made from the outline of the foot, and divided opposite the medio-tarsal joint into two parts, attached at the outer border by a hinge and manipulated by means of a screw. To this sole-plate is attached a steel trough, fitted to the inner side of the leg and divided obliquely opposite the ankle-joint by an oblique hinge, the direction of which is such as to allow pressure, exerted by means of a screw, to operate upon the varus deformity. The mechanism of the divided portion of the sole-plate has been improved by Dr. Roberts by substituting for the extension-bar employed by Shaffer a triple-thread screw. This apparatus is applied to the foot in the direction of an angle corresponding to the angle of deformity. The foot is then forced into a corrected position, which does not produce pain, and on several occasions during the day the contracted tissues are momentarily overstretched, held there for a few moments, and then relaxed. In this manner little by little is daily gained, until in time severe deformities are corrected. The writer would express his satisfaction in the use of this apparatus, and report several severe congenital cases cured by its use. A pushing or retraction force has also been employed by Shaffer.[1]

Elastic extension. Elastic force has been recommended by Barwell, Davis, Sayre, and others, in the treatment of this deformity. These appliances have been found very efficient in the correction of mild cases, but the writer prefers a non-elastic force. De F. Willard employed elastic traction without the use of plaster, by firmly lacing bands of felt or blanket around the foot and upper part of leg and connecting them with an elastic strap. (Fig. 263.)

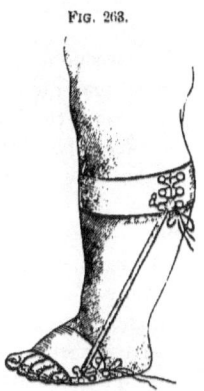

Fig. 263.

Willard's elastic club-foot appliance.

Tenotomy combined with extension and fixation. A combination of tenotomy with extension and fixation offers one of the quickest and best and surest methods of correcting these deformities. The tendons which will require division to correct the varus are the tibialis anticus, tibialis posticus, and plantar fascia, and in some cases the division of astragalo-scaphoid and calcaneo-cuboid ligaments. The part should at once be strongly manipulated until the over-corrected position remains with the application of but little force. It is then retained in this over-corrected position by means of a well-applied plaster-of-Paris bandage, or by means of other retentive apparatus.

The severer forms will require, in addition to these measures, the use of powerful correcting force and some of the cutting operations upon the tarsus.

Brisement forcé should be applied by one of the wrenches or apparatuses already described, in such a manner as to force the foot into a varus position.

Tarsectomy and tarsotomy are only necessary in the severest forms of club-foot, and since they expose the patient to considerable risk, their field of application should be limited. The operation which would be called for in these cases would be one of the three opera-

---

[1] Trans. Amer. Orthop. Assoc., vol. i. p. 291.

tions before given under the general head of tarsectomy and tarsotomy (*vide ante*), selecting the proper operation according to the severity of the individual case. If the over-correction has been thoroughly accomplished the danger of relapse will be greatly diminished, but in all cases it is necessary to strictly carry out the after-treatment, and in the majority of cases a retention walking-shoe is advisable for at least one year.

Treatment of Equinus.

The second division of the treatment of equino-varus, the correction of the equinus, will require resort to all the measures before given except the *brisement forcé*. In the milder cases strong flexion of the foot upon the leg, many times repeated, will in many instances overcome the deformity. In this, as in other deformities, what has been gained by these manual means may be retained by the use of fixation apparatus of different varieties. Elastic extension has in these cases been employed with great benefit. In the severer cases . tenotomy of the tendo Achillis, combined with extension and fixation, should be resorted to. The technique of this has already been given under the treatment of equinus proper. The cutting operations upon the tarsus which have been performed for the relief of the deformity (equino-varus) include all the methods introduced into orthopedic surgery except the one just given for the correction of varus, and for these the reader is referred to the subject under general treatment. In this connection the writer believes that it is in only exceptionally severe cases that cutting operations upon the tarsus should be resorted to. The order in which these operations should be performed is: (*a*) linear osteotomy of the neck of the astragalus; (*b*) removal of the astragalus. These two operations have established positions in orthopedic surgery, but the writer has never seen a case in which he considered anything beyond these operative procedures justifiable. Certainly, as Reeves remarks, "these are the days of *les folies chirurgicales*," and excision of the astragalus in infantile cases under a year is not justifiable. Even in cases in which this operation is justifiable, the surgeon should faithfully apply mechanical measures, and should resort to multiple tenotomy, open incision, and *brisement forcé* before resorting to this operation. The writer has, however, seen several cases which have been greatly benefited by this operation.

Treatment of Acquired Equino-varus.

Treatment of acquired equino-varus will require, in addition to the measures already given, a more frequent resort to rubber muscles to supplement the action of the weakened or paralyzed muscles.

Treatment of Equino-valgus.

The treatment of this compound variety will include the treatment prescribed for the two forms of which it is a compound. In the congenital forms and in the severer grades tenotomy will become necessary.

Treatment of Calcaneo-varus and Valgus.

The treatment of these compound varieties has already been given in sufficient detail in the individual forms of which these are the compounds.

CHAPTER XXVII.

CONGENITAL DISLOCATION OF THE HIP.

CONGENITAL dislocation of the hip is a displacement of the head of the femur, occurring before birth, due in most cases to arrest of growth or retarded development of the acetabulum, and tending, without treatment, to remain stationary or to grow worse. The deformity, in the strictest sense, is not a dislocation, since there is an arrest of development of the constituent parts of the joint, and some of the most important elements of a luxation, as the rupture of the capsule, are absent. The designation of "dysarthrosis congenita" (Von Ammon), and "congenital malposition" (Reeves) would therefore seem more appropriate, but the term congenital dislocation will be here retained on account of its general acceptance.

Frequency. While in itself not a common affection, congenital dislocation of the hip is the most frequent of all congenital dislocations. Its relative frequency is a little less than one in a hundred cases of surgical disease; thus, in 3100 cases of surgical disease treated at the Boston Children's Hospital, 24 were cases of congenital dislocation of the hip, all of which were in girls; 7 were on the right, 11 on the left side, and 6 were double. Of 19 cases treated by Hoffa, 14 were girls, 5 were boys; 12 were double and 7 were single.

Dollinger, in 859 cases of deformity, found there were 9 cases of this affection, or 1.1 per cent., but Hoffa in 1444 orthopedic cases found only 7, or 0.49 per cent. Chaussier, in 23,292 newborn children at the Paris Maternity, found only one case, whereas Parisé found it three times in 332 autopsies upon newborn children at the Hôpital des Enfants Trouvés.

Etiology. As in the study of the causes of all congenital affections, there is in this much that is speculative and purely theoretical. Dislocation of the hip differs from other congenital affections in that it is rarely associated with other deformities, the children being usually otherwise well formed and healthy. The writer has, however, recently seen an instance in which there was congenital deficiency of the femur, tibia, and other parts associated. The affection appears to be more frequently single than double, in the proportion of 191 to 122 cases, and girls are more often affected than boys. (Krönlein, Adams, Albert.) The relative number of boys and girls is well shown in the following table from Lovett[1] :

Reporter.	Number.	Boys.	Girls.
Drachmann . .	. 77	10	67
Pravaz . . .	. 107	11	96
Krönlein . . .	. 90	14	76
N. Y. Orth. Hosp. .	. 25	2	23
Boston Char. Hosp.	. 24	0	24
Prahl . . .	. 18	3	15
	341	40	301

The numerous theories which have been advanced may be considered under the following heads :

1. Theory of heredity.
2. Theory of mechanical intra-uterine pressure or traumatism.
3. Theory of pre-natal disease.
4. Theory of arrest or defect of development.

Traumatism has been ascribed as a cause, and the abnormally large proportion of breech deliveries in these cases has been cited as corroborative evidence. Adams reports seven breech presentations in forty-five cases. The necessary obstetrical operations employed during breech deliveries may, and frequently do, cause dislocation—the so-called *obstetric dislocations*—which, if unreduced, will later in life resemble in every respect congenital dislocation ; but such cases should not be classed as congenital, but as traumatic.

1. *Heredity* exerts a powerful and important influence over the occurrence of this deformity, and may explain, as pointed out by Vallette, the frequency of this affection in certain parts of France, a fact also referred to by Albert, who met with it exceptionally often in the Tyrol. As an etiological factor its influence can scarcely be doubted. Persons suffering from this deformity may leave several children similarly afflicted (Reeves, Bouvier, Caswell, Verneuil, Stadfeldt, Volkmann, Brodhurst), and individual hereditary cases are recorded by almost all writers. The case of Margaret Cardas, reported by Dupuytren, eight of whose relatives were similarly afflicted, is particularly interesting.

2. *The theory of mechanical intra-uterine pressure, or traumatism,* through deficient liquor amnii, compression, or the peculiar position of the fœtus *in utero*, or by a fall or blow, is as old as medicine itself.

---

[1] Diseases of the Hip-joint, 1891. p. 183.

Hippocrates himself averred that "infants in the very womb may have their joints dislocated by a fall, a blow, or compression."[1] Cruveilhier, Roser, and Lücke have supported the uterine compression view due to deficiency of the amniotic fluid, strong adduction being induced in this manner. Dupuytren believed the abnormal position to be one of strong flexion, causing, pressure on the lower and posterior part of the capsule.

External violence of the mother is believed by some to account for occasional cases. Thus, three cases were attributed to a fall in the seventh month by Kleeburg, Chatelain, and Zielewicz. The violent muscular movements of the fœtus itself have likewise been ascribed as a cause, and Chaussier quotes a case of congenital dislocation of the forearm in support of this theory.

The arguments against the pressure theory under the etiology of congenital club-foot apply with greatest force here, since congenital dislocation of the hip is very rarely associated with other deformity; there is no appreciable difference observed in the quantity of liquor amnii over previous births of normal children, and but few cases are recorded where the cause can be directly traced to external or internal traumatism to the fœtus. Moreover, external violence would produce fracture *in utero* rather than dislocation.

8. *The theory of pre-natal disease,* or the musculo-nervous theory, due to intra-uterine lesions identical with post-natal diseases, has been supported by a host of eminent authorities.

(*a*) According to some, the dislocation is spontaneous, being due to softening and looseness of the coxo-femoral ligaments (Sédillot, Stromeyer), to effusion and fungous synovitis (Broca, Verneuil), to hydrarthrosis (Parise), or to caries, arthritis, or other destructive joint inflammations (Morel, Lavallée, Albers, Guérin, Von Ammon).

(*b*) According to the majority it is due to primary muscular contraction, which is to be regarded as secondary and consecutive to a central nervous lesion. Carnochan, Jules Guérin, Adams, Mercer, Melicher, and Chaussier have all given it the weight of their indorsement, the last considering the condition one of intra-uterine convulsions.

(*c*) According to a few (Verneuil, Dalby, Reclus[2]), it is regarded as often the last stage of a paralysis and atrophy of the peri-trochanteric muscles, and the frequency of the dorsal iliac dislocations

---

[1] Paré, 1678, Book xvi. p. 347.   [2] Rev. mens. de Méd. et de Chir., 1878, p. 88.

over other forms is cited as confirmatory. In regard to this theory it may be remarked that the analogous deformities from which this theory was derived—club-foot, wry-neck, and scoliosis—are not now regarded as the result of primary muscular spasm; central cerebral and spinal lesions have not been demonstrated; and, as pointed out by Phelps, the anatomy of the hip-joint is such that muscular spasm, independent of other causes, could not produce dislocation. More-over, the dislocation is not always, as it should be, upward and back-ward, but may be, as pointed out by Phelps, upward and forward, or, according to Cornigan, downward and forward or downward and backward. While in anencephalous fœtal monsters the association of congenital dislocation of the hip, club-foot, and other deformities is confirmatory of this theory, many monstrosities with extensive demonstrable nervous lesions have no deformity of the extremities whatever.

4. *The theory of arrest or defect of development,* or the osseous theory, while it cannot be held accountable for all cases, has received the whole weight of modern authority as the most rational and scientific of all the theories. Proposed by Paletta, indorsed by Von Ammon, Dupuytren, Breschet, Schreger, and others, it has of late years been confirmed and established upon a scientific basis. Dollinger believed the cause to be a premature ossification of the Y-cartilage of the acetabulum, while Grawitz, from the examination of speci-mens and numerous and careful experiments, including micro-scopic examinations, seems to have demonstrated it to be due to arrest of development of the same. The same lack of development of the acetabulum is observed from non-use after luxation of the head of the femur, and it has been questioned whether this condition is primary and causative or secondary from non-use. The observations of Grawitz would seem to prove the former. The malformation may consist of the absence of a rim to the acetabulum, with or without displacement of the head of the femur, as in two specimens reported by Mr. Lockwood.[1] This theory makes congenital dislocation of the hip analogous with other deformities, explains its association with other malformations, and offers the best explanation that has been advanced, and while occasional cases may be due to traumatism, to mechanical or intra-uterine pressure, or to pre-natal disease, in the majority the cause will be found to be a primary arrest of develop-ment of the acetabulum.

---

[1] Trans. Path. Soc. London, 1887, vol. xxxiii. p. 208.

Pathology. The pathological changes, as pointed out by Gurlt, are of two kinds—those found in the affection itself and those produced by walking upon the deformed joint. In autopsies upon new-born children the acetabulum is undeveloped and narrowed, elongated, less concave than normal, and occasionally filled with fat and connective tissue (Paletta, Parisé). The head of the femur is occasionally deformed and atrophied, but is always slightly larger than the concavity of the acetabulum (Porto, Hovel, Cruveilhier). The

FIG. 264.

Dislocation upward and forward. (PHELPS.)

capsule and ligaments are unchanged, and the ligamentum teres is never wanting (Krönlein). The position of the head is usually upon the dorsum of the ilium, but cases are reported where it was dislocated on to the pubis, into the obturator foramen, and downward and backward. In cases which have walked the changes are more marked and progressive. The acetabulum is smaller, triangular, and filled with exostoses (Porto, Reeves). One or more depressions exist where the head of the femur has rested, and on the position of these new sockets depends the angle of the pelvis and the amount of lordosis, for if they form directly above the acetabulum the normal plane remains practically unchanged, but if they are much behind the pelvis is tilted and severe lordosis results. The head is generally flattened, and the neck is shortened or entirely wanting. The capsule

FIG. 265.

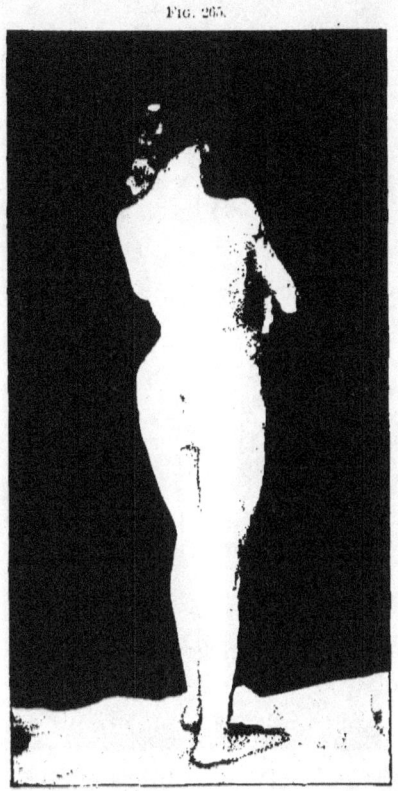

Double congenital dislocation of a mild grade, showing less lordosis and but little deformity. (LOVETT.)
(By permission of the Trustees of the Fiske Prize Fund.)

is relaxed, enlarged, and thickened, and in some cases is constricted into an hour-glass form. The ligamentum teres is flattened and thin, sometimes arising by two heads, and in others being entirely absent (Morgan, Bennett, Bowlby, Coudray). The gluteal muscles are contracted, and the unused muscles are atrophied and degenerated

(Bardeleben). In adult cases the changes are all advanced, exostoses fill the cavity of the acetabulum and form an elevated border above the new socket. The pelvis is suspended by the capsules, as pointed

Fig. 266.

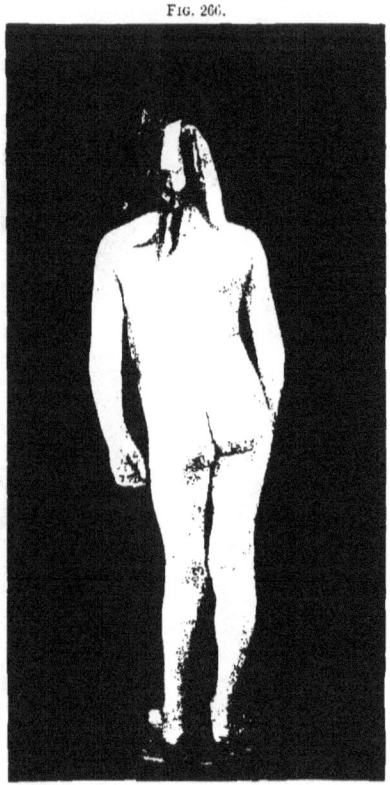

Unilateral congenital dislocation of the left hip, showing the natural position in standing. (LOVETT.)
(By permission of the Trustees of the Fiske Prize Fund.)

out by Volkmann, as the old-fashioned stage-coach was hung upon its leather springs. The pelvis becomes contracted above, expanded below, and the iliac bones are carried backward and upward—a condition which does not interfere with delivery (Lassmann, Bouvier).

Symptoms. The affection is characterized by the peculiar gait and the deformity. In bilateral dislocation the child walks very late and waddles with a peculiar, goose-like gait, with marked lordosis, flexion of the pelvis, and protrusion of the abdomen. The hips are prominent, the trochanters are conspicuous, and the lower extremities

FIG. 267.

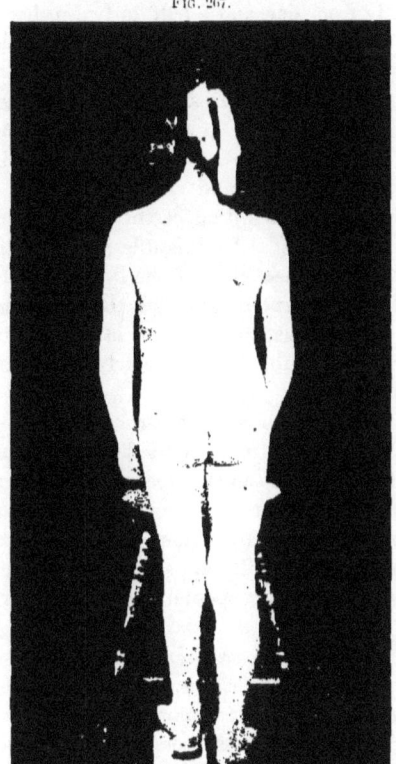

The same case as in Fig. 266, showing the corrected position. (LOVETT.)
(By permission of the Trustees of the Fiske Prize Fund.)

relatively short. In unilateral dislocation the gait becomes an exaggerated limp, the limb is shortened, the pelvis flexed and tilted, and scoliosis is present. The dislocated joints are excessively mobile and free from swelling or pain.

Diagnosis. In very young children the affection is frequently overlooked, but in adults the diagnosis can often be made at a glance. The diagnosis rests upon the history, the prominence of the trochanter, excessive mobility of the joint, width of the pelvis, and especially the position of the trochanter in its relation to Nélaton's line, drawn from the anterior superior spinous process of the ilium to the tuberosity of the ischium. Instead of being in line, the trochanter is usually from one and a half to two inches above. This affection must be distinguished from traumatic dislocations, hip-disease, bow-legs, and infantile paralysis. In traumatic dislocations the history of an obstetrical injury and the early fixation of the joint are distinguishing points. In hip-disease the spasm, atrophy, deformity, pain, abscess, and other symptoms are distinctive. In bow-legs the waddling walk and marked lordosis resemble this affection, but the normal condition of the hip is decisive, although both affections may coexist. In infantile paralysis the gait, laxity of the hip-joint, and the inequality of the limbs resemble dislocation, and where paralytic dislocation also exists, the condition is confusing ; but the history of paralysis, the cold and atrophied limb, the laxity of all the joints of the limb, and particularly the electrical reactions, serve to distinguish it.

Prognosis. The prognosis in older cases is unfavorable. In very young children, under modern methods, the outlook is more favorable than formerly, but much appears to depend upon the condition of the acetabulum and the time devoted to treatment. In older cases the newer operative methods also offer some hope. Without treatment the deformity appears either to remain stationary or grow somewhat worse, no spontaneous cures ever having been reported.

Treatment. *Treatment by extension and apparatus.* Cases have been reported in which benefit has resulted from the use of continuous prolonged extension by Pravaz senior, Pravaz junior, Guérin, and Adams, but in all of these cases there appears to have been some doubt as to the ultimate result. More recently, perfect cures have been reported by Buckminster Brown[1] and Bradford,[2] both having been accomplished by similar methods and apparatus. In the latter case the child was placed in a recumbent position, with a weight and pulley traction upon the limb for several months, so that the trochanter stood at nearly its normal level on both sides. Later, it was arranged that the child should sit up, wearing an appliance which exercised traction upon the femur at right angles to the trunk.

[1] Boston Med. and Surg. Journ., 1885, No. 23.

[2] Trans. Amer. Orthop. Assoc., 1891, p. 308.

FIG. 268.

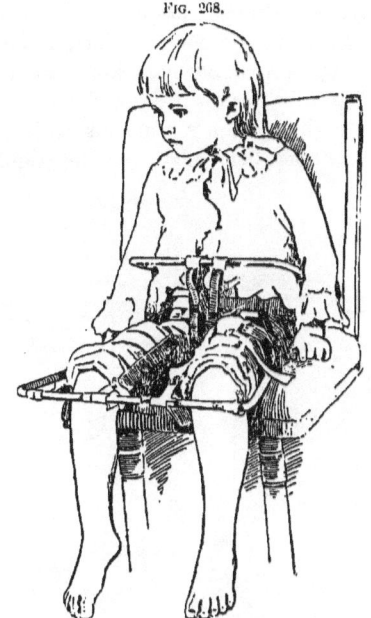

Traction appliance for congenital dislocation of hip. (BRADFORD.)

FIG. 269.

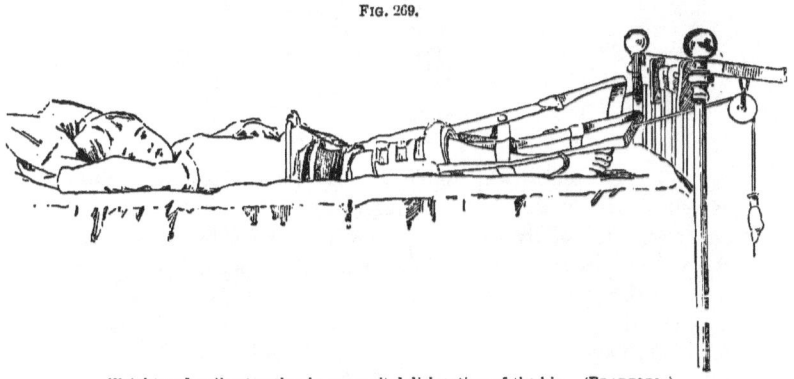

Weight-and-pulley traction in congenital dislocation of the hip. (BRADFORD.)

When the ligamentous traction of the fasciæ was overcome, passive motion was made with the limb in the normal position to facilitate the formation of a new socket in this position. At the end of two years the head of the femur not only remained in the nearly normal position in which it was placed, but it could not by any manual force be pushed out. The child was then suspended in an appliance consisting of a perineal sling, upon a trolley, so that the child could touch the feet to the floor, but could bear no weight, and finally a

Fig. 270.

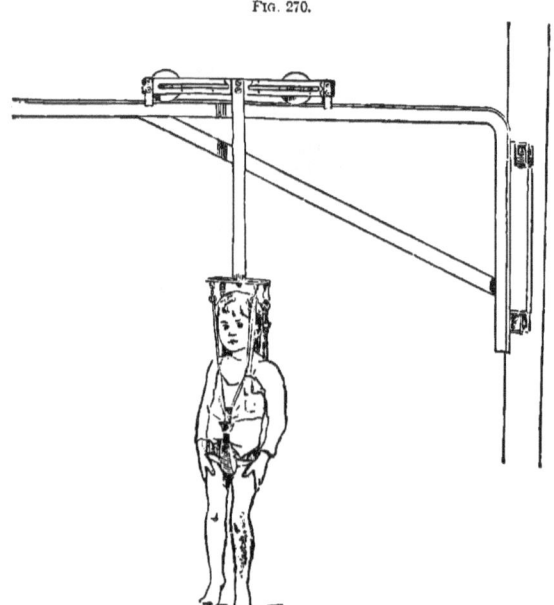

Trolley arrangement.   (BRADFORD.)

protective hip splint was applied. The entire apparatus employed is well shown in the accompanying illustrations. A unilateral case has been successfully treated by Post, of Boston, by anæsthetizing the patient, placing the head of the femur in the normal position, and fixing it there by plaster-of-Paris bandages, renewed at intervals for a year, and more recently Lovett has reported[1] improvement in a girl of thirteen by six months' bed extension and later an extension splint. Corset and pelvic bands of plaster-of-Paris, silicate, or felt, have been much advocated, especially by the Germans. These have their uses where nothing more radical or curative can be employed.

[1] Diseases of the Hip, 1891, p. 200.

Motta makes a plaster mould while the patient is suspended, and from this makes a poroplastic splint to be worn during the day, using a gaiter extension at night. The principle of the steel appliances generally employed is shown in the illustration. Modifications of the long extension splint have been employed by Ridlon, Phelps, and others.

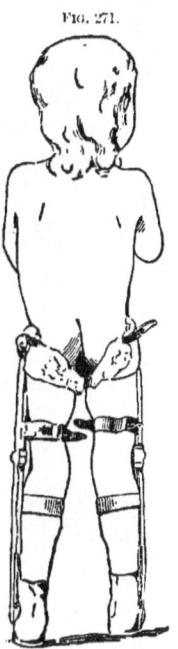

Fig. 271.

Operative Treatment. Many operations have been performed for the relief of this deformity. Guérin, Bouvier, Pravaz, Jr., Coolidge, Brodhurst, and Barwell performed subcutaneous tenotomy of the resisting muscles, with some apparent improvement in most of the cases. Hueter suggested removal of the head of the femur, and attempting by periosteal flaps to obtain bony union between the femur and pelvis, and König performed a similar operation, but the child died of scarlet fever before the results of the procedure could be ascertained. De Paoli enlarged the acetabulum, pushed down the head and nailed the femur in place, but the result was so bad that it was abandoned for resection.

Convalescent protection splint. (BRADFORD.)

Resection of the head of the femur. Resection of the head of the femur for congenital dislocation of the hip has been performed many times, and 27 cases are reported,[1] 17 unilateral, 7 bilateral, and 3 in which the side was not stated. Of the 7 bilateral cases, 3 walked badly with a stick, 4 walked quite well but were obliged to use a cane. Of the 17 unilateral cases, the result is not recorded in 1. Of the remaining, 1 can walk all day without fatigue (Motta), 2 can walk one and a half hours very well; in 5 the walk was in general improved, 2 require a splint, and 6 either limp or walk worse than before the operation (Battini, Langenbeck). The results were, in brief, 3 good, 5 moderate, and 8 bad. The objections to excision are a stiff joint and shortening from removal of the part of the epiphysis in a growing child. The results of resection are far from satisfactory. In bilateral dislocation the results are in many cases bad, and in unilat-

[1] Lovett: Diseases of the Hip, 1891. p. 207.

eral cases the results are unsatisfactory, and in neither do the results compare with the cures reported by traction methods.

Hoffa's operation. Of all the operative procedures proposed, that of Hoffa,[1] of Würzburg, deserves special mention. He would only resort to resection where osteitis is present in the joint, and believes

Fig. 272.

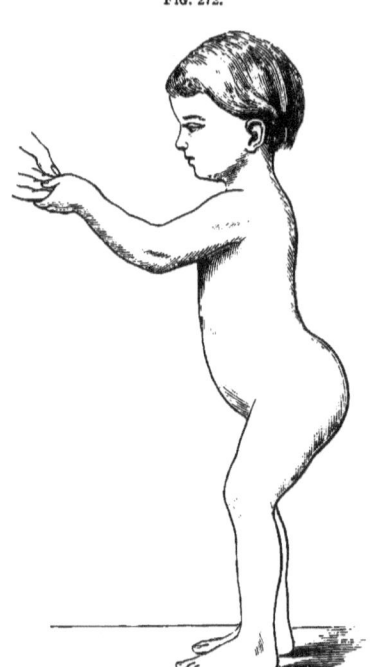

Double congenital luxation of hip : before operation. (HOFFA.)

in the thorough separation from the femur of all the shortened tissues. He makes a longitudinal incision down to the femur after Langenbeck's method, divides the fascia lata, rectus, etc., extirpates the capsule, enlarges the acetabulum with a heavy curette, reduces the dislocation by manipulation, and subperiosteally separates the muscles from the great trochanter. The flap detached from the bottom of the acetabulum is turned back and attached to the head

[1] Revue d'Orthopédie, March, 1891.

of the femur. The wound is drained, and the thigh, abducted and extended, is fixed in a plaster-of-Paris dressing. In the seven cases reported the results are superior to any set of cases so far reported, the movements in several being recorded as perfect. In cases where

FIG. 273.

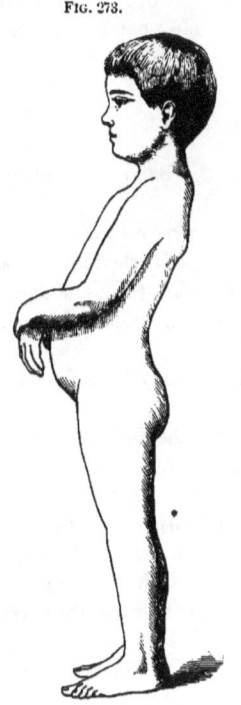

Double congenital luxation of hip: after operation. Side view. (HOFFA.)

FIG. 274.

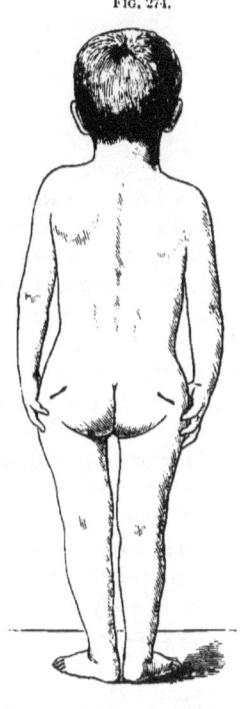

Double congenital luxation of hip: after operation. Back view. (HOFFA.)

no trace of an acetabulum exists, it has been proposed to overcome the difficulty by chiselling through the ilium, as Ogston is said to have done.

In conclusion, mechanical measures, under the most favorable circumstances, offer the best method of treatment in children, but are otherwise unsatisfactory. The ordinary operative methods, including multiple tenotomies, are generally unsatisfactory, and resection may cause a stiff joint and great shortening. Hoffa's operation, while promising much, may be said to be still on trial.

CHAPTER XXVIII.

PERVERTED DEVELOPMENT.

THE number of congenital deformities dependent upon perverted development is very large, but some of these have been omitted altogether; others belong more properly to works upon teratology, and only three need here be described—club-hand, deformities of fingers and toes, and deficiency of parts.

Club-hand.

Club-hand includes any deviation of the hand from its normal relation to forearm at the wrist.

Synonyms. French, *Main-bote*. German, *Klumphand*.

The deformity may be in the direction of—(1) flexion, or palmar, (2) extension, or dorsal, (3) adduction, or radial, (4) abduction, ulnar or cubital, or any combination of these, as radio-palmar, radio-dorsal, cubito-palmar, and cubito-dorsal.

The affection is generally congenital, but may be acquired. One or both hands may be affected.

The flexion forms are most common, and the deformity is usually associated with other malformations.

Etiology. The cause is obscure, but may be accounted for in congenital cases by the pressure theory of absence or deficiency of liquor amnii, and where there is associated deficiency of bones, by the theory of attachment of the amnion to the skin while these parts were in contact in early embryological life. A rare cause from cerebral injury has been reported by the writer.[1] The acquired variety may result from nerve irritation from bullets, tumors, bone, or as sequelæ of burns and scalds.

Symptoms. The hand is not rigidly held in the deformed position, but admits of a certain range of motion beyond which it is checked by the shortened muscles or bony changes. The diagnosis

---

[1] Medical News, May 12, 1888.

is made upon inspection. The possibilities of recovery without treatment are slight.

Treatment. In mild congenital cases, manipulation and the use of a pressure bandage will accomplish a cure, as in the case reported by Piéchaud, of a child with double club-hand of the ulnar variety, cured by the mother's manipulation in five months.

Plaster-of-Paris bandages have been successfully employed by L. A. Sayre. In the severer forms tenotomy will be indicated, the tendons requiring division being generally the palmaris longus, flexor carpi ulnaris, and flexor carpi radialis. The possibilities of ununited tendon after tenotomy is greater in this region than elsewhere, and in most cases where practicable the tendon should be lengthened by the modern method of open incision, identical with Willett's operation in the foot. A recent operation of this character upon a young woman suffering from monoplegia, in which the writer was associated, proved eminently successful.

In cases due to severe irritation the removal of the cause will accomplish a cure, and in paralytic cases the use of suitable mechanical appliances should be assisted by massage and electricity.

Deformities of the Fingers and Toes.

The congenital deformities of the fingers consist of six classes: supernumerary fingers and toes, congenital deficiencies, congenital union, hypertrophy, contraction, and tumors.

Supernumerary fingers and toes, known also as polydactylism, is not an uncommon inheritance in some families. It may recur in every generation, or one or more generations may escape this deformity. It may be unilateral, but as a rule is bilateral, there being usually one additional member on each hand and foot. This number may be greatly exceeded, as in the cases of Sâviard and Voigt, which had respectively ten and thirteen fingers and toes on each hand and foot, or the case of Bradford and Lovett, which had fifteen fingers and ten toes. The additional parts are usually added upon the ulnar side of the hand. Some are fully formed, but more often they are imperfect and associated with other deformity, especially congenital union, and Annandale[1] has classified four groups according to the degree of perfection.

---

[1] Malformation of the Fingers and Toes, 1866, p. 26.

The proper treatment consists in the removal of the supernumerary parts, and this can be accomplished with perfect safety at two or three months after birth. In exceptional cases the amputated part may be perfectly reproduced, as in the remarkable case of C. White,[1] where a supernumerary thumb was twice entirely reproduced.

Congenital deficiencies, either in the number or bulk of digits, are not common, and when they do occur are usually the result of amniotic inflammatory adhesion or amputation, not hereditary, but the result of maternal impression. All the fingers of one hand may be wanting, as in one of Annandale's cases.

The treatment is limited to the separation of webbed fingers, or plastic operations to restore deficient parts.

Congenital union, or webbed fingers and toes, scientifically known as syndactylism, is common, and may involve the union of the digits throughout their whole length or only the terminal phalanges.

FIG. 275.

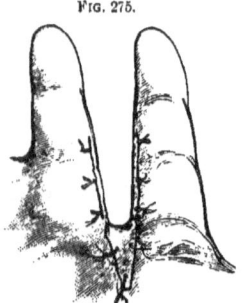

Dorsal flap brought through between fingers and stretched on the palmar side ; also wound closed on opposite side of fingers. (AGNEW.)

FIG. 276.

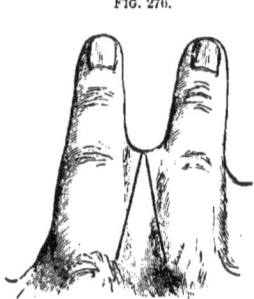

Form of incision for webbed fingers. (AGNEW.)

The united parts may be divided by a simple incision, and the parts held apart by strips of oiled lint until cicatrization is complete, but the plastic operations of Agnew[2] or Didot[3] are more satisfactory. Agnew's operation (Figs. 275 and 276) consists of a V-shaped flap cut from the dorsal surface of the base of the web, with the apex anterior, extending through one-half of the thickness of the band. This flap is dissected back, the remaining portions of the web slip longitudinally, the flap drawn through the cleft at the base of the

[1] C. White: "On the Regeneration of Animal Substances."
[2] Surgery, vol. iii. p. 371.
[3] Bull. de l'Acad. de Méd. de Belgique, March 23, 1850, lix. p. 351.

fingers, its apex stitched to the palmar surface of the wound, and its sides to the sides of the fingers.

Didot's operation (Figs. 277 and 278) consists of a palmar flap from one finger and a dorsal flap from the adjoining one, the flaps extending to the middle of the fingers. The remaining web is

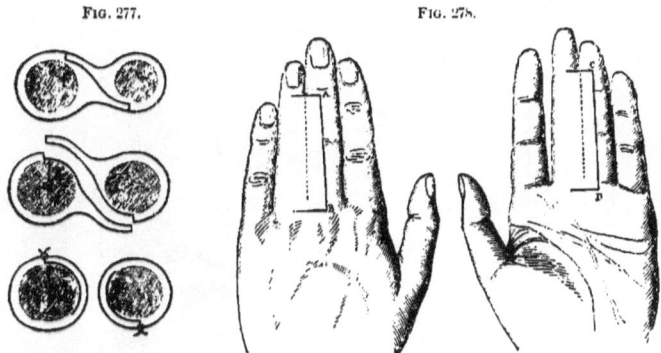

Fig. 277. Fig. 278.

Didot's operation for syndactylism, represented in cross section.
(Bradford and Lovett.)

Didot's operation for syndactylism.
(Bradford and Lovett.)

divided and the dorsal flap of one covers the palmar surface of the other. Separation of webbed toes is not so often demanded, since the deformity is easily concealed, but when necessary it should be performed by the same methods as for webbed fingers.

Hypertrophy of the fingers and toes usually affects but one or two digits. Some of the cases formerly classed under this head belong more properly under akromegaly. The treatment demanded is amputation.

Congenital contraction of the fingers and toes is, in some cases, hereditary, and is usually the result of defect or deficiency of the bones or contracture of the fascias and muscles.

Congenital tumors of the fingers and toes have been grouped by Annandale into four classes: (1) pedunculated growths or excrescences of the skin, (2) fatty growths, (3) fibrous growths, and (4) cartilaginous growths.

The treatment consists in removal of the growths, those of the first class being removed early, and those of the other groups being permitted to remain until the period of infancy is passed.

Congenital Deficiency of Parts.

Congenital deficiencies vary in extent from the absence of a single digit to the entire lack of one or more extremities. The total absence of both upper and lower extremities is an exceedingly rare malformation. In the systematic works of Förster, St. Hilaire, Tarnier,

Fig. 279.

Congenital deficiency of extremities. Position assumed when walking on hands. (G. E. Shoemaker.)

Ammon, Otto, Ahlfeld, and Little, reference is made to this deformity, and cases have been reported by Hare and Hardy. Entire absence of both upper extremities has been observed by Ramon, T. Smith, Curran, Gee, and the writer has observed two unrecorded cases, one an artist familiar to many for his reproductions of paintings in the modern gallery in Antwerp, Belgium, and the other an inmate of the Philadelphia Almshouse, the proud father of a large family. In such cases the feet acquire great tact and skill and fulfil the offices of the hands.

Entire absence of the lower extremities is very rare also, and the accompanying illustrations of Shoemaker's case[1] (Figs. 279 and 280) is an excellent example of this type. Partial deficiencies are much more common, and many cases of this character are recorded.

Fig. 280.

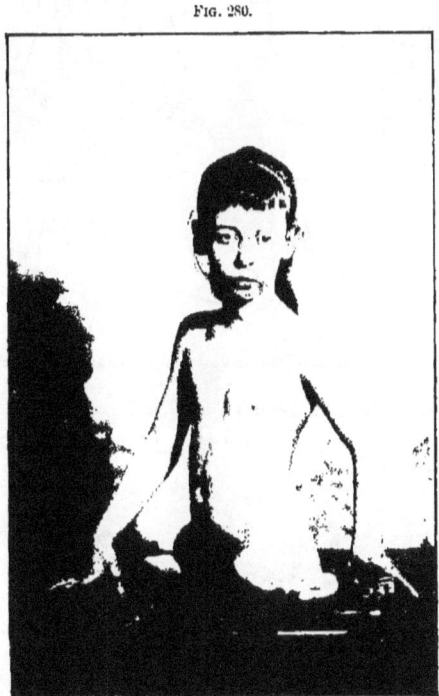

Congenital deficiency of extremities. Sitting position. (G. E. SHOEMAKER.)

St. Hilaire,[2] in his classic treatise, has made one simple division of anomalies—hemiteres—and three complex divisions, heterotaxis, hermaphrodism, and monstrosities, and has placed these deformities under Class V. of the first division as "anomalies by numerical diminution of the part."

[1] Trans. Coll. Phys. Philadelphia, 1892, p. 191. Internat. Med. Mag., March, 1893.
[2] Hist. gén. et part. des Anom. de l'Organ. chez l'Homme, 1832.

Förster[1] has given but three divisions : " 1. Complete or partial absence of an extremity.   2. Deformities so great that the parts are scarcely recognizable.   3. Diminution in size, in which the extremity

FIG. 281.

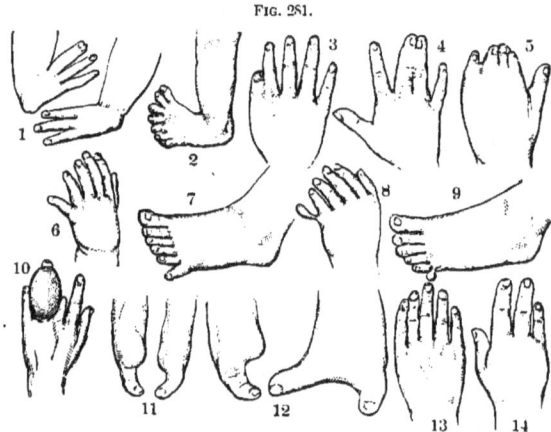

Types of different kinds of deformities of the hands and feet.   (BRYANT.)

is properly formed, but remains undeveloped "—and it is quite evident that they belong to the first.   Tarnier,[2] again, has adopted the plan of St. Hilaire.

FIG. 282.

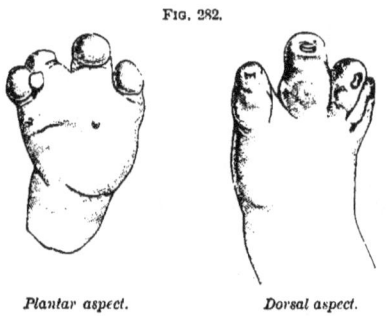

Plantar aspect.   *Dorsal aspect.*
Congenital hypertrophy of toes and foot.

Of absence of the humerus, partial or complete, no cases are found in literature, but of all the other bones the number is very large.

[1] Die Missbild. des Menschen, 1861.
[2] Traité de l'Art des Accouch., vol. ii.

Wenzel Gruber has contributed two communications on congenital absence of the radius, Malgaigne two cases, and Ehrlich, Parker, Reeves, Kaczande, Swaagman (double), Ledru (double), Hodge,

FIG. 283.

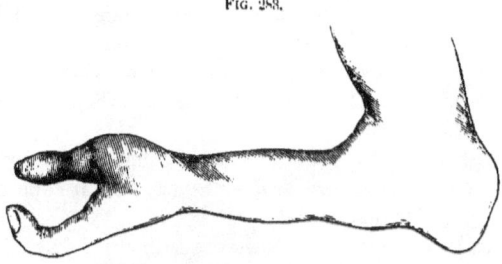

Congenital absence of ulna. (ROBERTS.)

Letolle, and Erichsen have each recorded cases of this defect. The absence of the ulna is much less frequent; Schnelle, Senftleben, and A. S. Roberts have each recorded cases. The writer enjoyed the pleasure of examining this last case. Ehrlich and Hirst have

FIG. 284.

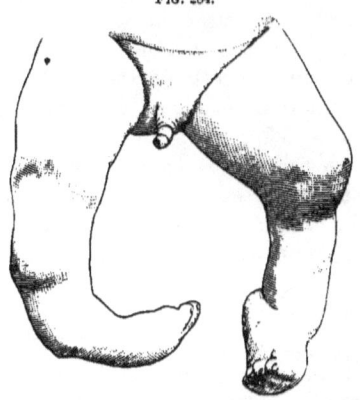

Author's case of double congenital absence of tibia.

recorded cases of rudimentary femora, Williams entire deficiency of both femora, and Meyersohn has collected eighteen cases of fibular defect, to which may be added a case by Little.[1] Congenital absence of the tibia has received more attention and study than any other

[1] Deformities of the Human Frame, 1850, p. 329.

similar deformity, and for a full description of this remarkable malformation the reader is referred to the case of the writer.[1] In all the systematic writers before referred to, not a single case of total absence of the tibia was recorded. In fact, not until 1861, when Billroth[2] recorded his case, was this anomaly recognized and described. Since then, as far as the literature is accessible, only eleven, including Billroth's, have been recorded; they are those of Albert,[3] first; Meyersohn,[4] Pauli,[5] Albert,[6] second; Parker,[7] Ehrlich[8] (three cases), Thümmel,[9] and Busachi.[10]

These cases attract the greatest interest from an orthopedic standpoint on account of the associated club-foot, and are the only ones demanding extended description.

The difficulty of ascertaining the cause with any degree of certainty at once becomes apparent when we appreciate the early period at which such defects must necessarily occur.

The popular theory of hereditary influence and maternal impression would find support in the cases of Parker and Thümmel, there being in the former a "remote family history on the mother's side of a similar deformity," and "the mother states that she had a presentiment while pregnant that the child would be deformed;" and in the latter the mother laid great stress on the fact that during the last half of her pregnancy she was shocked by a man who had a wound on his right leg, and attributed the deformity to this. The latter we are particularly inclined to doubt, as the deformity had doubtless existed many months before the shock which led to the coincidence occurred. That the deformity is not the result of inordinate muscular contraction through disease of the nervous centres[11] we are led to believe from the fact that the power required to produce such a result is out of all proportion to the development of the muscles at this early period.

With the opinion of Billroth that they are congenital luxations,

---

[1] Amer. Journ. Med. Sci., 1888, p. 145.

[2] Ueber einige durch Knochendefecte hervorgerufene Verkrümmungen des Fusses. Arch f. klin. Chir., i. S. 201.

[3] Implantation des Fib. bei angebor. Def. des ganz. Tibia. Wien. med. Presse, 1877.

[4] Meyersohn : Loc. cit.

[5] Ein Fall Klumpfuss durch Mangel des Diaphyse und der unteren Epiphyse des Tibia. Arch. f. klin. Chir., Bd. xxiv.

[6] Zwei seltene Fälle von Missbild. der Extremitäten. Wiener med. Blätter, 1880.

[7] Congenital Absence of Radius, etc. Pathol. Trans., Lond., xxxiii.

[8] Ehrlich : Loc. cit.

[9] Ein Fall von congen. Def. d. ganz. Tibia. Halle, 1886.

[10] Gior. della v. Accademia Medico.-chir. di Turin, 1887.

[11] Little : Loc. cit.

followed by a disappearance of the tibia, we cannot agree, first, because in many of the cases a large rudiment of the tibia remained, and, second, because such an explanation could not be applied to the simultaneous presence of defects of the upper extremity, such as occurred in some of the recorded cases.

We are inclined to accept the view of Hasse[1] and Pauli, or Dareste,[2] that at a very early period of intra-uterine life, when the amnion was in contact with the foetus, the membrane experienced a serous inflammation, resulting in a plastic exudate which led to an adhesion between the amnion and the integument ; that later by the increase of the liquor amnii these bands were gradually broken up, resulting in distortions, fractures, and amputations. In favor of this theory are the membranous attachments at the point of the rudiment of the tibia, and also in the case observed by Dr. Keihn,[3] a hole was found in the foetal membranes, exactly corresponding to the fragment of the tibia.

These bands may also be formed by the bursting of the amnion,[4] the chorion retaining the integrity of the ovum, and the rolling of the amnion on itself into bands, and the active movements of the child.

In the case of Meyersohn the process was more marked, the result being an amputation of all but the fragment of the tibia. That this was not the result of pressure of the umbilical cord is now generally accepted, because the cord can only cut down to but not through the bone, being harder than foetal tissues, but softer than bone ; and also in this case by the absence on the other side of the fibula and patella.

In reference to the marked club-foot present in all these cases, the absence of the internal condyle and the unopposed action of the adductors are sufficient explanation, to my mind, for this deformity. That they were not the result of deficiency of liquor amnii and intra-uterine pressure is clearly proven by the absence of pressure marks on other parts ; the rare association of club-hand (but one case[5]) ; the fact that no appreciable diminution in the quantity of liquor amnii was observed in these cases over previous or subsequent labors, but more particularly by the case[6] recently observed " of double equino-

[1] *Vide* Pauli : Loc. cit.
[2] Mém. sur les Anomalies des Membres, etc. Journ. de l'Anat. et de la Phys., 1882.
[3] Virch. Arch., Bd. c. 114.
[4] G. Braun : Oesterreich. Zeitschr. f. prakt. Heilkunde, 1865, Nos. 9 und 10. W. Lebedeff : Annales de Gynéc., Avril, 1878.
[5] Ehrlich (first case) : Loc. cit.
[6] A. Sydney Roberts : Clinical Lectures. Med. News, March 13 and 20, 1886.

varus in a twin, the other child showing no deformity whatever."
Aside from these, the theory of Berg,[1] and Parker and Shattuck, of
failure to rotate, would offer the most satisfactory explanation, supported as is is by embryological research.

FIG. 285.

Congenital deficiency of extremities. Permanent apparatus applied. (G. E. SHOEMAKER.)

The treatment will depend on the degree and inconvenience, and
the condition of the bones. Amputations at the knee were performed
in the cases of Billroth, Pauli (Carden's method), and Thümmel.
Albert preferred a more conservative method, and performed an
inter-condyloid resection of a wedge-shaped piece, so that the fibula
would come more directly in the line of support. Busachi, who was
inclined to follow the plan of Albert, thought that a tenotomy of the
tendo Achillis, with energetic massage long continued, would suffice

[1] Berg : Seguin's Archives, vol. viii. p. 226.

to bring the foot into an improved position, and under the existing conditions decided against an immediate operation.

In the writer's case massage was advised, with the use, later, of lateral shoes and supports. This seemed most advisable, from the fact that the case being bilateral nothing could be gained by operative interference, and further from the information derived from Parker's patient, in whom the deformity was bilateral, who at three years supported himself and walked on the fibulæ, but required the aid of a chair to steady himself.

INDEX.

28

LEA BROTHERS & CO'S

CLASSIFIED CATALOGUE

OF

MEDICAL AND SURGICAL

Publications.

N ASKING the attention of the profession to the works advertised in the follow-ing pages, the publishers would state that no pains are spared to secure a continuance of the confidence earned for the publications of the house by their careful selection and accuracy and finish of execution.

The printed prices are those at which books can generally be supplied by booksellers throughout the United States, who can readily procure for their customers any works not kept in stock. Where access to bookstores is not convenient books will be sent by mail by the publishers postpaid on receipt of the printed price, and as the limit of mailable weight has been removed, no difficulty will be experienced in obtaining through the post-office any work in this catalogue. No risks however are assumed either on the money or on the books, and no publications but our own are supplied, so that gentlemen will in most cases find it more convenient to deal with the nearest bookseller.

LEA BROTHERS & CO.

Nos. 706, 708 & 710 Sansom St., Philadelphia, June, 1894.

Practical Medical Periodicals.

THE AMERICAN JOURNAL OF THE MEDICAL SCIENCES, Monthly, $4.00 per annum.

THE MEDICAL NEWS, Weekly, $4.00 per annum.

} *To one address. post-paid,* **$7.50** *per annum.*

THE MEDICAL NEWS VISITING LIST (4 styles, see page 3), $1.25. With either or both above periodicals, in advance, 75c.

THE YEAR-BOOK OF TREATMENT (see page 16), $1.50. With either JOURNAL or NEWS, or both, 75c. Or JOURNAL, NEWS, VIS-ITING LIST AND YEAR-BOOK, in all $10.75, for $8.50 in advance.

Subscription Price Reduced to $4.00 Per Annum.

THE MEDICAL NEWS.

B Y KEEPING closely in touch with the needs of the active practitioner, THE News has achieved a reputation for utility so extensive as to render practicable its reduction in price from five to **Four Dollars** per annum. It is now by far the cheapest as well as the best large weekly medical journal published in America. Employing all the recognized resources of modern journalism, such as the cable, telegraph, resident correspondents, special reporters, etc., THE News supplies in the 28 quarto pages of each issue the latest and best information on subjects of importance and value to practitioners in all branches of medicine. The foremost writers, teachers and practitioners of the day furnish original articles, clinical lectures and notes

(Continued on next page.)

The New { Gray's Anatomy, 13th Edition. Just Ready. See page 5.
Dunglison's Dictionary, 21st Edition. Just Ready. See page 4.
National Dispensatory, 1894, 5th Edition. Just Ready. See page 11.

THE MEDICAL NEWS---Continued.

on practical advances; the latest methods in leading hospitals are constantly reported; a condensed summary of progress is gleaned each week from a large exchange list, comprising the best journals at home and abroad; a special department is assigned to abstracts requiring full treatment for proper presentation; editorial articles are secured from writers able to deal instructively with questions of the day; books are carefully reviewed; society proceedings are represented by the pith alone; regular correspondence is furnished from important medical centres, and minor matters of interest are grouped each week under news items. In a word THE MEDICAL NEWS is a crisp, fresh, weekly professional newspaper and as such occupies a well-marked sphere of usefulness, distinct from and complementary to the ideal monthly magazine, THE AMERICAN JOURNAL OF THE MEDICAL SCIENCES.

The American Journal
of the
Medical Sciences

Published Monthly
at $4.00
Per Annum.

THE AMERICAN JOURNAL entered with 1894 upon its seventy-fifth year, still maintaining the foremost place among the medical magazines of the world. A vigorous existence during two and a half generations of men amply proves that it has always adapted itself to meet fully the requirements of the time.

Being the medium chosen by the best minds of the profession during this period for the presentation of their ablest papers, THE AMERICAN JOURNAL has well earned the praise accorded it by an unquestioned authority—*"From this file alone, were all other publications of the press for the last fifty years destroyed, it would be possible to reproduce the great majority of the real contributions of the world to medical science during that period."* Original Articles, Reviews and Progress of the Medical Sciences constitute the three main departments of this ideal medical monthly.

COMMUTATION RATE.

Taken together, THE JOURNAL and THE NEWS afford to medical readers the advantages of the monthly magazine and the weekly newspaper. Thus all the benefits of medical periodical literature can be secured at the low figure of $7.50 per annum.

Subscribers can obtain, at the close of each volume, cloth covers for THE JOURNAL *(one annually), and for* THE NEWS *(one annually), free by mail, by remitting Ten Cents for* THE JOURNAL *cover, and Fifteen Cents for* THE NEWS *cover.*

The Medical News Visiting List for 1894

Is published in four styles, Weekly (dated for 30 patients); Monthly (undated, for 120 patients per month); Perpetual (undated, for 30 patients weekly per year); and Perpetual (undated, for 60 patients weekly per year). The 60-patient Perpetual consists of 256 pages of assorted blanks. The first three styles contain 32 pages of important data and 176 pages of assorted blanks. Each style is in one wallet-shaped book, leather-bound, with pocket, pencil, rubber, and catheter-scale. Price, each, $1.25.

This list is all that could be desired. It contains a vast amount of useful information, especially for emergencies, and gives good tables of doses and therapeutics.—*Canadian Practitioner.*

Its compactness and simplicity are such as to indicate that the highest point of perfection has been reached in works of this class.—*University Medical Magazine.*

The new issue maintains its previous reputation. It adapts itself to every style of book-keeping; there is space for all kinds of professional records; it is furnished with a ready reference thumb-letter index, and has a most valuable text.—*Medical Record.*

For convenience and elegance it is not surpassable.—*Obstetric Gazette.*

SPECIAL COMBINATIONS WITH THE VISITING LIST, see p. 1.

☞The safest mode of remittance is by bank check or postal money order, drawn to the order of the undersigned; where these are not accessible, remittances for subscriptions may be sent at the risk of the publishers by forwarding in *registered* letters addressed to the Publishers (see below).

The Medical News Physicians' Ledger.

Containing 300 pages of fine linen "ledger" paper ruled so that all the accounts of a large practice may be conveniently kept in it, either by single or double entry, for a long period. Strongly bound in leather, with cloth sides, and with a patent flexible back which permits it to lie perfectly flat when opened at any place. Price, $4.00.

Lea Brothers & Co., Publishers, 706, 708 & 710 Sansom Street, Philadelphia.

THE STUDENTS'
DICTIONARY OF MEDICINE
AND THE ALLIED SCIENCES,

COMPRISING THE PRONUNCIATION, DERIVATION AND FULL EXPLANATION OF MEDICAL
TERMS; TOGETHER WITH MUCH COLLATERAL DESCRIPTIVE MATTER,
NUMEROUS TABLES, ETC.

By ALEXANDER DUANE, M. D.,

Assistant Surgeon to the New York Ophthalmic and Aural Institute; Reviser of Medical Terms for
Webster's International Dictionary.

In one square octavo volume of 658 pages. Cloth, $4.25; half leather, $4.50; full
sheep, $5.00. *Just ready.*

Dr. Duane has spared no time, pains or expense
in his endeavor to bring before the profession,
and especially the students of medicine, a book embodying completeness and explicitness. The
vocabulary is abundant and its fulness is paral-
leled by the explanation accorded each word. It
also contains extensive tables. Each word is followed by its correct pronunciation, a new feature
in works of this kind, given by means of a simple
and obvious phonetic spelling. Derivation, the
greatest aid to memory, is fully treated of, and for
the convenience of those who do not understand
Greek, the English letters are substituted for
those of the Greek in giving the roots of the words
derived from that language. The author's experience as a lexicographer is fully attested by his
position as Reviser of Medical Terms for Webster's International Dictionary. We predict that
this will become a standard and favorite work of
its class.—*Medical Fortnightly*, March, 1894.

From A. L. LOOMIS, M. D, *Professor Pathology and*
Practice of Medicine, Medical Department, Univer-
sity City of New York, New York.
It seems to me entirely satisfactory for the purpose for which it is intended.

From J C. WILSON, M. D., *Professor of Medicine,*
Jefferson Medical College, Philadelphia.
It appears to be well suited to the purposes of
the medical student, being simple as regards deri-
vations and pronunciation, explicit yet sufficiently
comprehensive in definitions, and thoroughly up
to the times.

From JAMES T. WHITTAKER, M.D., *Professor Theory*
and Practice of Medicine, Medical College of Ohio,
Cincinnati, O.
I find it admirably adapted to the wants of students, and thoroughly modern in every particular
in which I have taken occasion to consult it. I
shall certainly recommend it to my class.

THE STUDENTS' QUIZ SERIES.

A NEW Series of Manuals, comprising all departments of medical science and practice,
and prepared to meet the needs of students and practitioners. Written by prominent medical teachers and specialists in New York, these volumes may be trusted as
authoritative and abreast of the day. Cast in the form of suggestive questions, and concise
and clear answers, the text will impress vividly upon the reader's memory the salient
points of his subject. To the student these volumes will be of the utmost service in preparing for examinations, and they will also be of great use to the practitioner in recalling
forgotten details, and in gaining the latest knowledge, whether in theory or in the actual
treatment of disease. Illustrations have been inserted wherever advisable. Bound in
limp cloth, and in size suitable for the hand and pocket, these volumes are assured of
enormous popularity, and are accordingly placed at an exceedingly low price in comparison with their value. For details of subjects and prices see below.

ANATOMY (*Double Number*)—By FRED J.
BROCKWAY, M. D., Assistant Demonstrator of
Anatomy, College of Physicians and Surgeons,
New York, and A. O'MALLEY, M. D., Instructor
in Surgery, New York Polyclinic. $1.75.

PHYSIOLOGY—By F. A. MANNING, M.D.,
Attending Surgeon, Manhattan Hosp., N. Y. $1.

CHEMISTRY AND PHYSICS—By JOSEPH
STRUTHERS, Ph. B., Columbia College School of
Mines, N.Y., and D. W. WARD, Ph. B., Columbia
College School of Mines, N. Y., and Chas. H.
Willmarth, M. S., N. Y. $1.

**HISTOLOGY, PATHOLOGY AND BAC-
TERIOLOGY**—By BENNETT S. BEACH, M. D.,
Lecturer on Histology, Pathology and Bacte-
riology, New York Polyclinic. $1.

**MATERIA MEDICA AND THERAPEU-
TICS**—By L. F. WARNER, M. D., Attending
Physician, St. Bartholomew's Disp., N.Y. $1.

**PRACTICE OF MEDICINE, INCLUDING
NERVOUS DISEASES**—By EDWIN T. DOU-
BLEDAY, M.D., Member N.Y. Pathological Soci-
ety, and J. D. NAGEL, M.D., Member N. Y.
County Medical Association. $1.

SURGERY (*Double Number*)—By BERN B. GAL-
LAUDET, M. D., Visiting Surgeon, Bellevue
Hospital, N.Y., and CHARLES DIXON JONES, M.D.,
Assistant Surgeon Out-Patient Department,
Presbyterian Hospital, N. Y. $1.75.

**GENITO-URINARY AND VENEREAL
DISEASES**—By CHARLES H. CHETWOOD, M.D.,
Visiting Surgeon, Demilt Dispensary, Dep. of
Surg. and Gen.-Urin. Dis., New York. $1.

DISEASES OF THE SKIN—By CHARLES C.
RANSOM, M. D., Assistant Dermatologist, Vanderbilt Clinic, New York. $1.

**DISEASES OF THE EYE, EAR, THROAT
AND NOSE**—By FRANK E. MILLER, M.D.,
Throat Surgeon, Vanderbilt Clinic, New York,
JAMES P. McEVOY, M.D., Throat Surgeon, Belle-
vue Hosp., Out-Patient Dep., New York, and
J. E. WEEKS, M. D., Lect. on Ophthal. and
Otol., Bellevue Hosp., Med. Col., N. Y. $1.

OBSTETRICS—By CHARLES W. HAYT, M.D.,
House Physician, Nursery and Child's Hospital, New York. $1.

GYNECOLOGY—By G. W. BRATENAHL, M. D.,
Assistant in Gynecology, Vanderbilt Clinic,
New York, and SINCLAIR TOUSEY, M. D., Assistant Surgeon, Out-Patient Department, Roose-
velt Hospital, New York. $1.

DISEASES OF CHILDREN—By C. A. RHODES,
M. D., Instructor in Diseases of Children, New
York Post-Graduate Medical College. $1.

Lea Brothers & Co., Publishers, 706, 708 & 710 Sansom Street, Philadelphia.

NEW (THIRTEENTH) EDITION. JUST READY.

GRAY'S ANATOMY

IN COLORS OR IN BLACK.

Anatomy, Descriptive and Surgical,

BY HENRY GRAY, F. R. S.,

LECTURER ON ANATOMY AT ST. GEORGE'S HOSPITAL, LONDON.

EDITED BY T. PICKERING PICK, F. R. C. S.,

Surgeon to and Lecturer on Anatomy at St. George's Hospital, London, Examiner in Anatomy, Royal College of Surgeons of England.

A new American from the thirteenth enlarged and improved London edition. In one imperial octavo volume of 1118 pages, with 636 large and elaborate engravings on wood. Price, with illustrations in colors, cloth, $7; leather, $8.

Price, with illustrations in black, cloth, $6; leather, $7.

SINCE 1857 **Gray's Anatomy** has been the standard work used by students of medicine and practitioners in all English-speaking races. So preëminent has it been among the many works on the subject that thirteen editions have been required to meet the demand. This opportunity for frequent revisions has been fully utilized and the work has thus been subjected to the careful scrutiny of many of the most distinguished anatomists of a generation, whereby a degree of completeness and accuracy has been secured which is not attainable in any other way. In no former revision has so much care been exercised as in the present to provide for the student all the assistance that a text-book can furnish. The engravings have always formed a distinguishing feature of this work, and in the present edition the series has been enriched and rendered complete by the addition of many new ones. The large scale on which the illustrations are drawn and the clearness of the execution render them of unequalled value in affording a grasp of the complex details of the subject. As heretofore the name of each part is printed upon it, thus conveying to the eye at once the position, extent and relations of each organ, vessel, muscle, bone or nerve with a clearness impossible when figures or lines of reference are employed. Distinctive colors have been utilized to give additional prominence to the attachments of muscles, the veins, arteries and nerves. For the sake of those who prefer not to pay the slight increase in cost necessitated by the use of colors, the volume is published also in black alone.

The illustrations thus constitute a complete and splendid series, which will greatly assist the student in forming a clear idea of Anatomy, and will also serve to refresh the memory of those who may find in the exigencies of practice the necessity of recalling the details of the dissecting room. Combining as it does a complete Atlas of Anatomy with a thorough treatise on systematic, descriptive and applied Anatomy, the work covers a more extended range of subjects than is customary in the ordinary text-books. It not only answers every need of the student in laying the groundwork of a thorough medical education, but owing to its application of anatomical details to the practice of medicine and surgery, it also furnishes an admirable work of reference for the active practitioner.

We always had a kindly regard for the illustrations in Gray, where each organ, tissue, artery, and nerve bear their respective names, and in this edition color has been worked to advantage in bringing out the relationship of vessel and nerve. Of late years, many works on anatomy have been introduced to the profession, but as a reference book for the practical everyday physician, and as a text-book for the student, we think it will be difficult to supplant Gray.—*Buffalo Med. and Surg. Jour.*, Jan., 1894.

No book deals with the complex subject so systematically, or presents the material in a way better fitted for the student to memorize.—*International Medical Magazine*, Dec. 1893.

It embraces the whole of human anatomy, and it particularly dwells on the practical or applied part of the subject, so that it forms a most useful, intelligible and practical treatise for the student and general practitioner.—*Dublin Journal of Medical Science*, December, 1893.

Gray's has been the unvarying standard for anatomical study by the vast majority of English-speaking medical students for so long that it would seem an anomaly to see a student acquire such knowledge from some other source.—*Medical Fortnightly*, February, 1894.

There has not been a medical student in English-speaking countries who has not incurred a profound obligation to the author of the best descriptive text-book on anatomy. He who builds his medical knowledge on Gray's Anatomy need have no fear for the security of the foundation. Many good text-books on anatomy have been sent forth since "Gray" first appeared on the field; but the unquestioned excellence of the book has proven to be a source of perennial prosperity. The ability of its editors and the enterprise of its publishers have made the numerous editions reflect accurately the status of contemporaneous anatomical science. The last edition contains those additions which progress has made necessary since the publication of its predecessor.—*New Orleans Medical and Surgical Journal*, Jan. 1894.

Few treatises can claim such a record of endurance of usefulness. To say that it still maintains its reputation is to give it the highest compliment within the range of the critic's privilege. It has grown and strengthened with the advancements and requirements of the modern knowledge of anatomy, enlarging its scope for newer adaptations in surgery and physiology.—*Medical Record*, March 3, 1894.

HOBLYN'S DICTIONARY OF MEDICINE. A Dictionary of the Terms Used in Medicine and the Collateral Sciences. By RICHARD D. HOBLYN, M. D. In one large royal 12mo. volume of 520 double-columned pages. Cloth, $1.50; leather, $2.00.

Lea Brothers & Co., Publishers, 706, 708 & 710 Sansom Street, Philadelphia.

HUMAN MONSTROSITIES

BY BARTON C. HIRST, M.D., AND GEORGE A. PIERSOL, M.D.

Professor of Obstetrics in the University of Pennsylvania.   *Professor of Anatomy and Embryology in the University of Pennsylvania.*

Magnificent folio, containing 220 pages of text, illustrated with engravings, and 39 full-page, photographic plates from nature. In four parts, price, each, $5. Complete work *just ready*. Limited edition, for sale by subscription only. Address the Publishers.

We have before us the fourth and last part of the latest and best work on human monstrosities. This completes one of the masterpieces of American medical literature. Typographically and from an artistic standpoint, the work is unexceptionable. In this last and final volume is presented the most complete bibliography of teratological literature extant. No library will be complete without this magnificent work.—*Journal of the American Medical Asso., May 6, 1893.*

Altogether, *Human Monstrosities* is a satisfactory production. It will take its place as a standard work on teratology in medical libraries, and it must always retain the honor of being the first of its kind written in the English language.—*The British Medical Journal, May 27, 1893.*

This work promises to be one for which a place must be found in the library of every anatomist, pathologist, obstetrician and teratologist. It is the joint production of an obstetrician, and an embryologist, and histologist, and this fact makes it certain that both the obstetric and anatomical sides of the subject will be fully represented and described. The book promises to be one of the greatest value to the English-speaking medical world.—*Edinburgh Medical Journal, April, 1892.*

Allen's System of Human Anatomy.

A System of Human Anatomy, Including Its Medical and Surgical Relations. For the use of Practitioners and Students of Medicine. By HARRISON ALLEN, M.D., Professor of Physiology in the University of Pennsylvania. With an Introductory Section on Histology by E. O. SHAKESPEARE, M.D., Ophthalmologist to the Philadelphia Hospital. Comprising 813 double-columned quarto pages, with 380 illustrations on 109 full page lithographic plates, many of which are in colors, and 241 engravings in the text. In six Sections, each in a portfolio. Price per Section, $3.50; also bound in one volume, cloth, $23.00; very handsome half Russia, raised bands and open back, $25.00. *For sale by subscription only. Address the Publishers.*

Holden's Landmarks, Medical and Surgical.

Landmarks, Medical and Surgical. By LUTHER HOLDEN, F. R. C. S., Surgeon to St. Bartholomew's Hospital, London. Second American from the third and revised English ed., with additions by W. W. KEEN, M.D., Professor of Artistic Anatomy in the Penna. Academy of Fine Arts. In one 12mo. volume of 148 pages. Cloth, $1.00.

Clarke & Lockwood's Dissector's Manual.

The Dissector's Manual. By W. B. CLARKE, F. R. C. S., and C. B. LOCKWOOD, F. R. C. S., Demonstrators of Anatomy at St. Bartholomew's Hospital Medical School, London. In one pocket-size 12mo. volume of 396 pages, with 49 illustrations. Limp cloth, red edges, $1.50. See *Students' Series of Manuals*, page 30.

Messrs. Clarke and Lockwood have written a book that can hardly be rivalled as a practical aid to the dissector. Their purpose, which is "how to describe the best way to display the anatomical structure," has been fully attained. They excel in a lucidity of demonstration and graphic terseness of expression, which only a long training and intimate association with students could have given. With such a guide as this, accompanied by so attractive a commentary as Treves' *Surgical Applied Anatomy* (same series), no student could fail to be deeply and absorbingly interested in the study of anatomy.—*New Orleans Medical and Surgical Journal, April, 1884.*

Treves' Surgical Applied Anatomy.

Surgical Applied Anatomy. By FREDERICK TREVES, F. R. C. S., Senior Demonstrator of Anatomy and Assistant Surgeon at the London Hospital. In one pocket-size 12mo. volume of 540 pages, with 61 illustrations. Limp cloth, red edges, $2.00. See *Students' Series of Manuals*, p. 30.

Bellamy's Surgical Anatomy.

The Student's Guide to Surgical Anatomy: Being a Description of the most Important Surgical Regions of the Human Body, and intended as an Introduction to Operative Surgery. By EDWARD BELLAMY, F. R. C. S., Senior Assistant-Surgeon to the Charing-Cross Hospital. In one 12mo. vol. of 300 pages, with 50 illus. Cloth, $2.25.

Wilson's Human Anatomy.

A System of Human Anatomy, General and Special. By ERASMUS WILSON, F. R. S. Edited by W. H. GOBRECHT, M.D., Professor of General and Surgical Anatomy in the Medical College of Ohio. In one large and handsome octavo volume of 616 pages, with 397 illustrations. Cloth, $4.00; leather, $5.00.

HARTSHORNE'S HANDBOOK OF ANATOMY AND PHYSIOLOGY. Second edition, revised. 12mo., 310 pages, 220 woodcuts. Cloth, $1.75.

HORNER'S SPECIAL ANATOMY AND HISTOL-OGY. Eighth edition. In two octavo volumes of 1007 pages, with 320 woodcuts. Cloth, $6.00.

CLELAND'S DIRECTORY FOR THE DISSECTION OF THE HUMAN BODY. 12mo., 178 pp. Cloth, $1.25.

Lea Brothers & Co., Publishers, 706, 708 & 710 Sansom Street, Philadelphia.

Draper's Medical Physics.

Medical Physics. A Text-book for Students and Practitioners of Medicine. By JOHN C. DRAPER, M. D., LL. D., Prof. of Chemistry in the Univ. of the City of New York. In one octavo vol. of 734 pages, with 376 woodcuts, mostly original. Cloth, $4.

No man in America was better fitted than Dr. Draper for the task he undertook and he has provided the student and practitioner of medicine with a volume at once readable and thorough. Even to the student who has some knowledge of physics this book is useful, as it shows him its applications to the profession that he has chosen. Dr. Draper, as an old teacher, knew well the diffi-

culties to be encountered in bringing his subject within the grasp of the average student, and that he has succeeded so well proves once more that the man to write for and examine students is the one who has taught and is teaching them. The book is well printed and fully illustrated, and in every way deserves grateful recognition.—*The Montreal Medical Journal*, July, 1890.

Reichert's Physiology.—Preparing.

A Text-Book on Physiology. By EDWARD T. REICHERT, M. D., Professor of Physiology in the University of Pennsylvania, Philadelphia. In one very handsome octavo volume of 800 pages, fully illustrated.

Power's Human Physiology.—Second Edition.

Human Physiology. By HENRY POWER, M. B., F. R. C. S., Examiner in Physiology, Royal College of Surgeons of England. Second edition. In one 12mo. vol. of 509 pp., with 68 illustrations. Cloth, $1.50. See *Students' Series of Manuals*, p. 30.

Robertson's Physiological Physics.

Physiological Physics. By J. McGREGOR ROBERTSON, M. A., M. B., Muirhead Demonstrator of Physiology, University of Glasgow. In one 12mo. volume of 537 pages, with 219 illus. Limp cloth, $2. See *Students' Series of Manuals*, page 30.

The title of this work sufficiently explains the nature of its contents. It is designed as a manual for the student of medicine, an auxiliary to his text-book in physiology, and it would be particularly useful as a guide to his laboratory experi-

ments. It will be found of great value to the practitioner. It is a carefully prepared book of reference, concise and accurate, and as such we heartily recommend it.—*Journal of the American Medical Association*, Dec. 6, 1884.

Dalton on the Circulation of the Blood.

Doctrines of the Circulation of the Blood. A History of Physiological Opinion and Discovery in regard to the Circulation of the Blood. By JOHN C. DALTON, M. D., Professor Emeritus of Physiology in the College of Physicians and Surgeons, New York. In one handsome 12mo. volume of 293 pages. Cloth, $2.

Dr. Dalton's work is the fruit of the deep research of a cultured mind, and from the busy practitioner it cannot fail to be a source of instruction. It will inspire him with a feeling of gratitude and admir-

ation for those plodding workers of olden times, who laid the foundation of the magnificent temple of medical science as it now stands.—*New Orleans Medical and Surgical Journal*, Aug. 1885.

Bell's Comparative Anatomy and Physiology.

Comparative Anatomy and Physiology. By F. JEFFREY BELL, M. A., Professor of Comparative Anatomy at King's College, London. In one 12mo. vol. of 561 pages, with 229 illustrations. Limp cloth, $2. See *Students' Series of Manuals*, page 30.

The manual is preëminently a student's book— clear and simple in language and arrangement. It is well and abundantly illustrated, and is readable and interesting. On the whole we consider

it the best work in existence in the English language to place in the hands of the medical student.—*Bristol Medico-Chirurgical Journal*, Mar. 1886.

Ellis' Demonstrations of Anatomy.—Eighth Edition.

Demonstrations of Anatomy. Being a Guide to the Knowledge of the Human Body by Dissection. By GEORGE VINER ELLIS, Emeritus Professor of Anatomy in University College, London. From the eighth and revised London edition. In one very handsome octavo volume of 716 pages, with 249 illus. Cloth, $4.25; leather, $5.25.

Roberts' Compend of Anatomy.

The Compend of Anatomy. For use in the dissecting-room and in preparing for examinations. By JOHN B. ROBERTS, A. M., M. D., Lecturer in Anatomy in the University of Pennsylvania. In one 16mo. vol. of 196 pages. Limp cloth, 75 cents.

WÖHLER'S OUTLINES OF ORGANIC CHEMISTRY. Edited by FITTIG. Translated by IRA REMSEN, M. D., Ph. D. In one 12mo. volume of 550 pages. Cloth, $3.
LEHMANN'S MANUAL OF CHEMICAL PHYSIOLOGY. In one octavo volume of 327 pages, with 41 illustrations. Cloth, $2.25.

CARPENTER'S HUMAN PHYSIOLOGY. Edited by HENRY POWER. In one octavo volume.
CARPENTER'S PRIZE ESSAY ON THE USE AND ABUSE OF ALCOHOLIC LIQUORS IN HEALTH AND DISEASE. With explanations of scientific words. Small 12mo. 178 pages. Cloth, 60 cents.

Lea Brothers & Co., Publishers, 706, 708 & 710 Sansom Street, Philadelphia.

Foster's Physiology.—New (5th) American Ed.   Just Ready.

Text-Book of Physiology. By MICHAEL FOSTER, M. D., F. R. S., Prelector in Physiology and Fellow of Trinity College, Cambridge, England. New (fifth) and enlarged American from the fifth and revised English edition, with notes and additions. In one handsome octavo vol. of 1083 pages, with 316 illus.   Cloth, $4.50; leather, $5.50.

This is the standard work on physiology, being most thorough and complete in all branches and details; moreover it contains considerable material which has never before been presented to the medical public. Evidence of its success is shown in the fact that it is now in i's fifth English and fifth American edition. In its high character, in the care which is shown in the statements and their verification, and in it i thorough dealing with physiological and histological problems, it is far ahead of any book of the class yet issued.—*The Medical Age*, December 26, 1893.

One cannot read a single chapter without being impressed with the care that the author has bestowed upon it. Apparently nothing that is known up to the present year concerning vital processes has escaped his painstaking attention. The details receive the fullest consideration. The additions which have been made to this last edition are caused by an effort to explain more fully and at greater length what seemed to be the most fundamental and important topics. The publishers have subjected it to the searching revision of one of the foremost American professors of physiology. We have nothing but words of the highest praise for the classical and thorough manner in which the work is written, as well as for the liberality of the publishers for selling such a large work, and one which must necessarily be very costly to produce, for an extremely moderate price.—*The Canada Medical Record*, March, 1894.

Dalton's Physiology.—Seventh Edition.

A Treatise on Human Physiology. Designed for the use of Students and Practitioners of Medicine. By JOHN C. DALTON, M. D., Professor of Physiology in the College of Physicians and Surgeons, New York, etc. Seventh edition, thoroughly revised and rewritten. In one very handsome octavo volume of 722 pages, with 252 beautiful engravings on wood.   Cloth, $5.00; leather, $6.00.

From the first appearance of the book it has been a favorite, owing as well to the author's renown as an oral teacher as to the charm of simplicity with which, as a writer, he always succeeds in investing even intricate subjects. It must be gratifying to him to observe the frequency with which his work, written for students and practitioners, is quoted by other writers on physiology. This fact attests its value, and, in great measure, its originality. It now needs no such seal of approbation, however, for the thousands who have studied it in its various editions have never been in any doubt as to its sterling worth.—*N. Y. Medical Journal*, Oct. 1882.

Professor Dalton's well-known and deservedly-appreciated work has long passed the stage at which it could be reviewed in the ordinary sense. The work is eminently one for the medical practitioner, since it treats most fully of those branches of physiology which have a direct bearing on the diagnosis and treatment of disease. The work is one which we can highly recommend to all our readers.—*Dublin Journal of Medical Science*, Feb.'83.

Chapman's Human Physiology.

A Treatise on Human Physiology. By HENRY C. CHAPMAN, M. D., Professor of Institutes of Medicine in the Jefferson Medical College of Philadelphia. In one octavo volume of 925 pages, with 605 engravings.   Cloth, $5.50; leather, $6.50.

It represents very fully the existing state of physiology. The present work has a special value to the student and practitioner as devoted more to the practical application of well-known truths which the advance of science has given to the profession in this department, which may be considered the foundation of rational medicine.—*Buffalo Medical and Surgical Journal*, Dec. 1887.

Matters which have a practical bearing on the practice of medicine are lucidly expressed; technical matters are given in minute detail; elaborate directions are stated for the guidance of students in the laboratory. In every respect the work fulfils its promise, whether as a complete treatise for the student or for the physician; for the former it is so complete that he need look no farther, and the latter will find entertainment and instruction in an admirable book of reference.—*North Carolina Medical Journal*, Nov. 1887.

Schofield's Elementary Physiology.

Elementary Physiology for Students. By ALFRED T. SCHOFIELD, M. D., Late House Physician London Hospital. In one 12mo volume of 380 pages, with 227 engravings and 2 colored plates containing 30 figures.   Cloth, $2.00.

Frankland & Japp's Inorganic Chemistry.

Inorganic Chemistry. By E. FRANKLAND, D. C. L., F. R. S., Professor of Chemistry in the Normal School of Science, London., and F. R. JAPP, F. I. C., Assistant Professor of Chemistry in the Normal School of Science, London. In one handsome octavo volume of 677 pages with 51 woodcuts and 2 plates.   Cloth, $3.75; leather, $4.75.

Clowes' Qualitative Analysis.—Third Edition.

An Elementary Treatise on Practical Chemistry and Qualitative Inorganic Analysis. Specially adapted for use in the Laboratories of Schools and Colleges and by Beginners. By FRANK CLOWES, D. Sc., London, Senior Science-Master at the High School, Newcastle-under Lyme, etc. Third American from the fourth and revised English edition. In one 12mo. vol. of 387 pages, with 55 illus. Cloth, $2.50.

CLASSEN'S ELEMENTARY QUANTITATIVE ANALYSIS. Translated, with notes and additions, by EDGAR F. SMITH, Ph. D., Assistant Professor of Chemistry in the Towne Scientific School, University of Penna. In one 12mo. volume of 324 pages, with 36 illus.   Cloth, $2.00.

Simon's Chemistry.—New (4th) Edition.

Manual of Chemistry. A Guide to Lectures and Laboratory work for Beginners in Chemistry. A Text-book, specially adapted for Students of Pharmacy and Medicine. By W. SIMON, Ph. D., M. D., Professor of Chemistry and Toxicology in the College of Physicians and Surgeons, Baltimore, and Professor of Chemistry in the Maryland College of Pharmacy. New (4th) edition. In one 8vo. vol. of 490 pp., with 44 woodcuts and 7 colored plates illustrating 56 of the most important chemical tests. Cloth, $3.25.

A work which rapidly passes to its fourth edition needs no further proof of having achieved a success. In the present case the claims to favor are obvious. Emanating from an experienced teacher of medical and pharmaceutical students the volume is closely adapted to their needs. This is shown not only by the careful selection and clear presentation of its subject matter, but by the colored plates of reactions, which form a unique feature. Every teacher will appreciate the saving of his own time, and the advantages accruing to the student from a permanent and accurate standard of comparison for tests depending on colors, and frequently upon their changes. To the practitioner, who is likely to be confronted at any time with important pathological or toxicological questions to be answered by the test tube, the volume will be of the utmost value. Such it has proved in the past, and the author has accordingly been enabled, through frequent and thorough revisions to keep his work constantly in touch with the progress of its science and the best methods of its presentation.—*Kansas City Medical Index*, May, 1893.

Fownes' Chemistry.—Twelfth Edition.

A Manual of Elementary Chemistry; Theoretical and Practical. By GEORGE FOWNES, Ph. D. Embodying WATTS' *Physical and Inorganic Chemistry*. New Amer.can, from the twelfth English edition. In one large royal 12mo. volume of 1061 pages, with 168 engravings and a colored plate. Cloth, $2.75; leather, $3.25.

Fownes' *Chemistry* has been a standard text-book upon chemistry for many years. Its merits are very fully known by chemists and physicians everywhere in this country and in England. As the science has advanced by the making of new discoveries, the work has been revised so as to keep it abreast of the times. It has steadily maintained its position as a text-book with medical students. In this work are treated fully: Heat, Light and Electricity, including Magnetism. The influence exerted by these forces in chemical action upon health and disease, etc., is of the most important kind, and should be familiar to every medical practitioner. We can commend the work as one of the very best text-books upon chemistry extant.—*Cincinnati Med. News*, Oct. '85.

Attfield's Chemistry.—Twelfth Edition.

Chemistry, General, Medical and Pharmaceutical; Including the Chemistry of the U. S. Pharmacopœia. A Manual of the General Principles of the Science, and their Application to Medicine and Pharmacy. By JOHN ATTFIELD, M. A., Ph. D., F. I. C., F. R. S., etc., Professor of Practical Chemistry to the Pharmaceutical Society of Great Britain, etc. A new American, from the twelfth English edition, specially revised by the Author for America. In one handsome royal 12mo. volume of 782 pages, with 88 illustrations. Cloth, $2.75; leather, $3.25.

Attfield's Chemistry is the most popular book among students of medicine and pharmacy. This popularity rests upon real merits. Attfield's work combines in the happiest manner a clear exposition of the theory of chemistry with the practical application of this knowledge to the everyday dealings of the physician and pharmacist. His book is precisely what the title claims for it. The admirable arrangement of the text enables a reader to get a good idea of chemistry without the aid of experiments, and again it is a good laboratory guide, and finally it contains such a mass of well-arranged information that it will always serve as a handy book of reference. He does not allow any unutilizable knowledge to slip into his book; his long years of experience have produced a work which is both scientific and practical, and which shuts out everything in the nature of a superfluity, and therein lies the secret of its success. This last edition shows the marks of the latest progress made in chemistry and chemical teaching.—*New Orleans Medical and Surgical Journal*, Nov. 1889.

Bloxam's Chemistry.—Fifth Edition.

Chemistry, Inorganic and Organic. By CHARLES L. BLOXAM, Professor of Chemistry in King's College, London. New American from the fifth London edition, thoroughly revised and much improved. In one very handsome octavo volume of 727 pages, with 292 illustrations. Cloth, $2.00; leather, $3.00.

Comment from us on this standard work is almost superfluous. It differs widely in scope and aim from that of Attfield, and in its way is equally beyond criticism. It adopts the most direct methods in stating the principles, hypotheses and facts of the science. Its language is so terse and lucid, and its arrangement of matter so logical in sequence that the student never has occasion to complain that chemistry is a hard study. Much attention is paid to experimental illustrations of chemical principles and phenomena, and the mode of conducting these experiments. The book maintains the position it has always held as one of the best manuals of general chemistry in the English language.—*Detroit Lancet*, Feb. 1884.

Luff's Manual of Chemistry.—Just Ready.

A Manual of Chemistry. For the use of students of medicine. By ARTHUR P. LUFF, M. D., B. Sc., Lecturer on Medical Jurisprudence and Toxicological Chemistry St. Mary's Hospital Medical School, London. In one 12mo. vol. of 522 pages, with 36 engravings. Cloth, $2.00. See *Students' Series of Manuals*, page 30.

Greene's Medical Chemistry.

A Manual of Medical Chemistry. For the use of Students. By WILLIAM H. GREENE, M. D., Demonstrator of Chemistry in the Medical Department of the University of Pennsylvania. In one 12mo. volume of 310 pages, with 74 illus. Cloth, $1.75.

Lea Brothers & Co., Publishers, 706, 708 & 710 Sansom Street, Philadelphia.

Vaughan & Novy on Ptomaines and Leucomaines.—2d Edition.

Ptomaines, Leucomaines and Bacterial Proteids; or the Chemical Factors in the Causation of Disease. By VICTOR C. VAUGHAN, Ph. D., M. D., Professor of Physiological and Pathological Chemistry, and Associate Professor of Therapeutics and Materia Medica in the University of Michigan, and FREDERICK G. NOVY, M. D., Instructor in Hygiene and Physiological Chemistry in the University of Michigan. New (second) edition. In one handsome 12mo. vol. of 389 pages. Cloth, $2.25.

This book is one that is of the greatest importance, and the modern physician who accepts bacterial pathology cannot have a complete knowledge of this subject unless he has carefully perused it. To the toxicologist the subject is alike of great import, as well as to the hygienist and sanitarian. It contains information which is not easily obtained elsewhere, and which is of a kind that no medical thinker should be without.—*The American Journal of the Medical Sciences*, April, 1892.

Remsen's Theoretical Chemistry.—New (4th) Edition.

Principles of Theoretical Chemistry, with special reference to the Constitution of Chemical Compounds. By IRA REMSEN, M. D., Ph. D., Professor of Chemistry in the Johns Hopkins University, Baltimore. Fourth and thoroughly revised edition. In one handsome royal 12mo. volume of 325 pages. Cloth, $2.00.

The fourth edition of Professor Remsen's well-known book comes again, enlarged and revised. Each edition has enhanced its value. We may say without hesitation that it is a standard work on the theory of chemistry, not excelled and scarcely equalled by any other in any language. Its translation into German and Italian speaks for its exalted position and the esteem in which it is held by the most prominent chemists. We claim for this little work a leading place in the chemical literature of this country.—*The American Journal of the Medical Sciences*, July, 1893.

Charles' Physiological and Pathological Chemistry.

The Elements of Physiological and Pathological Chemistry. A Handbook for Medical Students and Practitioners. Containing a general account of Nutrition, Foods and Digestion, and the Chemistry of the Tissues, Organs, Secretions and Excretions of the Body in Health and in Disease. Together with the methods for preparing or separating their chief constituents, as also for their examination in detail, and an outline syllabus of a practical course of instruction for students. By T. CRANSTOUN CHARLES, M. D., F. R. S., M. S., formerly Assistant Professor and Demonstrator of Chemistry and Chemical Physics, Queen's College, Belfast. In one handsome octavo volume of 463 pages, with 38 woodcuts and 1 colored plate. Cloth, $3.50.

Dr. Charles is fully impressed with the importance and practical reach of his subject, and he has treated it in a competent and instructive manner. We cannot recommend a better book than the present. In fact, it fills a gap in medical text-books, and that is a thing which can rarely be said nowadays. Dr. Charles has devoted much space to the elucidation of urinary mysteries. He does this with much detail, and yet in a practical and intelligible manner. In fact, the author has filled his book with many practical hints.—*Medical Record*, December 20, 1884.

Hoffmann and Powers' Medicinal Analysis.

A Manual of Chemical Analysis, as applied to the Examination of Medicinal Chemicals and their Preparations. Being a Guide for the Determination of their Identity and Quality, and for the Detection of Impurities and Adulterations. For the use of Pharmacists, Physicians, Druggists and Manufacturing Chemists, and Pharmaceutical and Medical Students. By FREDERICK HOFFMANN, A. M., Ph. D., Public Analyst to the State of New York, and FREDERICK B. POWER, Ph. D., Professor of Analytical Chemistry in the Philadelphia College of Pharmacy. Third edition, entirely rewritten and much enlarged. In one octavo volume of 621 pages, with 179 illustrations. Cloth, $4.25.

Parrish's Pharmacy.—Fifth Edition.

A Treatise on Pharmacy: Designed as a Text-book for the Student, and as a Guide for the Physician and Pharmaceutist. With many Formulæ and Prescriptions. By EDWARD PARRISH, late Professor of the Theory and Practice of Pharmacy in the Philadelphia College of Pharmacy. Fifth edition, thoroughly revised, by THOMAS S. WIEGAND, Ph. G. In one handsome octavo volume of 1093 pages, with 256 illustrations. Cloth, $5.00; leather, $6.00.

Caspari's Pharmacy.—Preparing.

A Text-Book on Pharmacy, for Students and Pharmacists. By CHARLES CASPARI, JR, Ph. G., Professor of the Theory and Practice of Pharmacy in the Maryland College of Pharmacy, Joint Editor of *The National Dispensatory* of 1894. In one very handsome octavo volume, richly illustrated.

Ralfe's Clinical Chemistry.

Clinical Chemistry. By CHARLES H. RALFE, M. D., F. R. C. P., Assistant Physician at the London Hospital. In one pocket-size 12mo. volume of 314 pages, with 16 illus. Limp cloth, red edges, $1.50. See *Students' Series of Manuals*, page 30.

Lea Brothers & Co., Publishers, 706, 708 & 710 Sansom Street, Philadelphia.

JUST READY—NEW AND THOROUGHLY REVISED EDITION.

The National Dispensatory.

Containing the Natural History, Chemistry, Pharmacy, Actions and Uses of Medicines, including those recognized in the Pharmacopœias of the United States, Great Britain and Germany, with numerous references to the French Codex. By ALFRED STILLÉ, M. D., LL. D., Professor Emeritus of the Theory and Practice of Medicine and of Clinical Medicine in the University of Pennsylvania, JOHN M. MAISCH, Phar. D., late Professor of Materia Medica and Botany in Philadelphia College of Pharmacy, Secretary to the American Pharmaceutical Association, CHARLES CASPARI, JR., Ph. G., Professor of Pharmacy in the Maryland College of Pharmacy, Baltimore, and HENRY C. C. MAISCH, Ph. G., Ph. D. New (fifth) edition, thoroughly revised in accordance with the new U. S. Pharmacopœia (Seventh Decennial Revision, 1894). In one magnificent imperial octavo volume of 1910 pages, with 320 engravings. Cloth, $7.25, leather, $8.00. With Ready Reference Thumb-letter Index, cloth, $7.75; leather, $8.50.

ON the first appearance of *The National Dispensatory* fifteen years ago it was at once recognized by the pharmaceutical and medical professions as satisfying the need for a work affording all necessary information upon its subject, with authoritative accuracy, and with a completeness and convenience attainable only by the exclusion of obsolete matter. Its success in filling this want is fully attested by the rapid demand for five editions, and the opportunity thus afforded has been well used in successive revisions, each placing it abreast of the day and maintaining the characteristics which had won for it a leading position.

Of all its issues the present embodies the results of the most exhaustive revision. The sweeping changes in the new United States Pharmacopœia are thoroughly incorporated, with official authorization of the Committee of Revision, and full use has been made of all valuable material in the latest issues of foreign Pharmacopœias. The volume is accordingly rich in pharmaceutical and chemical information, with data, formulas, tables, etc., gathered from all official sources, but this constitutes only a single department of its usefulness. As an encyclopædia of the latest and best therapeutical knowledge it deals not only with all official drugs, but also with all the new synthetic remedies of value and with the unofficial preparations now so largely in use. Pharmacists will appreciate its systematic descriptions of the materia medica, its clear explanations of chemical and pharmaceutical processes and tests, and its illustrations of important drugs and of the most improved apparatus. Physicians will readily perceive the indispensable assistance offered by its authoritative statements as to the efficacy of drugs in the light of the most recent medical advances. *Arranged alphabetically in the text, this information is placed most suggestively at command by the recommendations grouped under the various Diseases in the Therapeutical Index. Together with the General Index this covers more than one hundred treble-columned pages containing 25,000 references. The immensity of detail comprised in this single volume of 1900 pages is thus most forcibly indicated. Though the present edition contains far more matter than its predecessor it is maintained at the same price in view of the ever-increasing demand. Weights and Measures are given in both Ordinary and Metric Systems.

In brief the new edition of *The National Dispensatory* is presented to the medical and pharmaceutical professions as the equivalent of a whole library of pharmaceutical and therapeutic information; it is the standard of accuracy, the embodiment of completeness without inconvenient bulk, and a marvel of cheapness owing to the widespread demand for it as *the* authority.

The careful examination of this large volume will strike the reader with surprise at the great number of new articles added, and the amount of useful and accurate information regarding their properties, methods of preparation and therapeutical effects. The large number of new articles containing all the latest synthetic remedies and unofficial remedies, compass the entire range of available information in the line of the work. A number of very complete tables together with all the official re-agents and solutions for qualitative and quantitative tests, appear in the appendix. Altogether this work maintains its previous high reputation for accuracy, practical usefulness and encyclopædic scope, and is indispensable alike to the pharmacist and physician. Every druggist knows of it and uses it, and almost every physician properly consults it when desirous of settling all doubtful questions regarding the properties, preparation and uses of drugs.—*Medical Record*, April 7, 1894.

The descriptions of materia medica are clear, thorough and systematic, as are also the explanations of chemical and pharmaceutical processes

and tests. The therapeutical portion has been revised with equal care and the statements of the action and uses have been arranged not only alphabetically under the various drugs, but for practical medical usefulness have also been placed at the instant command of those seeking information in the treatment of special diseases by being arranged under the various diseases in a therapeutical index. The readiness with which any of the vast amount of information contained in this work is made available is indicated by the twenty-five thousand references in the two indexes at the end of the volume.—*Boston Medical and Surgical Journal*, April, 1894.

It is the official guide for the medical and pharmaceutical professions.—*Buffalo Medical and Surgical Journal*, March, 1894.

The book is recommended most highly as a book of reference for the physician and is invaluable to the druggist in his every-day work.—*The Therapeutic Gazette*, March, 1894.

This edition of the Dispensatory should be recognized as a national standard.—*The North American Practitioner*, March, 1894.

Lea Brothers & Co., Publishers, 706, 708 & 710 Sansom Street, Philadelphia.

A System of Practical Therapeutics

BY AMERICAN AND FOREIGN AUTHORS.

Edited by HOBART AMORY HARE, M. D.

Professor of Therapeutics and Materia Medica in the Jefferson Medical College of Philadelphia.

In a series of contributions by seventy-eight eminent authorities. In three large octavo volumes of 3544 pages, with 434 illustrations. Price, per volume: Cloth, $5.00; leather, $6.00; half Russia, $7.00. *For sale by subscription only. Address the Publishers. Full prospectus free to any address on application.*

The various divisions have been elaborated by men selected in view of their special fitness. In every case there is to be found a clear and concise description of the disease under consideration, corresponding with the most recent and well-established views of the subject, embracing apposite pictorial illustrations where these are necessary. In treating of the employment of remedies and therapeutical measures, the writers have been singularly happy in giving in a definite way the exact methods employed and the results obtained, both by themselves and others, so that one might venture with confidence to use remedies with which he was previously entirely unfamiliar. The practitioner could hardly desire a book on practical therapeutics which he could consult with more interest and profit.—*The North American Practitioner*, September, 1892.

The scope of this work is beyond that of any previous one on the subject. The goal, after all, is the treatment of disease, and a work which contributes to its successful management is to be looked upon as of vast use to humanity. It cannot be denied that therapeutic resources, whether the treatment be confined to the mere administration of drugs, or allowed its more extended application to the management of disease, have so greatly multiplied within the last few years as to render previous treatises of little value. Herein will be found the great value of Hare's encyclopedic work, which groups together within a single series of volumes the most modern methods known in the management of disease, and especially deal with important subjects comprehensively, which could not be done in a more limited treatise. We cannot commend Hare's *System of Practical Therapeutics* too highly; it stands out first and foremost as a work to be consulted by authors, teachers, and physicians, throughout the world.—*Buffalo Med. and Surg. Jour.*, Aug. 1892.

Hare's Text-Book of Practical Therapeutics.—New (3d) Ed.

A Text-Book of Practical Therapeutics; With Especial Reference to the Application of Remedial Measures to Disease and their Employment upon a Rational Basis. By HOBART AMORY HARE, M. D., Professor of Therapeutics and Materia Medica in the Jefferson Medical College of Philadelphia; Sec. of Convention for Revision of U. S. Pharmacopœia of 1890. With special chapters by DRS. G. E. DE SCHWEINITZ, EDWARD MARTIN, J. HOWARD REEVES and BARTON C. HIRST. New (3d) and revised edition. In one octavo volume of 689 pages. Cloth, $3.75; leather, $4.75.

The student of other works, has often, indeed, very often, longed for less of the abstract materia medica and more of the practical application of drugs to disease. In this work that want is filled. The drugs are arranged alphabetically, which enables one to find any name quickly, and, with the excellent index at the end of the volume, naught is left to be desired in the way of quick reference. Each drug, including all the newer remedies which have been proved to possess true merit, is considered in a rational and scientific manner. This work also presents us with nearly 250 pages of practical therapeutics, as applied to the individual diseases. The subjects are arranged alphabetically. It is in the chapter on *Diseases* that the student finds the rationale of therapeutics. This section is properly the complement of the former, in which each drug was presented with notes as to its usefulness in numerous diseases, while in the latter each disease is considered very fully from a therapeutical standpoint, the applications and special indications of the different remedies in the different phases of that particular affection being given. It is not a wonder that this work was quickly adopted by many colleges as a text-book and so liberally purchased as to necessitate the publication of a third edition within two years. The student will find its pages filled with the choicest of therapeutical lore, systematically arranged and clearly and forcibly presented; the practitioner will appreciate its rationality and its general utility as an elbow consultant. It contains, without question, the best exposition of modern therapeutics of any text-book with which we are acquainted.—*The Chicago Clinical Review*, March, 1893.

Maisch's Materia Medica.—Fifth Edition.

A Manual of Organic Materia Medica; Being a Guide to Materia Medica of the Vegetable and Animal Kingdoms. For the Use of Students, Druggists, Pharmacists and Physicians. By JOHN M. MAISCH, Phar. D., Prof. of Materia Medica and Botany in the Philadelphia College of Pharmacy. New (fifth) edition, thoroughly revised. In one very handsome 12mo. volume of 544 pages, with 270 engravings. Cloth, $3.00.

This is an excellent manual of organic materia medica, as are all the works that emanate from the skilful pen of such a successful teacher as John M. Maisch. The book speaks for itself in the most forcible language. In the edition before us which is the fifth one published within the comparatively short space of eight years (and this is the best proof of the great value of the work and the just favor with which it has been received and accepted), the original contents have been thoroughly revised and much good and new matter has been incorporated. We have nothing but praise for Professor Maisch's work. It presents no weak point, even for the most severe critic. The book fully sustains the wide and well-earned reputation of its popular author. In the special line of work of which it treats it is fully up to the most recent observations and investigations. After a careful perusal of the book, we do not hesitate to recommend Maisch's *Manual of Organic Materia Medica* as one of the best, if not the best work on the subject thus far published. Its usefulness cannot well be dispensed with, and students, druggists, pharmacists and physicians should all possess a copy of such a valuable book.—*Medical News*, December 31, 1892.

Edes' Therapeutics and Materia Medica.

A Text-Book of Therapeutics and Materia Medica. Intended for the Use of Students and Practitioners. By ROBERT T. EDES, M. D., Jackson Professor of Clinical Medicine in Harvard University. Octavo, 544 pp. Cloth, $3.50; leather, $4.50.

Flint's Practice of Medicine.—Sixth Edition.

A Treatise on the Principles and Practice of Medicine. Designed for the use of Students and Practitioners of Medicine. By AUSTIN FLINT, M. D., LL. D., Professor of the Principles and Practice of Medicine, and of Clinical Medicine in Bellevue Hospital Medical College, N. Y. Sixth edition, thoroughly revised and rewritten by the Author, assisted by WILLIAM H. WELCH, M. D., Professor of Pathology, Johns Hopkins University, Baltimore, and AUSTIN FLINT, JR., M. D., LL. D., Professor of Physiology, Bellevue Hospital Medical College, N. Y. In one very handsome octavo volume of 1160 pages, with illustrations. Cloth, $5.50; leather, $6.50.

No text-book on the principles and practice of medicine has ever met in this country with such general approval by medical students and practitioners as the work of Professor Flint. In all the medical colleges of the United States it is the favorite work upon Practice; and, as we have stated before in alluding to it, there is no other medical work that can be so generally found in the libraries of physicians. In every state and territory of this vast country the book that will be most likely to be found in the office of a medical man, whether in city, town, village, or at some cross-roads, is Flint's *Practice*. We make this statement to a considerable extent from personal observation, and it is the testimony also of others. An examination shows that very considerable changes have been made in the sixth edition. The work may undoubtedly be regarded as fairly representing the present state of the science of medicine, and as reflecting the views of those who exemplify in their practice the present stage of progress of medical art.—*Cincinnati Medical News*, Oct. 1886.

Hartshorne's Essentials of Practice.—Fifth Edition.

Essentials of the Principles and Practice of Medicine. A Handbook for Students and Practitioners. By HENRY HARTSHORNE, M. D., LL. D., lately Professor of Hygiene in the University of Pennsylvania. Fifth edition, thoroughly revised and rewritten. In one 12mo. vol. of 669 pages, with 144 illus. Cloth, $2.75; half leather, $3.

Farquharson's Therapeutics and Materia Medica.—4th Ed.

A Guide to Therapeutics and Materia Medica. By ROBERT FARQUHARSON, M. D., F. R. C. P., LL. D., Lecturer on Materia Medica at St. Mary's Hospital Medical School, London. Fourth American, from the fourth English edition. Enlarged and adapted to the U. S. Pharmacopœia. By FRANK WOODBURY, M. D., Professor of Materia Medica and Therapeutics and Clinical Medicine in the Medico-Chirurgical College of Philadelphia. In one handsome 12mo. vol. of 581 pp. Cloth, $2.50.

It may correctly be regarded as the most modern work of its kind. It is concise, yet complete. Containing an account of all remedies that have a place in the British and United States Pharma-copœias, as well as considering all non-official but important new drugs, it becomes in fact a miniature dispensatory.—*Pacific Medical Journal*, June, 1889.

Bruce's Materia Medica and Therapeutics.—Fourth Edition.

Materia Medica and Therapeutics. An Introduction to Rational Treatment. By J. MITCHELL BRUCE, M. D., F. R. C. P., Physician and Lecturer on Materia Medica and Therapeutics at Charing-Cross Hospital, London. Fifth edition. In one 12mo. volume of 591 pages. Cloth, $1.50. See *Students' Series of Manuals*, page 30.

The pharmacology and therapeutics of each drug are given with great fulness, and the indications for its rational employment in the practical treatment of disease are pointed out. The Materia Medica proper contains all that is necessary for a medical student to know at the present day. The third part of the book contains an outline of general therapeutics, each of the symptoms of the body being taken in turn, and the methods of treatment illustrated. A lengthy notice of a book so well known is unnecessary.—*Med. Chronicle*, May, 1891.

COHEN'S HANDBOOK OF APPLIED THERAPEUTICS. Being a Study of Principles Applicable and an Exposition of Methods Employed in the Management of the Sick. By SOLOMON SOLIS COHEN, M D., Professor of Clinical Medicine and Applied Therapeutics in the Philadelphia Polyclinic. In one large 12mo. volume, with illustrations. *Preparing.*

REYNOLDS' SYSTEM OF MEDICINE. Edited by J. RUSSELL REYNOLDS, M D., Professor of the Principles and Practice of Med. in University College, London. With notes and additions by HENRY HARTSHORNE, A. M., M. D., late Professor of Hygiene in the University of Pennsylvania. Three octavo volumes, containing 3056 double-columned pages, with 317 illustrations. Price per volume, cloth, $5.00; sheep, $6.00; half Russia, $6.50. *Subscription only.*

WATSON'S LECTURES ON THE PRINCIPLES AND PRACTICE OF PHYSIC. From the fifth English edition. Edited with additions, and 190 illustrations, by HENRY HARTSHORNE, A. M., M. D., late Professor of Hygiene in the University of Pennsylvania. In two large octavo volumes of 1840 pages. Cloth, $9.00; leather, $11.00.

FLINT'S PRACTICAL TREATISE ON THE DIAGNOSIS, PATHOLOGY AND TREATMENT OF DISEASES OF THE HEART. Second revised and enlarged edition. In one octavo volume of 550 pages, with a plate. Cloth, $4.

FLINT ON PHTHISIS: ITS MORBID ANATOMY, ETIOLOGY, SYMPTOMATIC EVENTS AND COMPLICATIONS, FATALITY AND PROGNOSIS, TREATMENT AND PHYSICAL DIAGNOSIS; in a series of Clinical Studies. In one octavo volume of 442 pages. Cloth, $3.50.

FLINT'S ESSAYS ON CONSERVATIVE MEDICINE AND KINDRED TOPICS. In one very handsome royal 12mo. volume of 210 pages. Cloth, $1.38.

A TREATISE ON FEVER. By ROBERT D. LYONS, K. C. C. In one 8vo. vol. of 354 pp. Cloth, $2.25.

LECTURES ON THE STUDY OF FEVER. By A. HUDSON, M. D., M. R. I. A. In one octavo volume of 308 pages. Cloth, $2.50.

LA ROCHE ON YELLOW FEVER, in its Historical, Pathological, Etiological and Therapeutical Relations. Two octavo vols., 1468 pp. Cloth, $7.00.

BRUNTON'S PHARMACOLOGY, THERAPEUTICS AND MATERIA MEDICA. Octavo, 1305 pages, 230 illustrations.

HERMANN'S EXPERIMENTAL PHARMACOLOGY. A Handbook of Methods for Determining the Physiological Action of Drugs. Translated, with the Author's permission, and with extensive additions, by R. M. SMITH, M. D. 12mo, 199 pages, with 32 illustrations. Cloth, $1.50.

STILLE'S THERAPEUTICS AND MATERIA MEDICA. A Systematic Treatise on the Action and Uses of Medicinal Agents, including their Description and History. Fourth edition, revised and enlarged. In two octavo volumes, containing 1936 pages. Cloth, $10.00; leather, $12.00.

Lea Brothers & Co., Publishers, 706, 708 & 710 Sansom Street, Philadelphia.

Lyman's Practice of Medicine.

The Principles and Practice of Medicine. For the Use of Medical Students and Practitioners. By HENRY M. LYMAN, M.D., Professor of the Principles and Practice of Medicine, Rush Medical College, Chicago. In one very handsome octavo volume of 925 pages, with 170 illustrations. Cloth, $4.75; leather, $5.75.

This is an excellent treatise on the practice of medicine, written by one who is not only familiar with his subject, but who has also learned through practical experience in teaching, what are the needs of the student, and how to present the facts to his mind in the most readily assimilable form. Although the book contains over nine hundred pages, there has been no space wasted by useless historical essays, prolonged discussions on debatable topics, or "padding" of any kind. Each subject is taken up in order, treated clearly but briefly, and dismissed when all has been said that need be said in order to give the reader a clean-cut picture of the disease under discussion. The reader is not confused by having presented to him a variety of different methods of treatment, among which he is left to choose the one most easy of execution, but the author describes the one which is, in his judgment, the best. This is as it should be.

What the student should be taught is the one most approved method of treatment. We have spoken of the work as one for the student, and this because the author occupies so prominent a position as a teacher, but we would not be understood that it is adapted only for students. There is many a practitioner of ten years' or more standing, who has been unable to follow the constant advances made in medical science, to whom this work will be of great use. He will find here each subject presented in its latest aspect, and only such theories mentioned as have been generally accepted by the highest authorities. The practical and busy man who wants to ascertain in a short time all the necessary facts concerning the pathology or treatment of any disease, will find here a safe and convenient guide.—*Medical Record*, October 22, 1892.

The Year-Book of Treatment for 1894.—Just Ready.

A Comprehensive and Critical Review for Practitioners of Medicine and Surgery. In one 12mo. vol. of 501 pages. Cloth, $1.50.

*** For special commutations with periodicals see pages 1 and 2.

This is the tenth year of this valuable work, which comes to hand with unerring regularity. It would be difficult indeed to imagine a book more nearly suited to the every-day needs of the medical practitioner or writer than this. The contributors to this volume are among the most prominent and well-known writers and teachers of the day, and their articles and opinions will be appreciated by all who are fortunate and wise

enough to secure them. It is the very book the busy practitioner needs. He can find anything pertaining to any subject in a moment's time, and he may rest assured that it is the most modern and reliable view now accepted. It, year by year, keeps him apprised of important advances in all branches of medicine, and presents them in a well-condensed and classified form.—*The Charlotte Medical Journal*, April, 1894.

The Year-Books of Treatment for 1891, 1892 and 1893.

12mos., 485 pages. Cloth, $1.50 each.

The Year-Books of Treatment for 1886 and 1887.

Similar to above. 12mos., 320–341 pages. Cloth, $1.25 each.

A System of Practical Medicine.

BY AMERICAN AUTHORS.

EDITED BY WILLIAM PEPPER, M. D., LL. D.,

PROVOST AND PROFESSOR OF THE THEORY AND PRACTICE OF MEDICINE AND OF CLINICAL MEDICINE IN THE UNIVERSITY OF PENNSYLVANIA.

The complete work, in five volumes, containing 5573 pages, with 198 illustrations, is now ready. Price per volume, cloth, $5; leather, $6; half Russia, $7. Subscription only.

* * The greatest distinctively American work on the practice of medicine, and, indeed, the superlative adjective would not be inappropriate were even all other productions placed in comparison. An examination of the five volumes is sufficient to convince one of the magnitude of the enterprise, and of the success which has attended its fulfilment.—*The Medical Age*, July 26, 1886.

The feeling of proud satisfaction with which the American profession sees this, its representative system of practical medicine issued to the medical world, is fully justified by the character of the work. The entire caste of the system is in keeping with the best thoughts of the leaders and followers of our home school of medicine, and the combination of the scientific study of disease and the practical application of exact and experimental knowledge to the treatment of human maladies, makes every one of us share in the pride that has welcomed Dr. Pepper's labors. Sheared of the prolixity that wearies the readers of the German school, the articles glean these same fields for all that is valuable. It is the outcome of American brains, and is marked throughout by much of the sturdy independence of thought and originality that is a national characteristic. Yet no where is there lack of study of the most advanced views of the day.—*N. C. Med. Jour.*, Sept. 1886.

Habershon on the Abdomen.

On the Diseases of the Abdomen; Comprising those of the Stomach, and other parts of the Alimentary Canal, Œsophagus, Cæcum, Intestines and Peritoneum. By S. O. HABERSHON, M.D., Senior Physician to and late Lecturer on Principles and Practice of Medicine at Guy's Hospital, London. Second American from third enlarged and revised English edition. In one handsome octavo vol. of 554 pages, with illus. Cloth, $3.50.

This valuable treatise on diseases of the stomach and abdomen will be found a cyclopædia of information, systematically arranged, on all diseases of the alimentary tract, from the mouth to the rectum. A fair proportion of each chapter is devoted to symptoms, pathology, and therapeutics. —*New York Medical Journal*, April, 1879.

Lea Brothers & Co., Publishers, 706, 708 & 710 Sansom Street, Philadelphia.

Musser's Medical Diagnosis.—Just Ready.

A Practical Treatise on Medical Diagnosis For the Use of Students and Practitioners. By JOHN H. MUSSER, M. D., Assistant Professor of Clinical Medicine, University of Pennsylvania, Philadelphia. In one very handsome octavo volume of 873 pages, with 162 illustrations. Cloth, $5; leather, $6.

The aim of the author has been to make the work eminently practical. Dr. Musser has succeeded in bringing together a large and valuable collection of clinical data drawn from his own extended experience and from exhaustive literary research, and has presented them in an unusually clear and concise manner. In brief, the book is thoroughly modern, readable and instructive, and, we believe, superior to any work of the kind before the profession.— *University Medical Magazine*, March, 1894.

Modern methods of medical teaching and study have rendered treatises like the present an absolute necessity. The present work is to be com-

mended alike for its logical arrangement, accurate observation and clearness of expression. The chapter on bacteriology is especially commendable, because it contains everything practically necessary for clinical work.— *Med. Rec.*, Apr. 21, '94.

The book should receive a hearty reception from students and medical men; it contains much information essential to good, scientific medical work. It is with pleasure that we can state that the work has been adopted as a text-book at the Johns Hopkins Medical School and Harvard University, and that it has met with marked approval in other teaching centres.— *International Medical Magazine*, April, 1894.

Flint on Auscultation and Percussion.—Fifth Edition.

A Manual of Auscultation and Percussion; Of the Physical Diagnosis of Diseases of the Lungs and Heart, and of Thoracic Aneurism. By AUSTIN FLINT, M. D., LL. D., Professor of the Principles and Practice of Medicine in Bellevue Hospital Medical College, New York. Fifth edition. Edited by James C. Wilson, M. D., Lecturer on Physical Diagnosis in the Jefferson Medical College, Philadelphia. In one handsome royal 12mo. volume of 274 pages, with 12 illustrations. Cloth, $1.75.

Whitla's Dictionary of Treatment.

A Dictionary of Treatment; or Therapeutic Index, including Medical and Surgical Therapeutics. By WILLIAM WHITLA, M. D., Professor of Materia Medica and Therapeutics in the Queen's College, Belfast. Revised and adapted to the United States Pharmacopœia. In one square, octavo vol. of 917 pp. Cloth, $4.00.

We have already dictionaries of medicine and dictionaries of surgery; Dr. Whitla now provides us with a dictionary of treatment. And reference to the volume shows that it really is what it professes to be. The several diseased conditions are arranged in alphabetical order, and the methods—medical, surgical, dietetic and climatic—by which they may be met, considered. On every page we find clear and detailed directions for treatment supported by the author's personal authority and experience whilst the recommendations of other competent observers are also critically examined. The book abounds with useful, practical hints and suggestions, and

the younger practitioner will find in it exactly the help he so often needs in the treatment both of those who are ill, and those who are ailing. At the same time the most experienced members of the profession may usefully consult its pages for the purpose of learning what is really trustworthy in the later therapeutic developments. The *Dictionary* is, in short, the recorded experience of a practical scientific therapeutist, who has carefully studied diseases and disorders at the bed-side and in the consulting-room, and has earnestly addressed himself to the cure and relief of his patients.— *The Glasgow Medical Journal*, April, 1892.

Fothergill's Handbook of Treatment.—Third Edition.

The Practitioner's Handbook of Treatment; Or, The Principles of Therapeutics. By J. MILNER FOTHERGILL, M. D., Edin., M. R. C. P., Lond., Physician to the City of London Hospital for Diseases of the Chest. Third edition. In one 8vo. volume of 661 pages. Cloth, $3.75; leather, $4.75.

This is a wonderful book. If there be such a thing as "medicine made easy," this is the work to accomplish this result.— *Va. Med. Month.*, June,'87.

To have a description of the normal physiological processes of an organ and of the methods of treatment of its morbid conditions brought together in a single chapter, and the relations between the two clearly stated, cannot fail to prove

a great convenience to many thoughtful but busy physicians. The practical value of the volume is greatly increased by the introduction of many prescriptions. That the profession appreciates that the author has undertaken an important work and has accomplished it is shown by the demand for this third edition.— *N. Y. Med. Jour.*, June 11,'87.

Broadbent on the Pulse.

The Pulse. By W. H. BROADBENT, M. D., F. R. C. P., Physician to and Lecturer on Medicine at St. Mary's Hospital, London. In one 12mo. volume of 312 pages. Cloth, $1.75. See *Series of Clinical Manuals*, page 30.

TANNER'S MANUAL OF CLINICAL MEDICINE AND PHYSICAL DIAGNOSIS. Third American from the second London edition. Revised and enlarged by TILBURY FOX, M. D. In one 12mo. volume of 362 pp. with illus. Cloth, $1.50.

DAVIS' CLINICAL LECTURES ON VARIOUS IMPORTANT DISEASES. By N. S. DAVIS, M. D. Edited by FRANK H. DAVIS, M. D. Second edition. 12mo. 287 pages. Cloth, $1.75.

TODD'S CLINICAL LECTURES ON CERTAIN ACUTE DISEASES. In one octavo volume of 320 pages. Cloth. $2.50.

FLINT'S PRACTICAL TREATISE ON THE PHYSICAL EXPLORATION OF THE CHEST

AND THE DIAGNOSIS OF DISEASES AFFECTING THE RESPIRATORY ORGANS. Second and revised edition. In one handsome octavo volume of 591 pages. Cloth, $4.50.

STURGES' INTRODUCTION TO THE STUDY OF CLINICAL MEDICINE. Being a Guide to the Investigation of Disease. In one handsome 12mo. volume of 127 pages. Cloth, $1.25.

WALSHE ON THE DISEASES OF THE HEART AND GREAT VESSELS. Third American edition. In 1 vol. 8vo., 416 pp. Cloth, $3.00.

HOLLAND'S MEDICAL NOTES AND REFLECTIONS. 1 vol. 8vo., pp. 493. Cloth, $3.50.

Yeo's Medical Treatment.—Just Ready.

A Manual of Medical Treatment or Clinical Therapeutics. By
I. BURNEY YEO, M. D., F. R. C. P., Prof. of Clinical Therapeutics in King's Coll., London. In two 12mo. volumes containing 1275 pages, with illustrations. Cloth, $5.50.

This work is devoted entirely to the treatment of disease, being the first we have ever seen of the kind. Only so much of the pathology and etiology of disease is introduced as is necessary to arrive at the rational indications, without which the administration of a drug can hardly be called scientific. Half a dozen choice formulæ by leading physicians are appended to each chapter. The index is so arranged that one can find disease and the various remedies at a glance. Without exaggeration, we can say that one could hardly read anything affording at the same time so much pleasure and profit as this elegantly written and beautifully printed book.—*The Canada Medical Record*, November, 1893.

In Dr. Yeo's book the study of the treatment of disease is approached, not from the side of the drug or remedy as in works on therapeutics, but "from the side of the disease." The various diseases are grouped together, a short account is given of the clinical history, course and pathology of each, and from a consideration thereof, indications for treatment are arrived at; then follows a full dis-

cussion of the best methods of carrying out these indications. Each section contains a number of prescriptions which the author has found most useful, and at the end of every chapter is added a selection of formulæ from the writings of various well-known physicians. The work is exceedingly practical, and the details of the various methods of treatment are always given. Full directions are given with regard to diet, mode of life, and general treatment, wh'ch are often as important as the treatment by drugs.—*Med. Chronicle*, January, 1894.

The discussion of the different ailments has a distinctly practical turn toward the main purpose of the book. Standard formulæ are introduced from eminent practitioners, and all the drugs of recognized value are grouped in the order of their importance. The dosage r-ceives careful attention, which is a feature that cannot be too highly commended. It cannot fail to be an exceedingly useful, suggestive and instructive work to the physician who wishes to be well up in the present advanced and scientific therapeutics of the day.—*Medical Record*, November 25, 1893.

Yeo on Food in Health and Disease.

Food in Health and Disease. By I. BURNEY YEO, M. D., F. R. C. P.,
Professor of Clinical Therapeutics in King's College, London. In one 12mo. volume of 590 pages. Cloth, $2 00. *See Series of Clinical Manuals*, page 30.

Dr. Yeo supplies in a compact form nearly all that the practitioner requires to know on the subject of diet. The work is divided into two parts—food in health and food in disease. Dr. Yeo has gathered together from all quarters an immense amount of useful information within a comparatively small

compass, and he has arranged and digested his materials with skill for the use of the practitioner. We have seldom seen a book which more thoroughly realizes the object for which it was written than this little work of Dr. Yeo.—*British Medical Journal*, Feb. 8, 1890.

Bartholow on Electricity in Medicine and Surgery.—3d Ed.

Medical Electricity. A Practical Treatise on the Applications of Electricity
to Medicine and Surgery. By ROBERTS BARTHOLOW, A. M., M. D., LL. D., Emeritus Professor of Materia Medica and General Therapeutics in the Jefferson Med. Coll. of Philadelphia, etc. Third edition. In one octavo volume of 308 pp., with 110 illus. Cloth, $2.50.

Bartholow on Cholera.—Just Ready.

Cholera: Its Causes, Symptoms, Pathology and Treatment. By
ROBERTS BARTHOLOW, M. D., LL. D., Emeritus Professor of Materia Medica, General Therapeutics and Hygiene in the Jefferson Medical College of Philadelphia. In one 12mo. volume of 127 pages, with 9 illustrations. Cloth, $1.25.

The most scientific work on cholera extant. Broad yet comprehensive, concise but explicit, it treats the subject in a way to invite but little criticism. The most valuable chapter is the one on treatment, which, considering the author's therapeutical experience, and the great improvements made in practice, is indeed, a contribution to medical literature worthy of more than passing notice — *The Medical Fortnightly*, July 15, 1893.

The author has sought to make a practical book in the smallest compass. The symptoms and

pathology of the disease are described separately in a brief and comprehensive manner. The final chapter, on the treatment of cholera, gives the prophylactic measures, including quarantine and the latest therapeutical methods in vogue in India, Europe and America. The volume is written in the author's usual pleasant style, and will satisfy the desire of any one that wishes to obtain the most recent information on the subject.—*The New York Medical Journal*, July 29, 1893.

Richardson's Preventive Medicine.

Preventive Medicine. By B. W. RICHARDSON, M. D., LL. D., F. R. S., Fellow of the Royal Coll. of Phys., London. In one 8vo. vol. of 729 pp. Cloth, $4; leather, $5.

There is perhaps no similar work written for the general public that contains such a complete, reliable and instructive collection of data upon the diseases common to the race, their origins, causes, and the measures for their prevention. The descriptions of diseases are clear, chaste and

scholarly, the discussion of the question of disease is comprehensive, masterly and fully abreast with the latest an l best knowledge on the subject, and the preventive measures advised are accurate, explicit and reliable.—*The American Journal of the Medical Sciences*, April, 1884.

SCHREIBER'S MANUAL OF TREATMENT BY MASSAGE AND METHODICAL MUSCLE EXERCISE. Translated by WALTER MENDELSON, M. D., of New York. In one 8vo. volume of 274 pp., with 117 engravings.

CHAMBERS' MANUAL OF DIET AND REGIMEN IN HEALTH AND SICKNESS. In one handsome octavo volume of 302 pp. Cloth, $2.75.

STILLÉ ON CHOLERA: Its Origin, History, Causation, Symptoms, Lesions, Prevention and Treatment. In one handsome 12mo. volume of 163 pages, with a chart. Cloth, $1.25.

PAVY'S TREATISE ON THE FUNCTION OF DIGESTION; its Disorders and their Treatment. From the second London edition. In one octavo volume of 238 pages. Cloth, $2.00.

Seiler on the Throat and Nose.—New (4th) Ed.

A Handbook of Diagnosis and Treatment of Diseases of the Throat, Nose and Naso-Pharynx. By CARL SEILER, M. D., Lecturer on Laryngoscopy in the University of Pennsylvania. New (4th) edition. In one handsome 12mo. volume of 414 pages, with 107 illustrations and 2 colored plates. Cloth, $2.25.

This little book is eminently practical, and will prove of interest not only to the specialist, but to the general practitioner as well. It deals with the subject in a clear and distinct manner, and the text is copiously illustrated with diagrams and colored plates. So little attention is paid ordinarily to the examination of the larynx that the need of such a book has long been felt. By consulting its pages anyone can learn the necessary manipulations, and, by a little practice, soon become expert in the use of the laryngeal mirror, a method of examination too often neglected. The anatomy of the larynx is explained with especial care, and the operative procedures for various diseases of the throat, tonsils, etc., are carefully explained. Approved methods of treatment are dealt with in a very satisfactory way, and all the most useful remedial agents are described.—

International Medical Magazine, November, 1893.

As a guide to the practitioner and a text-book for the student, it is unexcelled, being plain, accurate, comprehensive and pleasantly written.—Atlanta Medical and Surgical Journal, August, 1893.

It is needless to say that it is brought up to date in the fullest possible sense of the term. Rarely has any treatise on any specialty met with a more cordial reception than the one under consideration. A most generous recognition is given to the work of American laryngologists. The main feature of the present edition has been the expansion of that portion which deals with the diseases of the nose. The author is to be commended on the excellence of his work, and congratulated that a new edition has been so speedily called for.—Medical Record, November 25, 1893.

Browne on the Throat and Nose.—New (4th) Ed. Just Ready.

The Throat and Nose and Their Diseases. By LENNOX BROWNE, F. R. C. S., E., Senior Physician to the Central London Throat and Ear Hospital. Fourth and enlarged edition. In one imperial octavo volume of 751 pages, with 120 illustrations in color, and 235 engravings on wood. Cloth, $6.50.

Although quite complete enough for the use of specialists, it is at the same time so clear as to be of daily value to the general practitioner, who will find at the end of the volume a number of well-tried formulas most in vogue at the London hospitals for diseases of the throat.—The Canada Medical Record, November, 1893.

It is an admirable presentation of its subject in the light of the large clinical experience of a careful observer. It is a book that no specialist can afford not to have, and that the general physician can rely upon as a safe guide and practical adviser.—The Medical News, Oct. 14, 1893.

Tuke on the Influence of the Mind on the Body.

Illustrations of the Influence of the Mind upon the Body in Health and Disease. Designed to elucidate the Action of the Imagination. By DANIEL HACK TUKE, M. D., Joint Author of the Manual of Psychological Medicine, etc. New edition. Thoroughly revised and rewritten. In one 8vo. volume of 467 pages, with 2 colored plates. Cloth, $3.00.

It is impossible to peruse these interesting chapters without being convinced of the author's perfect sincerity, impartiality, and thorough mental grasp. Dr. Tuke has exhibited the requisite amount of scientific address on all occasions, and the more intricate the phenomena the more firmly has he adhered to a physiological and rational

method of interpretation. Guided by an enlightened deduction, the author has reclaimed for science a most interesting domain in psychology, previously abandoned to charlatans and empirics. This book, well conceived and well written, must commend itself to every thoughtful understanding.—New York Medical Journal, September 6, 1884.

Clouston on Mental Diseases.

Clinical Lectures on Mental Diseases. By THOMAS S. CLOUSTON, M. D., Lecturer on Mental Diseases in the University of Edinburgh. With an Appendix, containing an Abstract of the Statutes of the United States and of the Several States and Territories relating to the Custody of the Insane. By CHARLES F. FOLSOM, M. D., Ass't Professor of Mental Diseases, Med. Dep. of Harvard Univ. In one octavo volume of 541 pages, with eight lithographic plates, four of which are colored. Cloth, $4.

☞ Dr. Folsom's Abstract also separate, in one 8vo. vol. of 108 pages. Cloth, $1.50.

The descriptions of the diseases and cases are simple and practical, but true; and one sees as he reads that they are given by one perfectly familiar from daily observation with the cases and diseases he is speaking of. One feature of the book which commends it highly, and which is not to be found in any other work on mental diseases, is the hints

and descriptions given as to the practical management and care of the cases. We can heartily recommend it to the student and busy general practitioner. Dr. Folsom's work greatly increases the value of Dr. Clouston's book for the American practitioner.—Archives of Medicine, June, 1884.

Playfair on Nerve Prostration and Hysteria.

The Systematic Treatment of Nerve Prostration and Hysteria. By W. S. PLAYFAIR, M. D., F. R. C. P. In one 12mo. volume of 97 pages. Cloth, $1.00.

BROWNE ON KOCH'S REMEDY IN RELATION TO THROAT CONSUMPTION. In one octavo volume of 121 pages, with 45 illustrations, 4 of which are colored, and 17 charts. Cloth, $1.50.

FULLER ON DISEASES OF THE LUNGS AND AIR-PASSAGES. Their Pathology, Physical Diagnosis, Symptoms and Treatment. From the second and revised English edition. In one octavo volume of 475 pages. Cloth, $3.50.

SLADE ON DIPHTHERIA; its Nature and Treatment, with an account of the History of its Pre-

valence in various Countries. Second and revised edition. In one 12mo. vol., 158 pp. Cloth, $1.25.

SMITH ON CONSUMPTION; its Early and Remediable Stages. 1 vol. 8vo., 253 pp. Cloth, $2.25.

LA ROCHE ON PNEUMONIA. 1 vol. 8vo. of 490 pages. Cloth, $3.00.

WILLIAMS ON PULMONARY CONSUMPTION; its Nature, Varieties and Treatment. With an analysis of one thousand cases to exemplify its duration. In one 8vo. vol. of 303 pp. Cloth, $2.50.

Lea Brothers & Co., Publishers, 706, 708 & 710 Sansom Street, Philadelphia.

Gray on Nervous and Mental Diseases.

A Practical Treatise on Nervous and Mental Diseases. By LANDON CARTER GRAY, M. D., Professor of Diseases of the Mind and Nervous System in the New York Polyclinic. In one very handsome octavo volume of 681 pages, with 168 illustrations. Cloth, $4.50; leather, $5.50.

A book that will be welcomed by the many who desire a modern text-book on nervous diseases that is comprehensive and practical, and especially full in the details of the treatment of these affections that are so often matters of perplexity to the general practitioner. It will be found, on this account, to meet the wants of a large number perhaps better than would another equally meritorious text-book less full in this regard. Dr. Gray states in his preface, and it is evident to anyone perusing the work, that "especial care has been taken to make the therapeutical suggestions sufficiently detailed and precise to cover the varying stages, symptoms and complications of disease, as well as to inform the important indications afforded by differential diagnosis," and that "only that knowledge has been admitted to these pages which has stood the test of experience." Its style is clear and very readable, and the illustrations are numerous and excellent. A glossary of special terms is appended which will be found useful by the student. While it is intended as a text-book, not assuming any special knowledge on the part of its readers, the volume is full of valuable orig-

inal matter that renders it a desirable addition to the library of the specialist in nervous and mental diseases.—*American Jour. of Mental Sci.* Feb., 1893.

None but one of the foremost neurologists in this country could have succeeded in condensing within less than seven hundred pages a text-book on nervous and mental diseases. But the author's long experience as a teacher and his wide range of information regarding the literature of his subject have enabled him to discriminate nicely as to what knowledge is most needed by the student and practitioner. We wish that every medical college could have inscribed on its walls the following: "Medical nihilism is an error of youth and a confession of impotence, for Nature rarely afflicts man beyond the hope of relief. The therapeutist has enormous resources." Too often the instructor informs his pupils with exceeding care of the ætiology, pathology, and diagnosis of disease, and then says that it is incurable; he makes no effort to teach them what to do to alleviate, and Dr. Gray's optimism is in refreshing contrast to such instruction.—*New York Medical Journal,* March, 1894.

Ross on Diseases of the Nervous System.

A Handbook on Diseases of the Nervous System. By JAMES ROSS, M. D., F. R. C. P., LL. D., Senior Assistant Physician to the Manchester Royal Infirmary. In one octavo vol. of 725 pages, with 184 illus. Cloth, $4.50; leather, $5.50.

This admirable work is intended for students of medicine and for such medical men as have no time for lengthy treatises. In the present instance the duty of arranging the vast store of material at the disposal of the author, and of abridging the description of the different aspects of nervous diseases, has been performed with singular skill, and the result is a concise and philosophical guide to

the department of medicine of which it treats. Dr. Ross holds such a high scientific position that any writings which bear his name are naturally expected to have the impress of a powerful intellect. In every part this handbook merits the highest praise, and will no doubt be found of the greatest value to the student as well as to the practitioner.—*Edinburgh Medical Journal,* Jan. 1887.

Hamilton on Nervous Diseases.—Second Edition.

Nervous Diseases; Their Description and Treatment. By ALLEN McLANE HAMILTON, M. D., Attending Physician at the Hospital for Epileptics and Paralytics, Blackwell's Island, N. Y. Second edition, thoroughly revised and rewritten. In one octavo volume of 598 pages, with 72 illustrations. Cloth, $4.00.

Savage on Insanity and Allied Neuroses.

Insanity and Allied Neuroses, Practical and Clinical. By GEORGE H. SAVAGE, M. D., Lecturer on Mental Diseases at Guy's Hospital, London. In one 12mo. vol. of 551 pp., with 18 illus. Cloth, $2.00. See *Series of Clinical Manuals,* p. 30.

Klein's Histology.—Fourth Edition.

Elements of Histology. By E. KLEIN, M. D., F. R. S., Joint Lecturer on General Anatomy and Physiology in the Medical School of St. Bartholomew's Hospital, London. Fourth edition. In one 12mo. volume of 376 pages, with 194 illus. Limp cloth, $1.75. See *Students' Series of Manuals,* page 30.

The large number of editions through which Dr. Klein's little handbook of histology has run since its first appearance in 1883 is ample evidence that it is appreciated by the medical student and that it supplies a definite want. The clear and

concise manner in which it is written, the absence of debatable matter, of conflicting views, added to the convenient size of the book and its moderate price, will account for its undoubted success.—*Medical Chronicle,* Feb., 1890.

Schäfer's Histology.—Third Edition.

The Essentials of Histology. By EDWARD A. SCHÄFER, F. R. S., Jodrell Professor of Physiology in University College, London. New (third) edition. In one octavo volume of 311 pages, with 325 illustrations. Cloth, $3.00.

BLANDFORD ON INSANITY AND ITS TREATMENT. Lectures on the Treatment, Medical and Legal, of Insane Patients. In one very handsome octavo volume.
JONES' CLINICAL OBSERVATIONS ON FUNCTIONAL NERVOUS DISORDERS. Second

American Edition. In one handsome octavo volume of 340 pages. Cloth, $3.25.
PEPPER'S SURGICAL PATHOLOGY. In one pocket-size 12mo. volume of 511 pages, with 81 illustrations. Limp cloth, red edges, $2.00. See *Students' Series of Manuals,* page 30.

Gibbes' Practical Pathology and Morbid Histology.

Practical Pathology and Morbid Histology. By HENEAGE GIBBES, M. D., Professor of Pathology in the University of Michigan, Medical Department. In one very handsome 8vo. vol. of 314 pp., with 60 illus., mostly photographic.   Cloth, $2.75.

This is, in part, an expansion of the little work published by the author some years ago, and his acknowledged skill as a practical microscopist will give weight to his instructions. Indeed, in fulness of directions as to the modes of investigating morbid tissues the book leaves little to be desired. The work is throughout profusely illustrated with reproductions of micro photographs. We may say that the practical histologist will gain much useful information from the book.—*The London Lancet,* January 23, 1892.

The student of morbid histology and bacteriology has at his hand, in this neat volume of some three hundred pages, a most excellent guide and one which, unless he be a very advanced student, he cannot afford to be without. The work is divided into four parts, the first, that of practical pathology, containing clear and precise directions

In histological technique, showing how to repair the tissues for examination, cut, stain and mount sections, etc. The second part deals with bacteriology, with the different forms of cultivation, microscopic examinations of the bacteria, etc. The third part, which comprises more than half the book, treats of morbid histology. This part is illustrated with a great number of beautiful photo micrographs in which the microscopic field is reproduced with a distinctiveness that is really remarkable. The fourth part contains some very practical instruction on photography with the microscope. Works like this of Dr. Gibbes will soon popularize histology among the profession at large, whereas it is now to a large number of physicians almost a sealed book.—*Medical Record,* Oct. 17, 1891.

Abbott's Bacteriology.

The Principles of Bacteriology: a Practical Manual for Students and Physicians. By A. C. ABBOTT, M. D., First Assistant, Laboratory of Hygiene, University of Pennsylvania, Philadelphia. In one 12mo. vol. of 259 pp., with 32 illus. Cloth, $2.00.

To a person desiring to learn the technique of bacteriological work, we cannot recommend any work which will be more suitable than the one before us. Dr. Abbott has shown great judgment in the selection and arrangement of his material. The student who follows it closely

will be in a condition to carry forward the work for himself. Medical practitioners generally could read the work with profit, especially the chapters on sterilization and disinfection, and those on tuberculosis and diphtheria in the second part.— *The Canadian Practitioner,* Nov. 1, 1892.

Senn's Surgical Bacteriology.—Second Edition.

Surgical Bacteriology. By NICHOLAS SENN, M. D., Ph. D., Professor of Surgery in Rush Medical College, Chicago. New (second) edition. In one handsome octavo of 268 pp., with 13 plates, of which 10 are colored, and 9 engravings. Cloth, $2.00.

The book is really a systematic collection in the most concise form of such results as are published in current medical literature by the ablest workers in this field of surgical progress; and to these are added the author's own views and the results of his clinical experience and original investigations. The book is valuable to the student, but its chief value lies in the fact that such a compilation

makes it possible for the busy practitioner, whose time for reading is limited and whose sources of information are often few, to become conversant with the most modern and advanced ideas in surgical pathology, which have "laid the foundation for the wonderful achievements of modern surgery."—*Annals of Surgery,* March, 1892.

Green's Pathology and Morbid Anatomy.—Seventh Edition.

Pathology and Morbid Anatomy. By T. HENRY GREEN, M. D., Lecturer on Pathology and Morbid Anatomy at Charing-Cross Hospital Medical School, London. Sixth American from the seventh and revised English edition. Octavo, 539 pages, with 167 engravings. Cloth, $2.75.

The Pathology and Morbid Anatomy of Dr. Green is too well known by members of the medical profession to need any commendation. There is scarcely an intelligent physician anywhere who has not the work in his library, for it is almost an essential. In fact it is better adapted to the wants of general practitioners than any work of the kind with which we are acquainted. The works of German authors upon pathology, which have been

translated into English, are too abstruse for the physician. Dr. Green's work precisely meets his wishes. The cuts exhibit the appearances of pathological structures just as they are seen through the microscope. The fact that it is so generally employed as a text-book by medical students is evidence that we have not spoken too much in its favor.—*Cincinnati Medical News,* Oct. 1889.

Payne's General Pathology.

A Manual of General Pathology. Designed as an Introduction to the Practice of Medicine. By JOSEPH F. PAYNE, M. D., F. R. C. P., Senior Assistant Physician and Lecturer on Pathological Anatomy, St. Thomas' Hospital, London. Octavo of 524 pages, with 152 illustrations and a colored plate. Cloth, $3.50.

Coats' Treatise on Pathology.

A Treatise on Pathology. By JOSEPH COATS, M. D., F. F. P. S., Pathologist to the Glasgow Western Infirmary. In one very handsome octavo volume of 829 pages, with 339 beautiful illustrations. Cloth, $5.50; leather, $6.50.

Medical students as well as physicians, who desire a work for study or reference, that treats the subjects in the various departments in a very thorough manner, but without prolixity, will certainly give this one the preference to any with which we are acquainted. It sets forth the most recent discoveries, exhibits, in an interesting

manner, the changes from a normal condition effected in structures by disease, and points out the characteristics of various morbid agencies, so that they can be easily recognized. But, not limited to morbid anatomy, it explains fully how the functions of organs are disturbed by abnormal conditions.—*Cincinnati Medical News,* Oct. 1883.

Lea Brothers & Co., Publishers, 706, 708 & 710 Sansom Street, Philadelphia.

Wharton's Minor Surgery and Bandaging.—2d Ed. Just Ready.

Minor Surgery and Bandaging. By HENRY R. WHARTON, M. D., Demonstrator of Surgery in the University of Pennsylvania. In one 12mo. volume of 529 pages, with 416 engravings, many being photographic. Cloth, $3.00.

It is but little more than two years ago that we published a review notice of Wharton's first edition. At that time, we remarked that the book was one of the very best treatises on minor surgery that had been published, that it ought to be adopted as a text book on the subjects of which it treats, and that it contained more practical surgery within its limits and boundaries than any book of its kind we had ever seen. What was true of the first edition may be, with propriety, repeated and accentuated in regard to this second and revised edition. Its illustrations are to be specially commended, particularly those that relate to bandaging, most of which have been taken from photographs of applied bandages in the several localities of the body. The author has thoroughly revised that portion of the work relating to the aseptic and antiseptic methods of wound treatment, than which there is no more important subject in the whole domain of surgery. Much new matter has been added, which brings it abreast of the very latest knowledge on the subjects of which it treats.—*Buffalo Medical and Surgical Journal*, January, 1891.

Treves' Operative Surgery.—Two Volumes.

A Manual of Operative Surgery. By FREDERICK TREVES, F. R. C. S., Surgeon and Lecturer on Anatomy at the London Hospital. In two octavo volumes containing 1550 pages, with 422 engravings. Complete work, cloth, $9.00; leather, $11.00.

Mr. Treves in this admirable manual of operative surgery has in each instance practically assumed that operation has been decided upon and has then proceeded to give the various operative methods which may be employed, with a criticism of their comparative value and a detailed and careful description of each particular stage of their performance. Especial attention has been paid to the preparatory treatment of the patient and to the details of the after treatment of the case, and this is one of the most distinctive among the many excellent features of the book. We have no hesitation in declaring it the best work on the subject in the English language, and indeed, in many respects, the best in any language. It can-not fail to be of the greatest use both to practical surgeons and to those general practitioners who, owing to their isolation or to other circumstances, are forced to do much of their own operative work. We feel called upon to recommend the book so strongly for the excellent judgment displayed in the arduous task of selecting from among the thousands of varying procedures those most worthy of description; for the way in which the still more difficult task of choosing among the best of those has been accomplished; and for the simple, clear, straightforward manner in which the information thus gathered from all surgical literature has been conveyed to the reader.—*Annals of Surgery*, March, 1892.

Treves' Student's Handbook of Surgical Operations. In one square 12mo. volume of 508 pages, with 94 illustrations. Cloth, $2.50.

A Manual of Surgery. In Treatises by Various Authors, edited by FREDERICK TREVES, F. R. C. S. In three 12mo. volumes, containing 1866 pages, with 213 engravings. Price per set, cloth, $6.00. See *Students' Series of Manuals*, page 30.

We have here the opinions of thirty-three authors, in an encyclopædic form for easy and ready reference. The three volumes embrace every variety of surgical affections likely to be met with, the paragraphs are short and pithy, and the salient points and the beginnings of new subjects are always printed in extra-heavy type, so that a person may find whatever information he may be in need of at a moment's glance.—*Cincinnati Lancet-Clinic*, August 21, 1886.

Treves on Intestinal Obstruction. In one 12mo. volume of 522 pages, with 60 illus. Limp cloth, blue edges, $2.00. See *Series of Clinical Manuals*, page 30.

Erichsen's Science and Art of Surgery.—Eighth Edition.

The Science and Art of Surgery; Being a Treatise on Surgical Injuries, Diseases and Operations. By JOHN E. ERICHSEN, F. R. S., F. R. C. S., Professor of Surgery in University College, London, etc. From the eighth and enlarged English edition. In two large 8vo. vols. of 2316 pp., with 984 engravings on wood. Cloth, $9; leather, $11.

For many years this classic work has been made by preference of teachers the principal text-book on surgery for medical students, while through translations into the leading continental languages it may be said to guide the surgical teachings of the civilized world. No excellence of the former edition has been dropped and no discovery, device or improvement which has marked the progress of surgery during the last decade has been omitted. The illustrations are many and executed in the highest style of art.—*Louisville Medical News*, Feb. 14, 1885.

Bryant's Practice of Surgery.—Fourth Edition.

The Practice of Surgery. By THOMAS BRYANT, F. R. C. S., Surgeon and Lecturer on Surgery at Guy's Hospital, London. Fourth American from the fourth and revised English edition. In one large and very handsome imperial octavo volume of 1040 pages, with 727 illustrations. Cloth, $6.50; leather, $7.50.

The present edition is a thorough revision of those which preceded it, with much new matter added. Almost every topic in surgery is presented in such a form as to enable the busy practitioner to review any subject in every-day practice in a short time. In short, the work is eminently clear, logical and practical.—*Chicago Med. Jour. and Examiner*, Apr. 1886.

MILLER'S PRACTICE OF SURGERY. Fourth and revised American edition. In one large 8vo. vol. of 682 pp., with 364 illustrations. Cloth, $3.75.
MILLER'S PRINCIPLES OF SURGERY. Fourth American from the third Edinburgh ed. In one 8vo. vol. of 638 pages, with 240 illus. Cloth, $3.75.
HOLMES' SYSTEM OF SURGERY. THEORETICAL AND PRACTICAL. By Various Authors. Edited by TIMOTHY HOLMES, M. A. American edition, revised and re-edited by JOHN H. PACKARD, M. D. Three large octavo volumes, 3137 pages, 979 illustrations on wood and 13 lithographic plates. Per set, cloth, $18.00; leather, $21.00. *Subscription only.*

Lea Brothers & Co., Publishers, 706, 708 & 710 Sansom Street, Philadelphia

Smith's Operative Surgery.—Revised Edition.

The Principles and Practice of Operative Surgery. By STEPHEN SMITH, M. D., Professor of Clinical Surgery in the University of the City of New York. Second and thoroughly revised edition. In one very handsome octavo volume of 892 pages, with 1005 illustrations. Cloth, $4 00; leather, $5.00.

This excellent and very valuable book is one of the most satisfactory works on modern operative surgery yet published. The book is a compendium for the modern surgeon. The present edition is much enlarged, and the text has been thoroughly revised, so as to give the most improved methods in aseptic surgery, and the latest instruments known for operative work. It can be truly said that as a handbook for the student, a companion for the surgeon, and even as a book of reference for the physician not especially engaged in the practice of surgery, this volume will long hold a most conspicuous place, and seldom will its readers, no matter how unusual the subject, consult its pages in vain. Its compact form, excellent print, numerous illustrations, and especially its decidedly practical character, all combine to commend it.— *Boston Medical and Surgical Journal*, May 10, 1888.

Holmes' Treatise on Surgery.—Fifth Edition.

A Treatise on Surgery; Its Principles and Practice. By TIMOTHY HOLMES, M. A., Surgeon and Lecturer on Surgery at St. George's Hospital, London. From the fifth English edition, edited by T. PICKERING PICK, F. R. C. S. In one octavo volume of 997 pages, with 428 illustrations. Cloth, $6.00; leather, $7.00.

To the younger members of the profession and to others not acquainted with the book and its merits, we take pleasure in recommending it as a surgery complete, thorough, well-written, fully illustrated, modern, a work sufficiently voluminous for the surgeon specialist, adequately concise for the general practitioner, teaching those things that are necessary to be known for the successful prosecution of the surgeon's career, imparting nothing that in our present knowledge is considered unsafe, unscientific or inexpedient.— *Pacific Medical Journal*, July, 1889.

Hamilton on Fractures and Dislocations.—Eighth Edition.

A Practical Treatise on Fractures and Dislocations. By FRANK H. HAMILTON, M. D., LL. D., Surgeon to Bellevue Hospital, New York. New (8th) edition, revised and edited by STEPHEN SMITH, M. D., Prof. of Clinical Surgery in Univ. of City of N. Y. In one octavo volume of 832 pp., with 507 illus. Cloth, $5.50; leather, $6.50.

Its numerous editions are convincing proof if any is needed, of its value and popularity. It is preeminently the authority on fractures and dislocations, and universally quoted as such. In the new edition it has lost none of its former worth. The additions it has received by its recent revision make it a work thoroughly in accordance with modern practice, theoretically, mechanically, aseptically. The task of writing a complete treatise on a subject of such magnitude is no easy one. Dr. Smith has aimed to make the present volume a correct exponent of our knowledge of this department of surgery. The more one reads the more one is impressed with its completeness. The work has been accomplished, and has been done clearly, concisely, excellently well.— *Boston Medical and Surgical Journal*, May 26, 1892.

Stimson's Operative Surgery.—Second Edition.

A Manual of Operative Surgery. By LEWIS A. STIMSON, B. A., M. D., Professor of Clinical Surgery in the University of the City of New York. Second edition. In one royal 12mo. volume of 503 pages, with 342 illustrations. Cloth, $2.50.

The author knows the difficult art of condensation. Thus the manual serves as a work of reference, and at the same time as a handy guide. It teaches what it professes, the steps of operations. In this edition Dr. Stimson has sought to indicate the changes that have been effected in operative methods and procedures by the antiseptic system, and has added an account of many new operations and variations in the steps of older operations. We do not desire to extol this manual above many excellent standard British publications of the same class, still we believe that it contains much that is worthy of imitation.— *British Medical Journal*, Jan. 22, 1887.

Stimson on Fractures and Dislocations.

A Treatise on Fractures and Dislocations. By LEWIS A. STIMSON, M. D. In two handsome octavo volumes. Vol. I., FRACTURES, 582 pages, 360 illustrations. Vol. II., DISLOCATIONS, 540 pages, with 163 illustrations. Complete work, cloth, $5.50; leather, $7.50. Either volume separately, cloth, $3.00; leather, $4.00.

The appearance of the second volume marks the completion of the author's original plan of preparing a work which should present in the fullest manner all that is known on the cognate subjects of Fractures and Dislocations. The volume on Fractures assumed at once the position of authority on the subject, and its companion on Dislocations will no doubt be similarly received. This volume exhibits the surgery of Dislocations as it is taught and practised by the most eminent surgeons of the present time. Containing the results of such extended researches it must for a long time be regarded as an authority on all subjects pertaining to dislocations. Every practitioner of surgery will feel it incumbent on him to have it for constant reference.— *Cincinnati Medical News*, May, 1888.

Pick on Fractures and Dislocations.

Fractures and Dislocations. By T. PICKERING PICK, F. R. C. S., Surgeon to and Lecturer on Surgery at St. George's Hospital, London. In one 12mo. vol. of 530 pp., with 93 illus. Limp cloth, $2.00. See *Series of Clinical Manuals*, page 30.

Marsh on the Joints.

Diseases of the Joints. By HOWARD MARSH, F. R. C. S., Senior Assistant Surgeon to St. Bartholomew's Hospital, London. In one 12mo. volume of 468 pages, with 64 woodcuts and a colored plate. Cloth, $2.00. See *Series of Clinical Manuals*, page 30.

Politzer on Diseases of the Ear.—New Edition. Just Ready.

A Text-Book of Diseases of the Ear and Adjacent Organs. By DR. ADAM POLITZER, Imperial-Royal Professor of Aural Therapeutics in the University of Vienna, Chief of the Imperial·Royal University Clinic for Diseases of the Ear in the General Hospital, Vienna. Translated into English from the third and revised German edition. by OSCAR DODD, M. D., Clinical Instructor in Diseases of the Eye and Ear, College of Physicians and Surgeons, Chicago. Edited by SIR WILLIAM DALBY, F. R. C. S., M. B., Consulting Aural Surgeon to St. George's Hospital, London. In one large octavo volume of 748 pages, with 330 illustrations. Cloth, $5.50.

This standard treatise, by a recognized German authority, will be eagerly welcomed by those English readers unacquainted with the work in the original. Politzer's views on otological subjects are those of an advanced yet conservative surgeon. His advice is that of a careful observer of vast and varied experience. The illustrations accompanying the text are good.—*Medical Record,* April 21, 1894.

Field's Manual of Diseases of the Ear.—Just Ready.

A Manual of Diseases of the Ear. By GEORGE P. FIELD, M. R. C. S, Aural Surgeon and Lecturer on Aural Surgery in St Mary's Hospital Medical School, London. In one octavo of 391 pp., with 73 engravings and 21 colored plates. Cloth, $3.75.

The author's views are so plainly and forcibly expressed that the student and general practitioner of medicine cannot afford to be without their teaching and careful guidance if they would do the justice to their patients that the present advanced state of otology demands. Within the covers of this book will be found information sufficient to supply the needs of the student and practitioner of general medicine in practical matters pertaining to diseases of the ear.—*The Therapeutic Gazette,* January 15, 1894.

Burnett on the Ear.—Second Edition.

The Ear, Its Anatomy, Physiology and Diseases. A Practical Treatise for the use of Medical Students and Practitioners. By CHARLES H. BURNETT, A. M., M. D., Professor of Otology in the Philadelphia Polyclinic; President of the American Otological Society. Second edition. In one handsome octavo volume of 580 pages, with 107 illustrations. Cloth, $4.00; leather, $5.00.

Roberts on Urinary and Renal Diseases.—Fourth Edition.

A Practical Treatise on Urinary and Renal Diseases, including Urinary Deposits. By SIR WILLIAM ROBERTS, M. D., Lecturer on Medicine in the Manchester School of Medicine, etc. Fourth American from the fourth London edition. In one handsome octavo volume of 609 pages, with 81 illustrations. Cloth, $3.50.

It may be said to be the best book in print on the subject of which it treats.—*The American Journal of the Medical Sciences,* Jan. 1886.
It is an unrivalled exposition of everything which relates directly or indirectly to the diagnosis, prognosis and treatment of urinary diseases, and possesses a completeness not found elsewhere in our language in its account of the different affections.—*Manchester Med. Chron.,* July, '85.

Purdy on Bright's Disease and Allied Affections.

Bright's Disease and Allied Affections of the Kidneys. By CHARLES W. PURDY, M. D., Professor of Genito-Urinary and Renal Diseases in the Chicago Polyclinic. In one octavo vol. of 288 pages, with illustrations. Cloth, $2.00.

On treatment the writer is particularly strong, steering clear of generalities, and seldom omitting, what text-books usually do, the unimportant items which are all important to the general practitioner.—*The Manchester Medical Chronicle,* Oct. 1886.

The American System of Dentistry.

In Treatises by Various Authors. Edited by WILBUR F. LITCH, M. D., D. D. S., Professor of Prosthetic Dentistry, Materia Medica and Therapeutics in the Pennsylvania College of Dental Surgery. In three very handsome octavo volumes containing 3160 pages, with 1863 illustrations and 9 full-page plates. Per volume, cloth, $6; leather, $7; half Morocco, gilt top, $8. *For sale by subscription only.*

As an encyclopædia of Dentistry it has no superior. It should form a part of every dentist's library, as the information it contains is of the greatest value to all engaged in the practice of dentistry.—*American Jour. Dent. Sci.,* Sept. 1886.
A grand system, big enough and good enough and handsome enough for a monument (which doubtless it is), to mark an epoch in the history of dentistry. Dentists will be satisfied with it and proud of it—they must. It is sure to be precisely what the student needs to put him and keep him in the right track, while the profession at large will receive incalculable benefit from it.—*Odontographic Journal,* Jan. 1887.

Coleman's Dental Surgery.—American Edition.

A Manual of Dental Surgery and Pathology. By ALFRED COLEMAN L. R. C. P., F. R. C. S., Exam. L. D. S., Lecturer on Dental Surgery at St. Bartholomew's Hospital, London. Thoroughly revised and adapted to the use of American Students, by by THOMAS C. STELLWAGEN, M. A., M. D., D. D. S., Prof. of Physiology in the Philadelphia Dental College. Octavo volume of 412 pages, with 331 illustrations. Cloth, $3.25.

MORRIS ON SURGICAL DISEASES OF THE KIDNEY. By HENRY MORRIS, F. R. C. S., Surgeon to Middlesex Hospital, London. 12mo., 554 pp., with 40 woodcuts, and 6 colored plates. Limp cloth, $2.25. See *Series of Clinical Manuals,* p. 30.

Gross on Impotence, Sterility, etc.—Fourth Edition.

A Practical Treatise on Impotence, Sterility, and Allied Disorders of the Male Sexual Organs. By SAMUEL W. GROSS, A. M., M. D., LL. D., Professor of the Principles of Surgery and of Clinical Surgery in the Jefferson Medical College of Philadelphia. Fourth edition, thoroughly revised by F. R. STURGIS, M. D., Prof. of Diseases of the Genito-Urinary Organs and of Venereal Diseases, N. Y. Post Grad. Med. School. In one 8vo. vol. of 165 pages, with 18 illus. Cloth, $1.50.

Three editions of Professor Gross' valuable book have been exhausted, and still the demand is unsupplied. Dr. Sturgis has revised and added to the previous editions, and the new one appears more complete and more valuable than before. Four important and generally misunderstood subjects are treated—impotence, sterility, spermatorrhœa, and prostatorrhœa. The book is a practical one and in addition to the scientific and very interesting discussions on etiology, symptoms, etc., there are lines of treatment laid down that any practitioner can follow and which have met with success in the hands of author and editor.—*Medical Record*, Feb. 25, 1891.

Taylor on Venereal Diseases.—Sixth Edition. Preparing.

The Pathology and Treatment of Venereal Diseases. Including the results of recent investigations upon the subject. By ROBERT W. TAYLOR, A. M., M. D., Clinical Professor of Genito-Urinary Diseases in the College of Physicians and Surgeons, New York. Being the sixth edition of *Bumstead and Taylor*, rewritten by Dr. Taylor. Large 8vo. volume, about 900 pages, with about 150 engravings, as well as numerous chromo-lithographs. *In active preparation.* A notice of the previous edition is appended.

It is a splendid record of honest labor, wide research, just comparison, careful scrutiny and original experience, which will always be held as a high credit to American medical literature. This is not only the best work in the English language upon the subjects of which it treats, but also one which has no equal in other tongues for its clear, comprehensive and practical handling of its themes.—*Am. Jour. of the Med. Sciences*, Jan. 1884.

Culver & Hayden's Manual of Venereal Diseases.

A Manual of Venereal Diseases. By EVERETT M. CULVER, M. D., Pathologist and Assistant Attending Surgeon, Manhattan Hospital, New York, and JAMES R. HAYDEN, M. D., Chief of Clinic Venereal Department, College of Physicians and Surgeons, New York. In one 12mo. volume of 289 pages with 33 illus. Cloth, $1.75.

This book is a practical treatise, presenting in a condensed form the essential features of our present knowledge of the three venereal diseases, syphilis, chancroid and gonorrhea. We have examined this work carefully and have come to the conclusion that it is the most concise, direct and able treatise that has appeared on the subject of venereal diseases for the general practitioner to adopt as a guide. The general practitioner needs a few simple, concise and clearly presented laws, in the execution of which he cannot fail either to cure or prevent the ravages of the maladies in question and their direful results.—*Buffalo Medical and Surgical Journal*, May, 1892.

Cornil on Syphilis.

Syphilis, its Morbid Anatomy, Diagnosis and Treatment. By V. CORNIL, Professor to the Faculty of Medicine of Paris, and Physician to the Lourcine Hospital. Specially revised by the Author, and translated with notes and additions by J. HENRY C. SIMES, M. D., Demonstrator of Pathological Histology in the Univ. of Pa., and J. WILLIAM WHITE, M. D., Lecturer on Venereal Diseases, Univ. of Pa. In one handsome octavo volume of 461 pages, with 84 very beautiful illustrations. Cloth, $3.75.

Hutchinson on Syphilis.

Syphilis. By JONATHAN HUTCHINSON, F. R. S., F. R. C. S., Consulting Surgeon to the London Hospital. In one 12mo. volume of 542 pages, with 8 chromolithographs. Cloth, $2.25. See *Series of Clinical Manuals*, page 30.

Hardaway's Manual of Skin Diseases.

Manual of Skin Diseases. With Special Reference to Diagnosis and Treatment. For the use of Students and General Practitioners. By W. A. HARDAWAY, M. D., Professor of Skin Diseases in the Missouri Medical College. 12mo., 440 pp. Cloth, $3.00.

Dr. Hardaway's large experience as a teacher and writer has admirably fitted him for the difficult task of preparing a book which, while sufficiently elementary for the student is yet sufficiently thorough and comprehensive to serve as a book of reference for the general practitioner. It embraces all essential points connected with the diagnosis and treatment of diseases of the skin, and we have no hesitation in commending it as the best manual that has yet appeared in the department of medicine.—*Journal of Cutaneous and Genito-Urinary Diseases.*

Hyde on the Skin.—New (3d) Edition. Just Ready.

A Practical Treatise on Diseases of the Skin. For the use of Students and Practitioners. By J. NEVINS HYDE, A. M., M. D., Professor of Dermatology and Venereal Diseases in Rush Medical College, Chicago. Third edition. In one octavo volume of 802 pages, with 9 colored plates and 108 engravings. Cloth, $5.00; leather, $6.00.

Dr. Hyde is an experienced scholar as well as a competent author, and his former editions were received with approval by dermatologists as well as by those general practitioners who are interested in the study and treatment of diseases of the skin. The treatise is one that affords much satisfaction in that it is a safe guide for both students and practitioners, either general or special, and particularly is it adapt itself to the use of dermatologists.—Buffalo Medical and Surgical Journal, March, 1894.

The qualities that have contributed so much to its previous popularity still remain. The chief of these unquestionably are the standpoint of practical medicine from which it speaks and its wealth of therapeutical information. The writer knows no book in which one can seek more satisfactorily for information as to how to manage his patients with skin diseases. The present edition may be commended as being an exposition of the subject fully up to the present state of our knowledge.—The Chicago Clinical Review, April, 1894.

Taylor's Clinical Atlas of Venereal and Skin Diseases.

A Clinical Atlas of Venereal and Skin Diseases: Including Diagnosis, Prognosis and Treatment. By ROBERT W. TAYLOR, A. M., M. D., Clinical Professor of Genito-Urinary Diseases in the College of Physicians and Surgeons, New York; In eight large folio parts, and comprising 58 beautifully colored plates with 213 figures, and 431 pages of text with 85 engravings. Price per part, $2.50. Bound in one volume, half Russia, $27; half Turkey Morocco, $28. For sale by subscription only. Specimen plates sent on receipt of 10 cents. A full prospectus sent to any address on application.

It would be hard to use words which would perspicuously enough convey to the reader the great value of this Clinical Atlas. This Atlas is more complete even than an ordinary course of clinical lectures, for in no one college or hospital course is it at all probable that all of the diseases herein represented would be seen. It is also more serviceable to the majority of students than attendance upon clinical lectures, for most of the students who sit on remote seats in the lecture hall cannot see the subject as well as the office

student can examine these true-to-life chromo-lithographs. Comparing the text to a lecturer, it is more satisfactory in exactness and fulness than he would be likely to be in lecturing over a single case. Indeed, this Atlas is invaluable to the general practitioner, for it enables the eye of the physician to make diagnosis of a given case of skin manifestation by comparing the case with the picture in the Atlas, where will be found also the text of diagnosis, pathology, and full sections on treatment.—Virginia Medical Monthly, Dec. 1889.

Jackson's Ready-Reference Handbook of Skin Diseases.

The Ready-Reference Handbook of Diseases of the Skin. By GEORGE THOMAS JACKSON, M. D., Professor of Dermatology, Woman's Medical College of the New York Infirmary. In one 12mo. volume of 544 pages, with 50 illustrations and a colored plate. Cloth, $2.75.

Intended to serve as a reference book for the general practitioner, "no attempt has been made to discuss debatable questions," and "hence patnology and etiology do not receive as full consideration as symptomatology, diagnosis and treat-

ment." It treats in alphabetical order of the diseases of the skin and their management. This book seems to us the best of its class that has yet appeared.—Boston Medical and Surgical Journal, May 18, 1893.

Morris on the Skin.—Just Ready.

Diseases of the Skin. An Outline of the Principles and Practice of Dermatology. By MALCOLM MORRIS, F. R. C. S., Surgeon to the Skin Department, St. Mary's Hospital, London. In one 12mo. volume of 572 pages, with 19 chromo-lithographic figures and 17 engravings. Cloth, $3.50.

This admirable manual, written as it evidently is by a keen, clever specialist of exceptionally wide experience, most satisfactorily meets the requirements of the American practitioner of medicine, in that it gives him a clear, comprehensive picture of every skin-affection for the cure of

which there is any chance of his being called, and formulates for him a system of therapeutics in the following of which he can feel well assured of obtaining the best results.—The Therapeutic Gazette, March, 1894.

Pye-Smith on Diseases of the Skin.—Just Ready.

A Handbook of Diseases of the Skin. By P. H. PYE-SMITH, M. D., F. R. S, Physician to Guy's Hospital, London. In one octavo volume of 407 pages, with 26 illustrations, 18 of which are colored. Cloth, $2.00.

It is a plain, practical treatise on dermatology, written for the student and general practitioner by a general practitioner of broad experience in the special subject of which he writes. He simplifies the nomenclature, and succeeds in removing much of the difficulty. After reviewing the recent

advances made in this department of medicine, he pays a merited compliment to the "important contributions made by the newest school of dermatology, that of America."—Pittsburg Medical Review, June, 1893.

Jamieson on Diseases of the Skin.—Third Edition.

Diseases of the Skin. A Manual for Students and Practitioners. By W. ALLAN JAMIESON, M. D., Lecturer on Diseases of the Skin, School of Medicine, Edinburgh. Third edition, revised and enlarged. In one octavo volume of 656 pages, with woodcut and 9 double-page chromo lithographic illustrations. Cloth, $6.00.

The scope of the work is essentially clinical, little reference being made to pathology or disputed theories. Almost every subject is followed by illustrative cases. The pages are filled with interest to all those occupied with skin diseases. The

general practitioner will find the book of great value in matters of diagnosis and treatment. The latter is quite up to date, and the formulæ have been selected with care.—Medical Record, April 9, 1892.

The American Systems of Gynecology and Obstetrics.

Systems of Gynecology and Obstetrics, in Treatises by American Authors. Gynecology edited by MATTHEW D. MANN, A. M., M. D., Professor of Obstetrics and Gynecology in the Medical Department of the University of Buffalo; and Obstetrics edited by BARTON COOKE HIRST, M. D., Associate Professor of Obstetrics in the University of Pennsylvania, Philadelphia. In four very handsome octavo volumes, containing 3612 pages, 1092 engravings and 8 plates. Complete work *now ready.* Per volume: Cloth, $5.00; leather, $6.00; half Russia, $7.00. *For sale by subscription only. Address the Publishers.* Full descriptive circular free on application.

These volumes are the contributions of the most eminent gentlemen of this country in these departments of the profession. Each contributor presents a monograph upon his special topic, so that everything in the way of history, theory, methods, and results is presented to our fullest need. As a work of general reference, it will be found remarkably full and instructive in every direction of inquiry.—*The Obstetric Gazette*, September, 1889.

One is at a loss to know what to say of this volume, for fear that just and merited praise may be mistaken for flattery. The papers of Drs. Engelmann, Martin, Hirst, Jaggard and Reeve are incomparably beyond anything that can be found in obstetrical works.—*Journal of the American Medical Association*, Sept. 8, 1888.

In our notice of the "System of Practical Medicine by American Authors," we made the following statement:—"It is a work of which the profession in this country can feel proud. Written exclusively by American physicians who are acquainted with all the varieties of climate in the United States, the character of the soil, the manners and customs of the people, etc., it is peculiarly adapted to the wants of American practitioners of medicine, and it seems to us that every one of them would desire to have it." Every word thus expressed in regard to the "American System of Practical Medicine" is applicable to the "System of Gynecology by American Authors." It, like the other, has been written exclusively by American physicians who are acquainted with all the characteristics of American people, who are well informed in regard to the peculiarities of American women, their manners, customs, modes of living, etc. As every practising physician is called upon to treat diseases of females, and as they constitute a class to which the family physician must give attention, and cannot pass over to a specialist, we do not know of a work in any department of medicine that we should so strongly recommend medical men generally purchasing.—*Cincinnati Med. News*, July, 1887.

Emmet's Gynæcology.—Third Edition.

The Principles and Practice of Gynæcology; For the use of Students and Practitioners of Medicine. By THOMAS ADDIS EMMET, M. D., LL. D., Surgeon to the Woman's Hospital, New York, etc. Third edition, thoroughly revised. In one large and very handsome 8vo. vol. of 880 pp., with 150 illus. Cloth, $5; leather, $6.

We are in doubt whether to congratulate the author more than the profession upon the appearance of the third edition of this well-known work. Embodying, as it does, the life-long experience of one who has conspicuously distinguished himself as a bold and successful operator, and who has devoted so much attention to the specialty, we feel sure the profession will not fail to appreciate the privilege thus offered them of perusing the views and practice of the author. His earnestness of purpose and conscientiousness are manifest. He gives not only his individual experience but endeavors to represent the actual state of gynæcological science and art.—*British Medical Journal*, May 16, 1885.

Tait's Diseases of Women and Abdominal Surgery.

Diseases of Women and Abdominal Surgery. By LAWSON TAIT, F. R. C. S., Professor of Gynæcology in Queen's College, Birmingham, late President of the British Gynecological Society, Fellow American Gynæcological Society. In two octavo vols. Vol. I., 554 pp., 62 engravings and 3 plates. Cloth, $3. Vol. II., *preparing.*

The plan of the work does not indicate the regular system of a text book, and yet nearly everything of disease pertaining to the various organs receives a fair consideration. The description of diseased conditions is exceedingly clear, and the treatment, medical or surgical, is very satisfactory. Much of the text is abundantly illustrated with cases, which add value in showing the results of the suggested plans of treatment. We feel confident that few gynecologists of the country will fail to place the work in their libraries.—*The Obstetric Gazette*, March, 1890.

Edis on Diseases of Women.

The Diseases of Women. Including their Pathology, Causation, Symptoms, Diagnosis and Treatment. A Manual for Students and Practitioners. By ARTHUR W. EDIS, M. D., Lond., F. R. C. P., M. R. C. S., Assistant Obstetric Physician to Middlesex Hospital, late Physician to British Lying-in-Hospital. In one handsome octavo volume of 576 pages, with 148 illustrations. Cloth, $3.00; leather, $4.00.

The special qualities which are conspicuous are thoroughness in covering the whole ground, clearness of description and conciseness of statement. Another marked feature of the book is the attention paid to the details of many minor surgical operations and procedures, as, for instance, the use of tents, application of leeches, and use of hot water injections. These are among the more common methods of treatment, and yet very little is said about them in many of the text-books. The book is one to be warmly recommended especially to students and general practitioners, who need a concise but complete *résumé* of the whole subject. Specialists, too, will find many useful hints in its pages.—*Boston Med. and Surg. Journ.*, March 2, 1882.

HODGE ON DISEASES PECULIAR TO WOMEN. Including Displacements of the Uterus. Second edition, revised and enlarged. In one beautifully printed octavo volume of 519 pages, with original illustrations. Cloth, $4.50.

WEST'S LECTURES ON THE DISEASES OF WOMEN. Third American from the third London edition. In one octavo volume of 543 pages. Cloth, $3.75; leather, $4.75.

Thomas & Mundé on Diseases of Women.—Sixth Edition.

A Practical Treatise on the Diseases of Women. By T. GAILLARD THOMAS, M. D., LL. D., Emeritus Professor of Diseases of Women in the College of Physicians and Surgeons, New York, and PAUL F. MUNDÉ, M. D., Professor of Gynecology in the New York Polyclinic. New (sixth) edition, thoroughly revised and rewritten by DR. MUNDÉ. In one large and handsome octavo volume of 824 pages, with 347 illustrations, of which 201 are new. Cloth, $5.00; leather, $6.00.

The profession has sadly felt the want of a text-book on diseases of women, which should be comprehensive and at the same time not diffuse, systematically arranged so as to be easily grasped by the student of limited experience, and which should embrace the wonderful advances which have been made within the last two decades. Thomas' work fulfilled these conditions, and the announcement that a new edition was about to be issued, revised by so competent a writer as Dr. Mundé, was hailed with delight. Dr. Mundé brings to his work a most practical knowledge of the subjects of which he treats and an exceptional acquaintance with the world's literature of this important branch of medicine. The result is what is, perhaps, on the whole, the best practical treatise on the subject in the English language. It is, as we have said, the best text-book we know, and will be of especial value to the general practitioner as well as to the specialist. The illustrations are very satisfactory. Many of them are new and are particularly clear and attractive. The book will undoubtedly meet with a favorable reception from the profession.—*Boston Medical and Surgical Journal*, January 14, 1892.

Sutton on the Ovaries and Fallopian Tubes.

Surgical Diseases of the Ovaries and Fallopian Tubes, including Tubal Pregnancy. By J. BLAND SUTTON, F. R. C. S., Assistant Surgeon to the Middlesex Hospital, London. In one crown octavo volume of 544 pages, with 119 engravings and 5 colored plates. Cloth, $3.00.

This is not a book to be read and then shelved; it is one to be studied. It is not based upon hypotheses but upon facts. It makes pathology practical, and inculcates a practice based upon pathology. It is succinct, yet thorough; practical, yet scientific; conservative, yet bold. It is probably on the table of all gynecologists; but it is not for them alone; the general practitioner needs just such a book. It will be of immense service to him in the study of pelvic diseases, and will assuredly open his eyes to the progress made by conscientious, painstaking workers like Dr. Sutton in the field of pathology and differential diagnosis.—*International Medical Magazine*, September, 1892.

Davenport's Non-Surgical Gynæcology.—Second Edition.

Diseases of Women, a Manual of Non-Surgical Gynæcology. Designed especially for the Use of Students and General Practitioners. By FRANCIS H. DAVENPORT, M. D., Assistant in Gynecology in the Medical Department of Harvard University, Boston. New (second) edition. In one handsome 12mo. volume of 314 pages, with 107 illustrations. Cloth, $1.75.

Many valuable volumes already exist on the surgical aspects of gynecology, but scant attention has been paid in recent years to the non-surgical treatment of women's diseases. The present volume, dealing with nothing which has not stood the actual test of experience, and being concisely and clearly written, conveys a great amount of information in a convenient space.—*Annals of Gynæcology and Pædiatry*, June, 1893.

May's Manual of Diseases of Women.—Second Edition.

A Manual of the Diseases of Women. Being a concise and systematic exposition of the theory and practice of gynecology. By CHARLES H. MAY, M. D., late House Surgeon to Mount Sinai Hospital, New York. Second edition, edited by L. S. RAU, M. D., Attending Gynecologist at the Harlem Hospital, N. Y. In one 12mo. volume of 360 pages, with 31 illustrations. Cloth, $1.75.

This is a manual of gynecology in a very condensed form, and the fact that a second edition has been called for indicates that it has met with a favorable reception. It is intended, the author tells us, to aid the student who after having carefully perused larger works desires to review the subject, and he adds that it may be useful to the practitioner who wishes to refresh his memory rapidly but has not the time to consult larger works. We are much struck with the readiness and convenience with which one can refer to any subject contained in this volume. Carefully compiled indexes and ample illustrations also enrich the work. This manual will be found to fulfil its purposes very satisfactorily.—*The Physician and Surgeon*, June, 1890.

Duncan on Diseases of Women.

Clinical Lectures on the Diseases of Women; Delivered in Saint Bartholomew's Hospital. By J. MATTHEWS DUNCAN, M. D., LL. D., F. R. S. E., etc. In one octavo volume of 175 pages. Cloth, $1.50.

They are in every way worthy of their author; indeed, we look upon them as among the most valuable of his contributions. They are all upon matters of great interest to the general practitioner. Some of them deal with subjects that are not, as a rule, adequately handled in the text-books; others of them, while bearing upon topics that are usually treated of at length in such works, yet bear such a stamp of individuality that they deserve to be widely read.—*N. Y. Medical Journal*, March, 1880.

ASHWELL'S PRACTICAL TREATISE ON THE DISEASES PECULIAR TO WOMEN. Third American from the third and revised London edition. In one 8vo. vol., pp. 520. Cloth, $3.50.

Lea Brothers & Co., Publishers, 706, 708 & 710 Sansom Street, Philadelphia.

Taylor's Medical Jurisprudence.—New Edition.   Just Ready.

A Manual of Medical Jurisprudence. By ALFRED S. TAYLOR, M. D., Lecturer on Med. Jurisprudence and Chemistry in Guy's Hosp., London. New American from the 12th English edition. Thoroughly revised by CLARK BELL, Esq., of the New York Bar. In one octavo volume of 787 pages, with 56 illus. Cloth, $4.50; leather, $5.50.

This is a complete revision of all former American and English editions of this standard book. This edition contains a large amount of entirely new matter, many portions of the book having been rewritten by the editor. Many cases and authorities have been cited, and the citations brought down to the latest date. The book has long been a standard treatise on the subject of medical jurisprudence, and has gone through many editions—twelve English and eleven American. Mr. Clark Bell has enlarged and improved what already seemed complete, by bringing his many citations of cases down to meet the present law; and by adding much new matter he has furnished the medical profession and the bar with a valuable book of reference, one to be relied upon in daily practice, and quite up to the present needs, owing to its exhaustive character. It would seem that the book is indispensable to the library of both physician and lawyer, and particularly the legal practitioner whose duties take him into the criminal courts. The editor has given to two professions a reference-book to be relied upon. —*The American Journal of the Medical Sciences,* April, 1893.

No library is complete without Taylor's *Medical Jurisprudence,* as its authority is accepted and unquestioned by the courts.—*Buffalo Medical and Surgical Journal,* June, 1893.

There is no other work upon the subject which has been so uniformly recognized or so widely quoted and followed by courts in England and this country. This eleventh American edition is fully abreast with the most recent thought and knowledge. On the basis of his own researches, and of the investigations of scientists throughout the world, and of the decisions of our own courts, Mr. Bell has incorporated in it a wealth of practical suggestion and instructive illustration which cannot fail to strengthen the hold it has so long had upon the profession.—*The Criminal Law Magazine and Reporter,* January, 1893.

By the Same Author.

Poisons in Relation to Medical Jurisprudence and Medicine. Third American, from the third and revised English edition. In one large octavo volume of 788 pages. Cloth, $5.50; leather, $6.50.

Lea's Superstition and Force.—New Edition.   Just Ready.

Superstition and Force: Essays on The Wager of Law, The Wager of Battle, The Ordeal and Torture. By HENRY CHARLES LEA, LL. D., New (4th) edition, revised and enlarged. Royal 12mo., 629 pages. Cloth, $2.75.

Both abroad and at home the work has been accepted as a standard authority, and the author has endeavored by a complete revision and considerable additions to render it more worthy of the universal favor which has carried it to a fourth edition. The style is severe and simple, and yet delights with its elegance and reserved strength. The known erudition and fidelity of the author are guarantees that all possible original sources of information have been not only consulted but exhausted. The subject matter is handled in such an able and philosophic manner that to read and study it is a step toward liberal education. It is a comfort to read a book that is so thorough, well conceived and well done. We should like to see it made a text-book in our law schools and prescribed course for admission to the bar.—*Legal Intelligencer,* March 3, 1893.

A work as remarkable for the wealth of historical material treated as for the masterly style of the exposition.—*London Saturday Review,* Feb. 25, 1893.

By the same Author.

Chapters from the Religious History of Spain.—In one 12mo. volume of 522 pages. Cloth $2.50.

The width, depth and thoroughness of research which have earned Dr. Lea a high European place as the ablest historian the Inquisition has yet found are here applied to some side-issues of that great subject. We have only to say of this volume that it worthily complements the author's earlier studies in ecclesiastical history. His extensive and minute learning, much of it from inedited manuscripts in Mexico, appears on every page.—*London Antiquary,* Jan. 1891.

By the same Author.

The Formulary of the Papal Penitentiary. In one 8vo. volume of 221 pages, with a frontispiece. Cloth, $2.50. *Just Ready.*

By the Same Author.

Studies in Church History. The Rise of the Temporal Power—Benefit of Clergy—Excommunication—The Early Church and Slavery. Second and revised edition. In one royal octavo volume of 605 pages. Cloth, $2.50.

The author is preëminently a scholar; he takes up every topic allied with the leading theme and traces it out to the minutest detail with a wealth of knowledge and impartiality of treatment that compel admiration. The amount of information compressed into the book is extraordinary, and the profuse citation of authorities and references makes the work particularly valuable to the student who desires an exhaustive review from original sources. In no other single volume is the development of the primitive church traced with so much clearness and with so definite a perception of complex or conflicting forces.—*Boston Traveller.*

By the Same Author.

An Historical Sketch of Sacerdotal Celibacy in the Christian Church. Second edition, enlarged. In one octavo volume of 685 pages. Cloth, $4.50.

This subject has recently been treated with very great learning and with admirable impartiality by an American author, Mr. Henry C. Lea, in his *History of Sacerdotal Celibacy,* which is certainly one of the most valuable works that America has produced. Since the great history of Dean Milman, I know no work in English which has thrown more light on the moral condition of the Middle Ages, and none which is more fitted to dispel the gross illusions concerning that period which positive writers and writers of a certain ecclesiastical school have conspired to sustain.—*Lecky's History of European Morals,* Chap. V.

Lea Brothers & Co., Publishers, 706, 708 & 710 Sansom Street, Philadelphia.